Geochemistry and the Environment

VOLUME II
THE RELATION OF OTHER SELECTED TRACE ELEMENTS TO HEALTH AND DISEASE

A Report of the Workshop at Capon Springs and Farms
Capon Springs, West Virginia
May 6–12, 1973
under the Auspices of the
Subcommittee on the Geochemical Environment
in Relation to Health and Disease
U.S. National Committee for Geochemistry
Assembly of Mathematical and Physical Sciences
National Research Council

NATIONAL ACADEMY OF SCIENCES

WASHINGTON, D.C.   1977

Library of Congress Cataloging in Publication Data

U. S. National Committee for Geochemistry. Subcommittee on the Geochemical Environment in Relation to Health and Disease.
    Geochemistry and the environment.
    Includes bibliographies.
    CONTENTS: v. 1. The relation of selected trace elements to health and disease.—v. 2. The relation of other selected trace elements to health and disease.
    1. Environmental health—Congresses. 2. Geochemistry—Congresses. 3. Trace elements—Congresses. 4. Medical geography—Congresses. I. Title.
[DNLM: 1. Geology—Congresses. 2. Trace elements—Congresses. QU 130 G341 1972]

RA565.A1U53 1974 vol. 2   612'.01524'08s   [612'.01524]
ISBN 0-309-02223-1 (v. 1)   74-13310
ISBN 0-309-02548-6 (v. 2)

Available from
Printing and Publishing Office, National Academy of Sciences
2101 Constitution Avenue, Washington, D.C. 20418

Dedicated to

WILLIAM WALDEN RUBEY

BILL RUBEY, *former Chairman, Division of Earth Sciences, and former Chairman, National Research Council, and National Academy of Sciences Member since 1945, was instrumental in the establishment of the group that eventually became the Subcommittee on the Geochemical Environment in Relation to Health and Disease.*

He died of cancer April 12, 1974.

GENNARD MATRONE

"CHICK" MATRONE, *Head, Department of Biochemistry 1967–1975, and William Neal Reynolds Professor 1962–1967, Professor of Animal Nutrition 1952–1962, North Carolina State University, was Chairman of the Manganese Work Group and served on the Interactions of Trace Elements Work Group at the Capon Springs Workshop (Capon Springs, West Virginia) of the Subcommittee on the Geochemical Environment in Relation to Health and Disease, May 6–12, 1973.*

He died of a cardiovascular condition April 2, 1975, while traveling in Belgium.

PERRY ROBERT STOUT

PERRY STOUT, *Professor Emeritus at the University of California at Davis, was from 1950 until recently Professor of Soil Science. His interests spanned soil chemistry, plant nutrition, micronutrient elements, natural radioactivity of the biosphere, radioisotopes and rare stable isotopes as applied to soil and plant research, and physical instruments for agricultural research. He served as a member of the Work Group on the Consequences of Soil Imbalances at the Subcommittee on the Geochemical Environment in Relation to Health and Disease Capon Springs Workshop (Capon Springs, West Virginia), May 6–12, 1973.*

He died July 14, 1975, at the age of 66.

Preface

This volume, which continues the consideration of possible relationships between the geographic distribution of certain elements and patterns of disease, is the second of a triology, *Geochemistry and the Environment,* undertaken by the NAS–NRC Subcommittee on the Geochemical Environment in Relation to Health and Disease. The reports of these three workshops, at Asilomar, Capon Springs, and Captiva Island, are all outgrowths of the 1968 meetings of the American Association for the Advancement of Science (AAAS) in Dallas, where a few individuals resolved to further interdisciplinary consideration of ways to prove or disprove causal relationships between chemical elements in the environment and animal (including human) health. Since this original meeting, interest in the natural and man-made occurrence of trace substances in the environment and their possible effect on health has grown immensely. The Society for Environmental Geochemistry and Health has been established; a NAS–NRC subcommittee for furthering this interdisciplinary study has been set up; a number of National Science Foundation grants and contracts have been made on trace contaminants under the Research Applied to National Needs (RANN) program, and several other environmental and biomedical agencies of the federal government have become interested and have begun furnishing both intramural and extramural support for projects related to their missions. Several symposia have been held and papers published, which furnish background information that may be of interest to the readers of this volume. These include the following titles:

Cannon, H. L., and D. F. Davidson [eds.]. 1967. *Relation of Geology and Trace Elements to Nutrition.* (Symposium held at the Annual Meeting of the Geological Society of America, New York, 1963.) Geol. Soc. Am. Spec. Pap. No. 90. 67 pp. (Available at $4.25 from The Geological Society of America, 3300 Penrose Place, Boulder, Colorado 80301.)

Cannon, H. L., and H. C. Hopps [eds.]. 1971. *Environmental Geochemistry in Health and Disease.* (Symposium held at the AAAS meetings, Dallas, 1968.) Geol. Soc. Am. Mem. No. 123. 230 pp. (Available at $14 from The Geological Society of America, 3300 Penrose Place, Boulder, Colorado 80301.)

Hopps, H. C., and H. L. Cannon [eds.]. 1972. *Geochemical Environment in Relation to Health and Disease.* (Conference held by the New York Academy of Sciences, New York, 1971.) Annals of the New York Academy of Sciences, Vol. 199. 352 pp. (Available at $25 + $0.35 postage from the New York Academy of Sciences, 2 East 63d Street, New York 10021.)

Cannon, H. L., and H. C. Hopps [eds.]. 1972. *Geochemical Environment in Relation to Health and Disease.* (Symposium held at the AAAS meetings, Chicago, 1970.) Geol. Soc. Am. Spec. Pap. No. 140. 77 pp. (Available at $4.00 from The Geological Society of America, 3300 Penrose Place, Boulder, Colorado 80301.)

Hemphill, D. D. [ed.]. *Trace Substances in Environmental Health.* (Annual Conferences held by the University of Missouri Environmental Health Center, Columbia.) Vol. I (1967), $4.00; Vol. II (1969), $5.00; Vol. III

(1970), $5.00; Vol. IV (1971), $7.50; Vol. V (1972), $10.00; Vol. VI (1973), $10.00; Vol. VII (1973), $12.50; Vol. VIII (1974), $15.00; Vol. IX (1975), $18.00. (Available from Publications Office, 221 Whitten Hall, University of Missouri, Columbia 65201.)

TRACE ELEMENTS

Specialists differ with respect to the definition of trace elements. In general, geologists and geochemists think of trace elements as those other than the eight abundant rock-forming elements found in the earth's crust (oxygen, silicon, aluminum, iron, calcium, sodium, potassium, and magnesium), and mineralogists consider trace elements to be those nonessential components found in small quantities, ordinarily comprising less than 1 percent of a mineral. Most biomedical researchers, however, consider trace elements to be those that are ordinarily present in plant or animal tissue in concentrations less than 0.01 percent of the organism. Although most trace elements are universally recognized as such, some of the abundant rock-forming elements, such as silicon, are considered trace elements in biomedical circles. To help avoid mis-understanding, the term "trace elements" is being used in the biomedical sense in this series of reports.

In *Geochemistry and the Environment*, Volumes I and II, the emphasis is on the health aspects of trace elements in man's environment contributed from *natural* sources. As in Volume I, certain selected trace elements are discussed in individual chapters. Except for beryllium, which possesses a special toxic potential, the elements discussed in Volume II were selected because of their possible essentiality to man. Thus, some of the more toxic (as well as the more necessary) elements made prominent by man's pollution of his environment have not received major attention by the Subcommittee, because of its concentration on the geochemical factors.

In presenting relationships between trace elements and health and disease, certain aspects of recognized importance were necessarily omitted because of space and time limitations. One of these is the special role that microorganisms play in the movement of the elements from rocks to soils to plants to animals and humans. Microorganism populations are as dependent upon soil composition and the geochemistry of the soils as are plants and animals. This is an important subject in itself, which can only be suggested here.

Acknowledgments

The Capon Springs Workshop of the Subcommittee on the Geochemical Environment in Relation to Health and Disease was made possible by the Division of Environmental Systems and Resources of Research Applications, National Science Foundation, under Contract NSF-C310, Task Order 206. Subsequent support from the U.S. Bureau of Mines under Contract S0144074, from the National Cancer Institute under Contract N01-CP-45616, and from the Environmental Protection Agency and the U.S. Atomic Energy Commission (now the Energy Research and Development Administration) through transfers of funds under Contract NSF-C310, Task Order 206, greatly assisted with the completion of this publication and the ongoing work of the Subcommittee. The financial support provided by these agencies is gratefully acknowledged.

The Subcommittee acknowledges with special appreciation the efforts of the prepublication reviewers for other National Research Council components: for the Committee on Medical and Biologic Effects of Environmental Pollutants, Orville Levander and K. V. Rajagopalan; for the Environmental Studies Board, J. M. Wood and Robert Harriss; for the Advisory Center on Toxicology, Ralph C. Wands and his staff associates; for the Food and Nutrition Board, Harry P. Broquist; for the U.S. National Committee for Geochemistry, Larry A. Haskin, Rosemary J. Vidale, and Jon J. Connor; and for the Subcommittee on the Geochemical Environment in Relation to Health and Disease, the members plus Cornel Wohlberg and former member Paul M. Newberne. Their comments and suggestions, along with those of the Report Review Committee reviewers, have been particularly helpful.

The Subcommittee also wishes to thank the participants and the chairmen of the various work groups for their contribution of time, effort, and scientific expertise, without which this report would not have been possible. John O. Ludwigson served as Editorial Consultant. He and Muriel Y. Duggan edited the material from the Workshop deliberations in conjunction with the work group chairmen and the individual authors.

Contents

I

Introduction and Summary

THE WORKSHOP

The second Workshop of the Subcommittee on the Geochemical Environment in Relation to Health and Disease (GERHD) was held at Capon Springs, West Virginia, from May 6 to 12, 1973. Participants from many different disciplines were invited in order to attain the objective: to examine the various ways in which the geochemical environment may influence the processes of health and disease in man. The list of trace elements known to be essential for human health is growing; yet there is only one example in which a clear cause-and-effect relationship between a deficiency of an element in the environment (iodine) and a disease in man (iodine-deficiency goiter) has been established. This Workshop did not attempt to establish more examples of such cause–effect relationships. Instead, its purpose was to examine the existing knowledge and to identify possible research approaches through which such relationships could be established in the future.

To this end, the participants in the Workshop critically evaluated our present knowledge of eight individual trace elements that seemed likely to be related to health and disease. They examined the behavior of these elements in rocks, soil, and water; their migration through the food chain to humans; and the biological effects of under- or overexposure. For some of the elements, the working groups were able to warn of the potential for over- or underexposure and to recommend research to clearly define, and possibly solve, the problems. For other trace elements this was impossible, as shown by the large gaps in our knowledge identified by the working groups.

The same is true for the subjects of the plenary sessions. These were difficult and complex, but it is only through recognition of the immense complexities of problems such as those found in the interactions of trace elements, the nature of soil imbalances, or the interpretation of analytical survey data, that oversimplified and futile research approaches can be avoided. This report of the second GERHD Workshop, representing a consensus of scientists from many disciplines, is an attempt to contribute to planning of future research that will permit an improved relationship between man and the geochemical environment.

This volume is accordingly organized into two sections: "Trace Elements Related to Health and Disease" (Chapters II–IX) and "Geochemical Environment and Man" (Chapters X–XVI), reflecting the special emphasis of the work groups and the plenary sessions, respectively. Brief specific introductions set the tone and scope of each section.

TRACE ELEMENTS RELATED TO HEALTH AND DISEASE

The relationship of trace elements in the environment to human health and disease may be examined in at least two ways. One of these is the attempt to prove a cause–effect relationship through a more thorough knowledge of the distribution, availability, and interaction of given elements and their health effects. The other approach relies on the development of empirical correlations between the geochemical environment and disease processes through geochemical and epidemiologic studies in which large amounts of data can be used to establish hypotheses helpful to further research.

Both of these approaches were thoroughly discussed at the Workshop. As expected, the discussions did not result in solutions, but they pointed out the very complex interactions of individual trace elements, not only in the environment but in the human organism as well, which must be kept in mind in the design of future studies. They also summarized the problems and preliminary results of ongoing studies of the relationship of environmental factors to health and disease in man. In seeking to identify the primary obstacle to further progress, these deliberations outlined the advantages of a National Environmental Specimen Index System (Chapter XVI).

The Elements

The sessions that were concerned with our present knowledge of eight elements—beryllium, magnesium, manganese, nickel, silicon, strontium, tin, and vanadium—can be summarized as follows: It is known that severe under- or overexposure to most of these elements results in disease in experimental animals. Except for beryllium, experimental deficiencies have been induced, although some of these findings need confirmation by independent investigators and in different animal species. While magnesium deficiency is not an uncommon problem in livestock, and manganese deficiency has been described under practical conditions in poultry, deficiencies of the remaining five elements are not yet known to occur naturally. On the other hand, no intensive effort has been made to detect such deficiencies in humans or animals; therefore, their existence, at least in marginal form, cannot be excluded.

Geographic variations of the concentrations of the elements in rocks, soils, and water are known for most of those discussed. Some of these variations are reflected in the elemental concentrations in plants and, in some cases, in human foods. However, it is not known to what degree these changes in trace element concentrations in foods are reflected in the total human intake of these elements, and whether deviations from the optimum affect health. Of particular importance in this regard is the exact assessment of the relative importance of our own activities on the environment. For example, to what degree do modern agricultural practices reinforce or override environmental influences on the trace element content of foods? Equally important, the influence of man-made changes of foods (for example, the appearance of "food analogues") and of rapidly changing nutritional habits on the trace element exposure of humans, and their relation to the natural environmental background, are poorly known and need to be examined.

Perhaps the most serious obstacle to future progress is our lack of knowledge of the optimal requirements of humans for individual trace elements. Whereas such requirements are reasonably well known for experimental animals, human needs can only be grossly extrapolated from the animal experiments. Even for nutrients like magnesium, whose function and essentiality have been long known, there is still considerable controversy as to the intake required for maintenance of optimal human health. Because of these gaps in our knowledge, it is impossible at the present time to establish cause–effect relationships between the geochemical environment and human health and disease. The great need for proving such relationships is evident, in view of the many diseases without known etiology and with proved or suspected geographical distribution patterns. The considerable amount of knowledge accumulated in the fields of rock, soil, and life sciences, and discussed in these sessions, strongly suggests that the establishment of cause–effect relationships is feasible if future intensive research can provide the links now missing.

GEOCHEMICAL ENVIRONMENT AND MAN

While the establishment of cause–effect relationships between the geochemical environment and disease is the ultimate scientific goal, this approach requires considerable effort and time and cannot be expected to produce end results within the immediate future. It is possible, however, to obtain useful preliminary information suggestive of environmental influences on health and disease processes. For example, outstanding features of the geochemical environment can be mathematically correlated with the incidence of certain diseases.

This has been attempted, and is discussed and proposed in this report for isolated population groups living in an unusual geochemical environment, such as certain Indian tribes in the southwestern United States (Chapter XII). The disease patterns in these groups and their deviations from the patterns of the general population are well known (for example, the very high incidence of diabetes, obesity, and cholelithiasis in the Pima Indians and the much lower occurrence of cardiovascular disease in this group than in the general population). Many of the geochemical characteristics of the area inhabited by these tribes are also well known, but a correlation between environment and disease has not yet been made. Correlations can be attempted on a wider geographic basis, as in the Missouri study (Chapter XIII), or by carefully coordinated studies on a worldwide basis, as in the World Health Organization program on trace elements in cardiovascular diseases (Chapter XIV). Yet, it must be realized in the interpretation of both today's preliminary results and future more complete data, that even the most perfect correlation between two variables does not establish a cause–effect relationship. However, a close correlation can serve to alert the researcher to a more thorough study of some environmental factors that may appear more promising than others that do not correlate with disease at all.

A session on the presentation of data in the form of

maps and the compatibility of different mapping systems served as an introduction to the focus of the third Workshop; it delineated an approach that would greatly facilitate the interpretation and correlation of geochemical and health data (Chapter XV). Accordingly, the geographic distribution of disease and environmental factors, depicted on maps, has since been examined in greater detail at the Workshop held in 1974 at Captiva Island, Florida.

As was true for the first Workshop at Asilomar, almost all sessions pointed out the existing deficiencies in trace element analyses. These deficiencies are not important where high concentrations are measured, or when pronounced over- or underexposures are of concern. They become, however, limiting to the progress of research in which very low concentrations or small deviations from a norm have to be accurately analyzed. This is particularly true for the "new" trace elements. For example, an assessment of human vanadium intake is impossible as long as different methods produce analytical results for one food that differ by three orders of magnitude. The trace element program of the World Health Organization, and particularly the coordination of the analytical aspects of this program by the International Atomic Energy Agency, has done much to alert investigators to the many improvements of analytical procedures required if a meaningful comparison of samples coming from different environments is to be made. These considerations and the conclusions reached by the Asilomar Workshop led to the concept of a National Environmental Specimen Index System (Chapter XVI). Such a system, designed and operated according to strict criteria, would not only serve as a guide to much-needed reference materials for analytical chemists, but may also become an invaluable index to archive specimens to help define the influence of human activities on the natural environmental background of trace elements.

WALTER MERTZ
Workshop Chairman

PART ONE

Trace Elements Related to Health and Disease

Eight chapters of *Geochemistry and the Environment, Volume I,* reported on the relation of the following trace elements to health and disease: Fluorine; Iodine; Chromium; Lithium; Cadmium, Zinc, and Lead; Selenium; Tellurium; and Copper and Molybdenum. The other five chapters provided an Overview; Analytical Methods; Sampling, Sample Preparations, and Storage for Analysis; Experimental Design and Epidemiological Considerations; and a Summary of Needs and Priorities.

In the same manner, *Geochemistry and the Environment, Volume II,* is concerned with eight additional trace elements (Chapters II through IX) that have been critically evaluated with respect to our present knowledge of their relation to health and disease, emphasizing *natural* environmental sources. They appear in alphabetical order: Beryllium, Magnesium, Manganese, Nickel, Silicon, Strontium, Tin, and Vanadium.

II

Beryllium

WALLACE R. GRIFFITTS, *Chairman*

William H. Allaway, David H. Groth

A strong contrast is evident between the apparent lack of effect on human health of beryllium in natural materials and the highly toxic effects of many beryllium-containing products of industry. This contrast results from the differences in both the beryllium content and its chemical form in these two classes of materials. Occupational exposure to high levels of beryllium-rich dust is associated not only with respiratory illness but with the development of lung cancer as well. There is also the possibility that beryllium in natural materials may act as a sensitizer to other metals, but this subject has been little studied.

GEOCHEMISTRY

Rocks

Beryllium is widespread in the rocks of the earth's crust and in soils derived from them. In igneous rocks, the beryllium content generally increases with increasing contents of silica and alkalies and with decreasing contents of iron, calcium, and magnesium. Thus, peridotites, with silica contents of less than 50 percent, contain much less than 0.5 ppm of beryllium; basalts and gabbros, with around 60 percent silica, commonly contain several tenths of a part per million, rarely as much as 1 ppm; and granites, with 70–75 percent silica, average about 5 ppm beryllium. Extrusive volcanic rocks generally contain less beryllium than do intrusive rocks with the same silica content (i.e., those that have crystallized at depth). Fi-

nally, in a very general way, the beryllium content of silicic rocks varies directly with the fluorine content—a condition most evident in glassy volcanic rocks.

Within the common igneous rocks, beryllium is dispersed through the common rock-forming minerals, replacing silicon or aluminum in tetrahedral coordination in the crystal structures. In granitic rocks, plagioclase and hornblende contain most of the beryllium, while potassium feldspars and micas contain the rest. In basic rocks, plagioclase, hornblende, and diopside contain more beryllium than associated minerals. Independent beryllium minerals crystallize only rarely, in granitic pegmatites (Wedepohl, 1969).

Beryllium-bearing minerals, unstable at the earth's surface, weather to clays, with loss of alkalies and calcium. These clays inherit nearly all of the beryllium of the original minerals and usually contain a few parts per million—amounts similar to those of the original fresh rock. Such clays and quartz are eroded, sorted, and redeposited, ultimately to become shales and sandstones. Most shales contain 2–5 ppm of beryllium, while most sandstones contain less than 1 ppm. Limestones contain very little beryllium—much less than 1 ppm.

Metamorphism has generally little effect on the beryllium content of rocks. Most schists and gneisses contain a few parts per million of beryllium, inherited from the original sedimentary and igneous rocks. As in igneous rocks, the beryllium is contained largely in plagioclase, hornblende, and diopside.

Soil

Clays and residual minerals contain most of the beryllium of the original rocks; thus, the soils composed of these minerals generally contain a few parts per million of beryllium. About 850 soils collected from localities throughout the conterminous United States were found by Shacklette *et al.* (1971) to contain from less than 1 to 7 ppm beryllium, averaging about 1 ppm. Several major regional variations have been reported, the most prominent of these being where values of 1.5 and 2 ppm predominate in southwestern and north-central Nevada, and in the mountains of central and southwestern Colorado, where values of 2–3 ppm prevail. In rather small areas in which rocks contain unusually high levels of beryllium, the soils may also be relatively high in beryllium. Thus, the granite in parts of the Pike's Peak batholith in Colorado contains 10 ppm, an amount that persists in the soils derived from the granite. Also, soils of the beryllium district in the Lost River Valley, Alaska, contain from less than 1 to 300 ppm of beryllium and average about 60 ppm. Such areas of beryllium-rich soils are small and are located in sparsely settled regions away from important areas of food production.

Water

Very little beryllium is released to groundwater during weathering, because the small amount that escapes prompt capture by growing clay minerals is generally sorbed by the surfaces of mineral grains. Under some circumstances, however, beryllium does remain in solution long enough to migrate at least short distances. For example, in the Red Desert of Wyoming, volcanic ash-bearing sediments released beryllium as the glass shards altered, and the beryllium was adsorbed by ferric hydroxide (limonite) precipitated in the vicinity. Even so, the total amount of beryllium moved was very small; the precipitates are small in amount and contain only a few parts per million of beryllium. Inasmuch as the metal is in solution at least briefly, it presumably could be assimilated by plants and may, therefore, be of more importance than the larger amount of beryllium that is fixed in the rocks. In a study of many water samples from the western United States, beryllium was detected in only three highly acid mine waters. In the eastern United States and in Siberia, surface water contains 0.1–0.9 ppb beryllium (Wedepohl, 1969). Pacific Ocean water contains 2–9 ppb of beryllium, about half of it dissolved and half in suspended particulate matter. Beryllium is so promptly adsorbed from neutral water by mineral and glass surfaces that it can exist in solution only as complexes, of which the most important in nature are probably fluoride, chlorocarbonate, and organic complexes.

INDUSTRIAL PRODUCTION AND USAGE

The amount of beryllium used annually in the United States is less than 300 tons; the total used in the 40-yr history of the industry in this country is about 9,000 tons. Industrial production before 1968 was from the beryllium aluminosilicate, beryl, which contains about 4.8 percent beryllium. Since 1968, a hydrous beryllium silicate, bertrandite, has also become an important source. The beryl ores are processed in one of two ways: They may be finely ground, sintered with sodium fluorosilicate, and the resulting beryllium fluoride (BeF_2) leached with water; or the lump ore can be fused, quenched, annealed, and the beryllium sulfate ($BeSO_4$) leached with sulfuric acid. In either case, the resulting solutions are purified, and beryllium hydroxide is precipitated. This may be converted to beryllium oxide (BeO) for use in refractory or insulating materials, converted to a variety of beryllium compounds, or reduced to metal. The reduction of beryllium oxide to metal can be done in the presence of another metal to produce alloys or alone to produce beryllium metal. Beryllium metal generally is then broken into small particles that are pressed into masses with various desired shapes through powder–metallurgical methods. Alloys, beryllium metal, and beryllium oxide account for all but a small amount of the beryllium used.

Although large amounts of manufacturing plant scrap are recycled, little beryllium is reclaimed from the ultimate products, because of the small size and, commonly, low beryllium content of most products. Hair springs for watches or instruments, base insulators for transistors, and spring clips are more typical products than the more spectacular massive heat shields for space capsules and parts for nuclear reactors.

The small items are usually either discarded with other solid wastes or salvaged for the copper in the alloy. This may represent by far the most widespread accidental addition of beryllium to the human environment.

PLANTS

The ash of many coals has long been known to contain beryllium, in a few places in amounts of 100 ppm. The beryllium is mainly in the organic matter of the coal, which strongly indicates that it was a constituent of the precursor plants.

Modern plants have also been found to contain beryllium. Hickory (various species of genus *Carya*) is the best beryllium accumulator found so far, containing as much as 1 ppm (dry weight) of beryllium (30 ppm in plant ash). Dogwood and other broad-leaved trees and shrubs contain more than conifers, which seldom contain 0.1 ppm (less than 1 ppm in the ash). Most studies have shown that leaves contain more beryllium than either twigs or fruit, but some desert shrubs in the Southwest contain more beryllium in their twigs than in their leaves. Romney and Childress (1965) showed that the roots of plants grown hydroponically contain more beryllium than the higher parts, but this has not been established for plants grown in soil. Tundra plants on the Seward Peninsula, Alaska, tend to concentrate beryllium from soils that contain no more than 20 ppm, yielding ash that exceeds the soil in metal content; however, when the soil contains more than

50 ppm beryllium, the plant ash contains less beryllium than the soil (Sainsbury *et al.*, 1968).

Work currently in progress shows that lodgepole pine, Englemann spruce, and Douglas fir in Colorado and Idaho contain small amounts of beryllium. Their ash contains somewhat less than 1 ppm, much less than the 5 ppm in the soil in which the trees grow. Beryllium is also detectable in liquids condensed from fluids transpired by these trees (Curtin *et al.*, 1974).

Toxic Effects

Romney and Childress (1965) found that increasing the levels of beryllium in culture solutions from 0 to 16 ppm resulted in a depression of more than 50 percent in the yield of peas, soybeans, alfalfa, and lettuce. The beryllium concentration in the roots was substantially higher than that in the aboveground portions of the plants, and the beryllium in the foliage was at a higher level than that in the fruit.

The addition of beryllium to three different soils, up to amounts equivalent to 16 percent of the cation-exchange capacity, also caused yield depression in plants, but the drop in yields was less marked than in the culture-solution experiments. The concentrations of beryllium in the air-dried tops of plants grown in nutrient solutions containing 16 ppm beryllium ranged from 27 ppm in alfalfa to 75 ppm in peas. The air-dried tops of ladino clover reached beryllium concentrations ranging from 18 to 22 ppm when grown in Aiken soil (a dominantly kaolinitic clay soil series named for a type location in California) treated with beryllium up to 16 percent of its cation-exchange capacity. When similar additions, in terms of cation-exchange capacity, were made to Vina soil (a strongly montmorillonitic and kaolinitic clay soil series also named for a type location in California), ladino clover tops reached 5–7 ppm beryllium. Increasing the levels of beryllium added to culture solutions and to soils tended to depress calcium uptake and increase phosphorus uptake by the plants. It is evident from this work that beryllium in water-soluble forms is toxic to plants and also that the soils tested fix added beryllium in phosphate or other forms that are less available to plants.

The forms of beryllium in plants and their digestibility by animals have not yet been determined. There is no evidence at present that beryllium is moving from soils into food plants in the United States in amounts that are detrimental to plants, animals, or people. However, the increasing industrial use of beryllium poses the possibility that beryllium waste or scrap will find its way to agricultural soils.

EFFECTS ON HEALTH

In the early 1940's, it was observed that several cases of pulmonary granulomatosis (formation of epithelioid cell tubercles) occurred in workers manufacturing fluorescent light tubes (Spencer, 1962). Later, it was found that the beryllium in the fluorescent powder used to coat the inside of the lamps induced the disease. Several cases of acute beryllium poisoning (berylliosis) also developed, some of which resulted in death. These cases exhibited clinical and pathological characteristics similar to those of other types of chemically induced pneumonitis. The chronic forms of the berylliosis usually occurred only after 1 or more years of exposure and frequently not until years after any known exposure to beryllium. This disease pattern is unique to beryllium exposure and continues to be the main feature that puzzles investigators in this field.

In 1948, the U.S. Atomic Energy Commission established guidelines for permissible levels of exposure to beryllium and instituted a Threshold Limit Value (TLV) of 2 μg/m³ of air. This means that over an 8-h working day the concentration of beryllium in the air in the worker's breathing-zone area should not exceed an average of 2 μg/m³. It was found that when no precautions were observed (before 1948) levels as high as 1,000 μg/m³ of air were common. In the years after 1948, some factories were able to succeed in keeping the beryllium levels near the TLV; in others, the levels were closer to 20–40 μg/m³.

Mancuso (1970) reported that workers in the beryllium extraction industry who worked between 1940 and 1948 and who had developed an occupational respiratory illness during employment (presumably acute berylliosis) had a higher incidence of lung cancer than the general population. It can be reasonably inferred from this that workers who received high exposures to beryllium and survived ran a higher risk of developing lung cancer.

The maximum permissible level of beryllium content in air is not likely to be met often in exposure to natural materials, except in extremely dusty places where the few parts per million of beryllium contained in airborne clay and silt might total 2 μg/m³. Beryllium disease is not induced by exposure to material with such low beryllium content.

There is good evidence that chronic berylliosis is an immunologic disease (Resnick *et al.*, 1970; Deodhar *et al.*, 1973). Inhalation sensitization studies in animals need to be done, however, to confirm that observation and to establish doses required for sensitization.

Although inhalation of high concentrations of some compounds of beryllium, such as the sulfate or fluoride, can produce acute beryllium disease, and other compounds, such as the hydroxide, can produce chronic beryllium disease, there is no evidence to indicate any public health hazard from the ingestion of the amounts of beryllium found in food or water.

RECOMMENDATIONS

Research is needed on the following questions:

1. What is the capacity of different soils to convert various forms of added beryllium to insoluble forms that are not available to plants?

2. What other elements enhance or depress the uptake and translocation of beryllium by plants?

3. What are the chemical forms of beryllium in plants,

and to what extent is the beryllium in food-plants assimilated by animals and people?

4. What are the thresholds of toxicity to animals and man of various forms of dietary beryllium?

5. What is the extent of sensitization and activation effects of beryllium, and what are possible protective reactions that may reduce the reactions to other materials?

REFERENCES

Curtin, G. C., H. D. King, and E. L. Mosier. 1974. Movement of elements into the atmosphere from coniferous trees in subalpine forests of Colorado and Idaho. J. Geochem. Explor. 3:245–263.

Deodhar, S. D., B. Barna, and H. S. Van Ordstrand. 1973. A study of the immunologic aspects of chronic berylliosis. Chest 63:309–313.

Mancuso, T. F. 1970. Relation of duration of employment and prior respiratory illness to respiratory cancer among beryllium workers. Environ. Res. 3(3):251–275.

Resnick, H., M. Roche, and W. K. C. Morgan. 1970. Immunoglobulin concentrations in berylliosis. Am. Rev. Resp. Dis. 101:504–510.

Romney, E. M., and J. D. Childress. 1965. Effects of beryllium in plants and soil. Soil Sci. 100(1):210–217.

Sainsbury, C. L., J. C. Hamilton, and C. Huffman, Jr. 1968. Geochemical cycle of selected trace elements in the tin-tungsten–beryllium district, western Seward Peninsula, Alaska—A reconnaissance study. U.S. Geol. Surv. Bull. 1242-F. U.S. Government Printing Office, Washington, D.C. 42 pp.

Shacklette, H. T., J. C. Hamilton, J. G. Boerngen, and J. M. Bowles. 1971. Elemental composition of surficial materials in the conterminous United States. U.S. Geol. Surv. Prof. Pap. 574-D. U.S. Government Printing Office, Washington, D.C. 71 pp.

Spencer, H. 1962. Pathology of the lung. Macmillan Co., New York. 850 pp.

Wedepohl, K. H. [ed.] 1969. Handbook of geochemistry, Vol. II–I, Sect. 4, Beryllium. Springer–Verlag, New York–Heidelberg–Berlin.

III

Magnesium

WARREN E. C. WACKER, *Chairman*

*Jerry K. Aikawa, George K. Davis,
Donald J. Horvath, Willard L. Lindsay*

Magnesium, an essential nutrient for all organisms, is a component of all types of chlorophyll. There is an average of 2 percent magnesium in rocks (Clark, 1924) and 0.3 percent in soils. The amount of magnesium in soil solution is important in governing the magnesium available to plants and may be important to understanding magnesium deficiencies in plants and animals. Magnesium deficiencies in plants vary from region to region and from species to species. In addition, high concentrations of potassium, calcium, and nitrogen in the soil can induce hypomagnesemia in animals.

The magnesium content of healthy plants, especially grasses and forbs, may not be sufficient to sustain normal animal functions, such as lactation (the water supply should not be overlooked as a significant source of magnesium). Thus, there is a disparity between plant growth requirements and animal requirements, especially under conditions of intensive animal production. The relative availability of magnesium is a crucial factor in the development of such deficiency. Although the total magnesium intake may be sufficient, a variety of conditioning factors may limit its availability to animals. The most susceptible animals usually are lactating ruminants grazing on grass. Magnesium-deficiency disease in animals is now found in many parts of the temperate zones, although there are regional differences. Magnesium deficiency is expected to increase as intensive farming increases.

Beyond the recommended allowances of magnesium (300 mg/day for adult women and 350 mg for adult men),

the nutritional value of magnesium supplementation has not been established.

Magnesium deficiency in humans, as in other animals, is conditioned. Populations vulnerable to magnesium deficiency are alcoholics, patients with intestinal malabsorption (especially those with protein–calorie malnutrition), patients fed intravenously with low-magnesium fluids, those with certain endocrine disturbances, individuals performing heavy labor (particularly in hot climates), and patients receiving such drugs as the newer potent diuretics. There is no known regional distribution of human magnesium deficiency, but, as there are areas in which animal magnesium deficiency occurs, it is possible that indigenous human populations also could be subjected to decreased magnesium intake.

Hypermagnesemia occurs in humans largely as a result of severe kidney disease.

GEOCHEMISTRY

Rocks and minerals usually contain a higher percentage of magnesium than do soils, which reflects the loss of magnesium during weathering. For example, Bowen (1966) reports the following magnesium compositions: igneous rocks, 2.33 percent; shales, 1.5 percent; sandstones, 1.1 percent; limestones, 0.3 percent; and dolomites, 20 percent. This contrasts with 0.3 percent for soils. Seawater contains approximately 1,350 ppm, compared to an average of 4.1 ppm for fresh waters.

11

Soil Solubility Relations

During weathering, magnesium may be precipitated in clay minerals, which play a significant role in governing the solubility of magnesium in soil. The Mg^{2+} ion often replaces Al^{3+} and leaves a net negative charge of -1 for each substitution in the mineral lattice. Magnesium montmorillonites reflect this type of substitution. Again, the substitution of Mg^{2+} for Fe^{2+} is common, but here there is no alteration in charge associated with the exchange. A major portion of soil magnesium thus consists of weathered primary minerals and secondary aluminosilicate minerals in which Mg^{2+} is substituted for Al^{3+}.

In soil solution, magnesium ions react with water, forming $MgOH^+$, which does not significantly affect the pH range of soils. The level of soluble magnesium in soils is generally in the range of 10^{-4} to 10^{-3} M (mole, mass in grams equivalent to its formula mass), slightly less than that of Ca^{2+}. These levels may result in a significant removal of both Mg^{2+} and Ca^{2+} under high rainfall conditions.

Mg^{2+} is held as an exchangeable cation by negatively charged silicate minerals and generally comprises from 5 to 30 percent of the total exchange capacity. Thus, exchangeable magnesium constitutes the major reserve of magnesium readily available to plants. Magnesium deficiencies usually occur when exchangeable magnesium drops below 5 percent of the total cation-exchange capacity (Chapman, 1966). This happens mainly in coarse-textured soils in areas of high rainfall. Excessive leaching favors the removal of magnesium from the exchange site and the development of acid soils. Figure 1 illustrates the lower magnesium content of soils at 8 in. depth in the high rainfall areas of the Southeast than in soils at 8 in. depth of the arid West. A familiar agricultural practice to replenish exchangeable magnesium in soil is to add dolomite $[CaMg(CO_3)_2]$ where magnesium comprises less than 5–10 percent of the total exchangeable cation. A guideline often used is that the ratio of magnesium to potassium should be greater than 2 (Tourtelot, 1971).

ANALYTICAL METHODS

Atomic-absorption spectroscopy provides a satisfactory method for measurement of magnesium in biological samples, provided adequate precautions are taken to assure the elimination of various anion and cation interferences. Many values for magnesium in biological materials obtained prior to the mid-1960's should be used with caution.

BIOCHEMISTRY

The size of a cation is important in determining its solubility and ligand binding properties. Mg^{2+} has an ionic radius two-thirds that of Ca^{2+} and Na^+ and half that of K^+. Because of its small size and relatively large charge, Mg^{2+} is the least easily polarized metal ion, but the best polarizer of other molecules in biological systems. In general, cations form stable complexes with anions or ligands of similar "hardness." Ligands such as Mg^{2+} contain highly electronegative electron donors; Mg^{2+} is most stably complexed by phosphate and carboxylate anions or by the lone pair of electrons of nitrogen. Because of its greater polarizing ability, Mg^{2+} has a larger hydration energy than Ca^{2+} and magnesium salts of long-chain organic acids are, for this reason, more soluble than the analogous calcium salts (Mildvan, 1970; Hughes, 1972; Vernon and Wacker, 1977).

Magnesium is the most abundant intracellular divalent cation in both plants and animals. In animal tissues it varies in concentration from 61 to 365 mg/kg (Williams and Wacker, 1967; Wacker and Parisi, 1968). Since the discovery by Erdtman (1927) that magnesium is an activator of mammalian alkaline phosphatase, many other enzymes have been found to be similarly activated, including most of those utilizing ATP (adenosine triphosphate) or catalyzing the transfer of phosphate (Wacker, 1969).

Since ATP is required in the energy processes of many biological systems, including membrane transport, amino acid activation, acetate activation, succinate activation, protein synthesis, nucleic acid synthesis, fat synthesis, coenzyme synthesis, nerve impulse generation and transmission, muscle contraction, and oxidative phosphorylation, the function of magnesium may extend to all such processes.

A divalent cation (presumably magnesium) is needed for most of the reactions leading to the synthesis of protein. An activating enzyme, aminoacyl t-RNA (transfer-RNA) synthetase, specific for each of the amino acids, catalyzes the formation of an amino acid adenylate, a mixed anhydride between the carboxyl of the amino acid and the 5'-phosphate of adenosine monophosphate (AMP) in the presence of magnesium derived from ATP. A specific t-RNA then reacts with the enzyme-bound aminoacyl adenylate, forming an ester between the carboxyl of the amino acid and either the 2'- or 3'-OH of the ribose moiety of the terminal adenosine of t-RNA. The amino acid-charged t-RNA then reacts with the ribosome-messenger RNA (m-RNA) complex, "the polysome," where the peptidyl t-RNA chain is growing. The t-RNA of the previous amino acid to be incorporated is still bound to the polysome, and it is displaced by a nucleophilic attack on the ester bond brought about by the incoming α-amino group of the aminoacyl t-RNA. An additional peptide bond is thereby formed. The growth of the peptide chains from the α-amino to the carboxyl terminal proceeds in this manner. The details of these processes are the subject of reviews by Novelli (1967) and Schweet and Heintz (1966).

Aminoacyl t-RNA synthetases require magnesium (or another divalent cation) for the overall reaction leading to the charging of a specific t-RNA with its amino acid. In all the systems examined, magnesium (or another divalent cation) is required for the formation of the aminoacyl

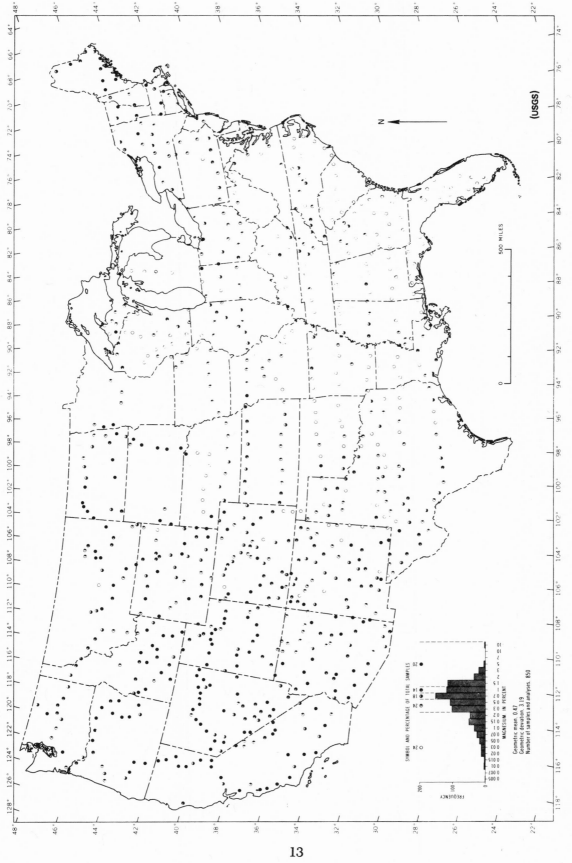

FIGURE 1 Magnesium content of surficial materials (Shacklette *et al.*, 1971).

(USGS)

500 MILES

SYMBOL AND PERCENTAGE OF TOTAL SAMPLES

MAGNESIUM, IN PERCENT

Geometric mean. 0.47
Geometric deviation. 3.19
Number of samples and analyses. 850

FREQUENCY

13

adenylate. The second step, the transfer of the activated amino acid to *t*-RNA, probably does not require a divalent cation (Allende *et al.*, 1964; Lagerkvist and Waldenstrom, 1965). The amount of magnesium required in the overall reaction catalyzed by the aminoacyl synthetases varies according to the specific amino acid being activated (Makman and Cantoni, 1965; Ravel *et al.*, 1965; Novelli, 1967). The magnesium concentration yielding optimal activity may vary from 1 to 30 times that of ATP.

A number of other reactions result in the formation of enzyme-bound acyl–adenylate intermediates analogous to the aminoacyl intermediate. These include enzymes that activate acetate and succinate (Webster, 1965, 1967; Gibson *et al.*, 1967). They, too, require magnesium as an activator.

Magnesium maintains the structure of ribosomal particles in a number of different species (Chao, 1957; Bolkens *et al.*, 1958; Tissieres and Watson, 1958; Tso *et al.*, 1958; Hall and Doty, 1959; Tissières *et al.*, 1959; Petermann, 1960). *E. coli* ribosomes, when suspended in a solution containing 10^{-4} *M* magnesium, exist only as 30-S (sedimentation constant) and 50-S subunits; increasing the magnesium concentration to 10^{-3} *M* results in the formation of 70-S ribosomes by the association of a 30-S and 50-S subunit. Increasing the concentration still further results in the formation of 100-S dimers of 70-S ribosomes (Tissières *et al.*, 1959). Removal of residual magnesium with EDTA from 30-S and 50-S ribosomal particles produces irreversible degradation of the ribosomal subunits. Magnesium is bound to the phosphate groups of ribosomal RNA (Goldberg, 1966; Chai and Carr, 1967).

When *E. coli* is incubated in a medium deficient in magnesium, the ribosomes disappear. After 6 h of starvation, only small numbers of 70-S ribosomes remain (McCarthy, 1962). Similar results have been obtained with magnesium-starved yeast (Beck *et al.*, 1967) and *Aerobacter aerogenes* (Kennell and Kotoulas, 1967; Marchesi and Kennell, 1967). Magnesium is apparently required for the maintenance of ribosomal structure both *in vivo* and *in vitro*.

Magnesium ions are required for the attachment of messenger RNA to ribosomes (Brenner *et al.*, 1961; Gros *et al.*, 1961; Takanami and Okamoto, 1963; Moore, 1966; Das *et al.*, 1967). Magnesium is also necessary in the final step of protein synthesis, the interaction between the ribosome-messenger RNA, and the *t*-RNA molecule charged with a specific amino acid (Gordon and Lipmann, 1967; Salas *et al.*, 1967). The fidelity of the RNA code has been shown to be affected by magnesium concentration; coding ambiguities increase as the magnesium concentration varies in several *in vitro* amino acid incorporating systems (So and Davie, 1964; Szer and Ochoa, 1964; Marshall *et al.*, 1967).

Alkaline phosphatase was the first enzyme shown to be activated by magnesium (Erdtman, 1927). Alkaline phosphatases have a wide substrate range, hydrolyzing organic orthophosphate of almost any type. Mammalian alkaline phosphatases also hydrolyze both inorganic and organic pyrophosphates (Cox *et al.*, 1967; Eaton and Moss, 1967a, 1967b, 1967c; Fernley and Walker, 1967). Unlike the orthophosphatase, activity in the pyrophosphatase is inhibited by magnesium at the optimal pH range of the orthophosphatase reaction (Eaton and Moss, 1967a, 1967b; Fernley and Walker, 1967).

Membrane-bound ATPases (adenosine triphosphatases) are considered to play decisive roles in several major biological systems, including mitochondrial oxidative phosphorylation, photosynthesis, active ion transport, and muscle contraction (Pulman and Schatz, 1967; Albers, 1967; Gibbs, 1967). While ATPases are usually present in cells in bound form, several have been isolated and purified in soluble form and all have an *in vitro* requirement for magnesium ions. A soluble ATPase (F_1) was first isolated from mitochondria by Pulman and coworkers (1960). This enzyme, F_1, when added to submitochondrial particles restores coupling of phosphorylation to electron transport (Penefsky *et al.*, 1960). A similar enzyme, isolated from bovine heart mitochondria (Selwyn, 1967), has an absolute requirement for a divalent cation, which can be satisfied by magnesium, cobalt, zinc, iron, manganese, cadmium, nickel, and calcium. However, the highest activity is obtained with magnesium. A phosphatase closely related to F_1, the chloroplast coupling factor (CF_1), catalyzes photphosphorylation and a magnesium-activated ATPase (Nelson *et al.*, 1972). Several short reviews of the various ATPases have recently been published. (Guidotti, 1976; Postma and Van Dam, 1976).

A similar magnesium-dependent ATPase has been obtained in soluble form from yeast mitochondria (Schatz *et al.*, 1967). An ATPase that requires magnesium for activity and binding has been isolated from the membranes of *Streptococcus fecalis* (Abrams and Baren, 1967). Magnesium ions also serve to bind the enzyme to the bacterial membrane. Magnesium does not increase the number of binding sites on the membrane, but does increase the strength of attachment of the enzyme to the membrane. Complete binding in the absence of magnesium can only be obtained with large excesses of enzyme. However, when magnesium is present, binding occurs at a low enzyme concentration. It may be that magnesium forms an ionic linkage between the acidic enzyme and the polyanionic lipid, present in the bacterial membrane (Abrams and Baren, 1968).

Another magnesium-activated ATPase is associated with cell membranes in a variety of tissues; it can also be activated by sodium and potassium. It has been suggested that active transport of these alkali metals is mediated by ATPase (Albers, 1967; Marchesi and Palade, 1967).

Enolase from yeast and mammals catalyzes the dehydration of phosphoglyceric acid to phosphoenolpyruvate. Yeast-derived enolase was shown by Warburg and Christian (1941) to require magnesium for catalysis activity; enolase from muscle has since been found to have a similar requirement. Early studies by equilibrium dialysis suggested that the active form of the enzyme was a metalloprotein complex (Malstrom, 1956). However, kinetic studies demonstrated that the Michaelis constant (K_m) for magnesium was identical for both the forward and backward reactions of enolase. Because the binding constants of magnesium for phosphoglyceric acid and phosphoenolpyruvate differ, the K_m values would be expected to differ if a magnesium–protein complex were the true substrate involved in this reaction. Identical K_m values would indicate that the enzyme–magnesium complex is the active catalytic moiety (Wold and Ballou, 1957a, 1957b). Cohn and her collaborators, however, have demonstrated that an enzyme–metal complex (presumably magnesium) is, indeed, the active species (Cohn and Leigh, 1962; Cohn, 1963).

These investigators have also carried out a series of studies, using nuclear magnetic resonance (NMR) techniques, which have provided fundamental understanding of the role of divalent metal ions in enzymatic reactions that utilize ATP.

The normal activator of creatine kinase and pyruvate kinase, based on the intracellular abundance of divalent cations, is probably magnesium (see Cohn, 1963). However, manganese also activates both, and the use of manganese in studies of metal ion activation offers a distinct advantage because of its paramagnetic character. As a result of developments in magnetic resonance techniques, it is possible to utilize these paramagnetic properties of manganese to study its interaction with these two enzymes, their substrates (creatine phosphate and phosphoenolpyruvate), and ATP (a nucleotide). Such studies have demonstrated that, in some instances, the metal–nucleotide complex is the true substrate of the reaction, whereas in others the metal–enzyme complex is the active species. Examples of each type of complex (i.e., involving creatine kinase and pyruvate kinase) are considered. In these experiments manganese is assumed to act like magnesium and is used because of its magnetic properties. For the details of the procedure see Cohn (1963).

The metal–nucleotide is the active species of creatine kinase, as indicated by kinetic studies and confirmed by NMR, because the binding of manganese to either creatine kinase or ATP alone enhances the measured nuclear magnetic resonance relaxation rate only slightly. However, when the enzyme, ATP, and manganese are all present, a greater than eightfold enhancement occurs. Thus, it appears that a ternary complex between enzyme, ATP, and manganese has formed.

Electron paramagnetic resonance (EPR) has been used in combination with NMR to demonstrate that manganese in the ADP (adenosine diphosphate)–manganese complex does not serve as a bridge atom between the nucleotide and the creatine kinase enzyme. The line widths of the EPR spectrum of Mn^{2+}, which are a measure of the relaxation time of the electron spin, are altered on addition of ATP, but do not change on addition of creatine kinase instead. Addition of the enzyme to Mn–ADP causes a change in the relative rotational motion of water and manganese in the complex, but causes no change in the type of motion that determines the electron spin relaxation rate. Manganese must, therefore, be in the same

relation to ADP in both the binary and ternary complexes (Cohn, 1963).

In a further study of the interaction of Mn–ADP and Mn–ATP with creatine kinase, the data were compared to those obtained by kinetic measurements (O'Sullivan and Cohn, 1966). The dissociation constants of creatine kinase–Mn–ADP and creatine kinase–Mn–ATP were found to agree with those derived from kinetic studies of manganese activation, and with those for magnesium previously obtained by kinetic and thermodynamic methods (Kuby *et al.*, 1962; Morrison and O'Sullivan, 1965). Thus, manganese and magnesium are the same kinetically.

In pyruvate kinase, unlike creatine kinase, proton-relaxation-rate studies demonstrate that manganese binds directly to the enzyme (Mildvan and Cohn, 1965). In contrast to creatine kinase, the addition of manganese to pyruvate kinase alone results in a marked enhancement of the proton relaxation rate. Comparison of the activator constant determined by kinetics, with the dissociation of manganese from the enzyme as determined by direct binding studies, demonstrates that pyruvate kinase functions as a metal–enzyme complex.

Calcium is a competitive inhibitor of pyruvate kinase. Direct binding studies using NMR have shown that calcium competes with manganese for magnesium-activated enzymes that function as metal–enzyme complexes, but, conversely, calcium is often an activator of systems that use a metal–nucleotide as substrate. Kinetic studies on β-methyl aspartase indicate that inhibition or activation by divalent metals is a function of the ionic radius of the particular metal (Bright, 1967).

Pyruvate kinase has a dual enzymatic specificity, being able to function also as a fluorokinase, converting fluoride ion in the presence of ATP to fluorophosphate (Mildvan *et al.*, 1967). Taking advantage of the magnetic resonance of fluorine, a complex has been demonstrated between fluorophosphate and the enzyme in which manganese serves as a bridge. Because the distance of the fluorine or fluorophosphate from manganese can be estimated by this technique, these studies demonstrate that fluorophosphate is bound to manganese through an oxygen bond and not through a direct fluoromanganese bond. Fluorophosphate is a competitive inhibitor of phosphoenolpyruvate; therefore, by analogy, the same type of binding by phosphoenolpyruvate to the enzyme can be assumed in accord with the previous data of Mildvan and Cohn (1966). Moreover, both phosphoenolpyruvate and fluorophosphate reduce the enhanced proton relaxation rate of water to the same extent, again suggesting a similar structure.

While progress is being made toward understanding the role of magnesium in isolated enzymatic systems, the actual intracellular function of magnesium in physiological terms can still only be judged by extrapolation. The function of magnesium as an activator of intracellular enzymes, for example, provides a potential mechanism through which control of intracellular processes could be mediated (Bygrave, 1967; Kerson *et al.*, 1967; Williams and Wacker, 1967; Wacker and Williams, 1968). Recent experiments by Rubin (1975) lend support to the idea that the concentration of free magnesium ions plays a major role in "coordinate control" of cellular activity. These studies, using chick embryo fibroblasts, show that DNA synthesis and a number of other reactions are activated by magnesium ions. All of the reactions where magnesium appears to function in this way are rate-limiting steps and are transphosphorylation reactions.

Chlorophyll, because of its role in photosynthesis, is undoubtedly the most important biological compound containing magnesium. A clear-cut chemical participation of chlorophyll in photosynthesis has yet to be defined. Although chlorophyll is known to serve as the primary absorbent of the radiant energy of sunlight, the mechanism of energy transfer from the excited chlorophyll molecule remains a matter of speculation and controversy. One proposal suggests that each absorbed quantum of light may split a C—O or an O—H bond; another theory postulates that the energy is conveyed to a reactive species.

MAGNESIUM IN HUMANS AND ANIMALS

Intake

The average adult American ingests 240–480 mg of magnesium daily. An intake of about 200 mg may meet daily nutritive requirements for an average adult, provided that the individual remains in positive magnesium balance. The recommended allowances are 300 mg/day for women and 350 mg for men; for a child, the requirement is 150 mg (Coussons, 1969). Greater importance of magnesium in childhood is suggested by the relative ease with which deficiency states are produced experimentally in young animals as compared with adult animals (Coussons, 1969). Schroeder et al. (1969) called attention to the theoretical relation of dietary magnesium deficiency to serious chronic diseases, including atherosclerosis.

Some common foods, ranked in order of decreasing mean concentrations (wet weight) of magnesium, are: nuts, 1,968 mg/kg; cereals, 802; seafoods, 352; meats, 267; legumes, 243; vegetables, 170; dairy products, 158; fruits, 73; refined sugars, 61; and fats, 7. Two major sources of caloric energy, fats and refined sugars, are virtually devoid of magnesium (Schroeder et al., 1969).

Absorption

In normal individuals on regular diets, the average daily absorption of magnesium from the gastrointestinal tract is 1.70 mg/kg (or 153 mg for a 90-kg man), an amount approximately 40 percent of the size of the extracellular pool. Although no single factor appears to play a dominant role in absorption as does vitamin D in calcium absorption, the absorption of magnesium is influenced by the load presented to the intestinal mucosa (Aikawa, 1959; Graham et al., 1960). Absorption begins within an hour of ingestion and continues at a steady rate for about 2–8 h.

Absorption from the small intestine is fairly uniform, but little or no magnesium is absorbed from the large bowel (Graham et al., 1960).

Evidence from a variety of animals suggests that the small intestine is the main site of magnesium absorption, but that the pattern of absorption varies with the species studied (Graham et al., 1960; Field, 1961). Absorption from the large intestine is negligible in the rabbit (Aikawa, 1959). In male albino rats, more than 79 percent of the total absorption of ^{28}Mg takes place in the colon, and secretion of endogenous magnesium occurs predominantly in the proximal gut (Chutkow, 1964). Both magnesium and calcium are bound to phosphate and to nonphosphate binding material of an unknown nature in the ileal contents of ruminating calves (Smith and McAllan, 1966) and, hence, are rendered nonultrafiltrable.

There appears to be an interrelation between the absorption of magnesium and of calcium in the proximal part of the small intestine in the rat (Alcock and MacIntyre, 1962). The suggestion has been made that there is a common mechanism for transporting calcium and magnesium across the intestinal wall (MacIntyre, 1960; Hendrix et al., 1963).

The concentration of *ionic* magnesium in the intestinal contents appears to be the main factor controlling the amount absorbed in a given period of time (Smith and McAllan, 1966).

CONSERVATION AND EXCRETION

Mendel and Benedict (1909), in a review of the early literature, showed quite clearly that rapid renal excretion of magnesium followed the subcutaneous injection of various magnesium salts, whereas intestinal excretion was minimal. Hirschfelder and Haury (1934) reported that in seven normal adults 40–44 percent of an injected dose of magnesium appeared in the urine within 24 h. Tibbetts and Aub (1937) reported that healthy individuals taking 595–899 mg/day excreted 498–802 mg, of which slightly more than half was in the stools. Smith et al. (1939) studied the excretion of magnesium in dogs after the intravenous administration of $MgSO_4$ and concluded that the magnesium distributed itself throughout the extracellular fluid during the first 3–4 h; during subsequent hours, some of the ion appeared to be segregated from the extracellular fluid and not excreted.

A tracer dose of ^{28}Mg was administered orally to 26 humans; fecal excretion within 120 h accounted for 60–88 percent of the administered dose (Aikawa et al., 1958). The concentration of radioactivity in plasma peaked at 4 h. When ^{28}Mg was injected intravenously into a normal human subject, only 1.8 percent of the radioactivity was recovered in the stool within 72 h (Aikawa et al., 1960). In humans, fecal magnesium thus appears to be primarily magnesium from material that is not absorbed by the body.

The diffusible magnesium in plasma is filtered by the glomeruli and is reabsorbed by the renal tubules, probably by an active process, although the control mechanisms are not known. There is evidence that magnesium may be secreted by the renal tubule (Forster and Berglund, 1956).

Magnesium is also excreted in sweat (Consolazio *et al.*, 1963). Following exposure of men to high temperature for several days, from 10 to 15 percent of the total body output of magnesium is recovered in sweat, and, under extreme conditions, up to 25 percent may be recovered.

Magnesium clearance by the kidney, corrected for protein binding, increases as a linear function of plasma magnesium concentration, but approaches a steady rate at high plasma levels of magnesium. It is primarily the ionic fraction of magnesium in plasma that appears in the glomerular filtrate. The excretion of magnesium may be greater than normal in renal diseases associated with heavy proteinuria. There normally appears to be almost maximal tubular reabsorption of magnesium (Chesley and Tepper, 1958).

Magnesium deficiency in humans with healthy kidneys is rare even with low-magnesium diets. Renal mechanisms are apparently efficient enough to conserve all but about 12 mg/day, and fecal losses are minimal (Barnes *et al.*, 1958). The kidney is the major excretory pathway for magnesium once this is absorbed into the body (Mendel and Benedict, 1909). In subjects on a normal diet, renal excretion amounts to a widely varying proportion of the daily ingested magnesium. The mean daily excretion of magnesium in the urine of 12 normal men on an unrestricted diet was 162 ± 42.5 mg/day (Wacker and Vallee, 1958) and in 8 other normal subjects was found to range between 72.9 and 437.4 mg (Aikawa *et al.*, 1958).

Metabolic-balance studies in 27 subjects showed a positive correlation between the level of dietary intake and the magnesium excretion in both the urine and the feces (Heaton, 1969). When dietary intake of magnesium is increased or decreased, urinary excretion of magnesium increases or decreases with no significant change in the plasma level of magnesium. Magnesium is retained by the kidney, however, in response to a restriction in the dietary intake (Fitzgerald and Fourman, 1956; Barnes *et al.*, 1958). Diurnal variations in the urinary excretion of magnesium, as well as that of calcium, sodium, and creatinine, have been demonstrated (Briscoe and Ragan, 1966) with a reduction in excretion at night. The mechanism of excretion of magnesium by the mammalian kidney is still unclear. It could involve glomerular filtration and partial reabsorption of the filtered material by the renal tubules; or the filtered material could be completely reabsorbed and the excreted magnesium appear by tubular secretion, as is believed to occur with potassium. Tubular secretion of magnesium undoubtedly occurs in the aglomerular fish (Berglund and Forster, 1958), but stop–flow studies with radioactive magnesium in dogs have produced conflicting evidence about secretion of magnesium by the tubules (Ginn *et al.*, 1959; Murdaugh and Robinson, 1960). In rabbits, the renal excretion of magnesium appears to be essentially glomerular (Raynaud, 1962).

The amount of magnesium filtered at the glomerulus in an adult human is about 117 mg/h, assuming a glomerular filtration rate of 130 ml/min, a total plasma magnesium concentration of 19.4 mg/l, and an ultrafiltrable fraction comprising 75 percent of the total. The mean rate of magnesium excretion in the urine (about 4 mg/h), therefore, represents only 3.5 percent of the filtered load. The whole range of excretion observed under physiologic conditions in humans can be explained if the tubular reabsorption of magnesium varies between 91 and 99 percent of the amount filtered at the glomerulus. In rats (Averill and Heaton, 1966), sheep (Wilson, 1960), and cattle (Storry and Rook, 1962), there is evidence for the existence of a renal threshold for excretion of magnesium at a value close to the lower limit of the normal blood level. There is reduction in net tubular reabsorption of magnesium above a total serum magnesium concentration of 14.6–17.0 mg/l; this could be due to either a decrease in the maximum capacity for tubular reabsorption or an increase in tubular secretion of magnesium.

The possibility of secretion of magnesium by the renal tubules in humans has been investigated under conditions of magnesium loading (Heaton, 1969). At serum concentrations above 75.3 mg/l, the amount excreted was more than twice the filtered load, thus demonstrating tubular secretion of magnesium beyond any likely experimental error. In other species, the data indicating that magnesium is actively secreted by the tubule are equivocal or negative (Massry and Coburn, 1973).

Until recently, all the available evidence observed in rats has been consistent with a mechanism for magnesium excretion that involves reabsorption of the filtered material, with the excreted magnesium derived chiefly by tubular secretion. This secretion only appears to start when the magnesium concentration in serum exceeds a critical value close to the lower limit of the normal range. However, studies with the standard stop–flow techniques to measure kidney functions *in vitro* did not show magnesium secretion in acutely magnesium-loaded rats undergoing mannitol or sulfate diuresis (Alfredson and Walser, 1970).

In dogs (Massry *et al.*, 1969), magnesium excretion, like sodium and calcium excretion, is determined by filtration and reabsorption alone, without evidence for tubular secretion. There is a maximal tubular reabsorptive capacity for magnesium of approximately 139.7 mg/min/kg body weight. The parathyroid hormone may directly increase tubular reabsorption of magnesium.

METABOLIC STUDIES

Tracer Studies

The introduction in 1957 of the radioactive isotope of magnesium, ^{28}Mg, for clinical studies made possible determination of the "exchangeable" magnesium pool in human subjects. When nine normal subjects were given intravenous infusions of 145.8–364.5 mg of magnesium tagged with ^{28}Mg, the material was rapidly cleared from the extracellular fluid (Aikawa *et al.*, 1960). After about 18 h, the specific activities in plasma and urine showed only a slight gradual increase, suggesting that the infused material had equilibrated with the stable magnesium in a somewhat labile pool and that further exchange was occurring very slowly in an even less labile pool. The concentration of radioactivity in plasma and urine was too low to follow after 38 h. The size of the labile pool in normal subjects ranged between 1,640 and 4,824 mg (32–64 mg/kg of body weight). Since the body content of magnesium is estimated to be as much as 364.5 mg/kg, only about 15 percent or less of the total body content of magnesium is measured by the ^{28}Mg exchange technique.

The results of the external survey and the tissue analyses suggest that the labile pool of magnesium is contained primarily in connective tissue, skin, and the soft tissues of the abdominal cavity (such as the liver and intestine) and that the magnesium in bone, muscle, and red cells exchanges very slowly.

In another study, Silver *et al.* (1960) followed the turnover of magnesium for periods up to 90 h after intravenous injections of ^{28}Mg. Even at 90 h, only one-third of the body's magnesium had reached equilibrium with the isotope. Graphic analysis of urinary ^{28}Mg curves in terms of exponential components yielded a slow component with a half-time of 13–38 h, which accounted for 10–15 percent of the injected dose, and two more rapid components with half-times of 1 h and 3 h, respectively, which accounted for 15–25 percent of the injected dose. The large fraction of magnesium remaining—about 25–50 percent of the body's total—had a turnover rate of less than 2 percent per day.

The reactivity of the skeleton, as measured by isotopic exchange, declines with age (Breibart *et al.*, 1960). The exchange of ^{28}Mg, expressed as bone/serum specific activity, is more rapid in younger animals than in older ones. ^{28}Mg accumulates in the bones of young rats about twice as fast as in the bones of adult rats (Lengemann, 1959), and the exchange of ^{28}Mg in cortical bone occurs much more rapidly in young rats than in old ones. The stable magnesium content of bone increases with age and varies inversely with the water content of bone. ^{28}Mg studies in lambs indicate that the magnesium reserve in bone is mobilized during dietary magnesium deficiency (McAleese *et al.*, 1961).

^{28}Mg Compartmental Analysis

Avioli and Berman (1966) used a combination of metabolic balance and ^{28}Mg turnover techniques to develop a mathematical model for magnesium metabolism in humans.

Multicompartmental analysis indicates that in humans there are at least three rapidly exchangeable magnesium pools with differing rates of turnover: Pools 1 and 2 have a relatively fast turnover and together approximate extracellular fluid in distribution, and pool 3, an intracellular pool containing over 80 percent of the exchanging mag-

nesium, has a turnover rate one-half that of the most rapid pool. A fourth pool, which is not rapidly exchangeable, probably accounts for most of the whole-body magnesium (Wallach *et al.*, 1966). Only 15 percent of whole-body magnesium, or an average of 43.01 mg/kg body weight, is accounted for by relatively rapid exchange processes (Avioli and Berman, 1966).

Parathyroid Hormone

There is considerable evidence for the hypothesis that the parathyroid hormone may help to control the concentration of plasma magnesium through a feedback mechanism (MacIntyre *et al.*, 1963; Heaton, 1965; Gill *et al.*, 1967).

Symptomatic magnesium deficiency in rats is accompanied by hypercalcemia and hypophosphatemia, provided the parathyroid glands are intact. The concentration of ionic calcium in plasma also is elevated. In the absence of parathyroid glands, magnesium-deficient rats do not develop hypercalcemia or hypophosphatemia. Moreover, these latter rats develop a lower concentration of ionized calcium in plasma than that observed in parathyroidectomized rats on a normal diet (Gitelman *et al.*, 1968a).

If parathyroid regulation is influenced by the concentration of magnesium in plasma, in a manner similar to calcium, hypermagnesemia should inhibit parathyroid gland activity. This hypothesis was tested in intact and in previously parathyroidectomized rats, which were nephrectomized to eliminate the urinary excretion of calcium as a variable. Isotonic magnesium chloride was administered subcutaneously to the experimental animals, and normal saline solution was administered to the controls. A significant decrease in the concentration of ionic calcium in the plasma was observed in the magnesium-treated animals with intact parathyroid glands. In contrast, magnesium-treated parathyroidectomized animals failed to develop a significant change in the concentration of ionic calcium in comparison with saline-treated parathyroidectomized controls, suggesting that hypermagnesemia may inhibit parathyroid gland activity. Similar results were found in goats and in a sheep by perfusion of the isolated parathyroid gland with whole blood of varying magnesium concentration (Buckle *et al.*, 1968). Sherwood *et al.* (1970), using an organ culture system of normal bovine parathyroid tissue, provided direct evidence that the release of parathyroid hormone is inversely proportional to both the calcium and the magnesium ion concentrations. All of these results are consistent with the hypothesis that the parathyroid regulatory mechanism involved in calcium homeostasis is modified by alterations in the concentration of plasma magnesium (Gitelman *et al.*, 1968b).

Relation of Extracellular Magnesium to Bone

Magnesium deficiency in the rat lowers the magnesium concentration in bone (Martindale and Heaton, 1965).

The observation of a direct relation between the magnesium concentrations in the plasma and in the femur of magnesium-deficient rats and calves supports the view that the skeleton provides the magnesium reserve in the body and suggests that an equilibrium exists between the magnesium of the plasma and the bone. This equilibrium is apparently independent of enzymatic activity and must, therefore, be physical–chemical in nature. The dependency of this equilibrium on the concentration of magnesium in both the medium and the bone suggests that the relation between bone and extracellular fluid magnesium is analogous to the ionization of a poorly dissociated salt, with the magnesium in bone corresponding to the undissociated salt.

DEFICIENCY IN ANIMALS

"Grass tetany" or hypomagnesemia in ruminant animals is an example of the influence of the geochemical environment on the health and disease of animals. In addition to environmental deficiencies, agricultural practices as well as the type of animal management system used can have a bearing on this disease (Amos *et al.*, 1975).

The nervous signs, clonic muscle responses, and death resulting from acute magnesium deficiency were described first in grazing, lactating cattle in the Netherlands in this century. The syndrome was called "grass staggers" and has since been recognized in cattle or sheep in most temperate areas. A related syndrome called "transit tetany" is occasionally seen in horses (Underwood, 1966). Although the muscle spasms are a pathognomonic sign, tetany is not primarily a disorder of muscle (Todd and Horvath, 1970). One of the more extensive recent reviews of grass tetany is that of Grunes *et al.*(1970).

In terms of animal losses, grass tetany ranks high as a cause of fatalities in cattle and sheep, with fewer losses in goats. Van der Molen (1964) reported that in 1963–1964 the fatal cases amounted to 2.7 percent of all dairy cows in the Netherlands. Allcroft and Burns (1968) noted that 1.1 percent of all dairy and beef cattle in Scotland had grass tetany each year; in Great Britain as a whole, 0.5 percent of all dairy cows were affected. Butler (1963) reported that, in the 41 percent of the herds in which grass tetany was observed, the overall frequency of disease was 8.7 percent, and in some herds as high as 10.9 percent mortality occurred among animals that showed clinical signs of grass tetany.

Herd (1966) reported grass tetany in Australian beef and dairy cattle and in ewes, with overall mortality rates of about 1 percent. Grass tetany is also a problem in many parts of the United States. Hjerpe (1964) reported that, in the previous winter in California, between 4,000 and 6,000 head of cattle died of grass tetany. In some herds, 20 percent of the cattle died. In West Virginia, however, Horvath (1959) reported an incidence rate of less than 1 percent. Almost all tetany is found to occur in ruminants that graze on pastures under a temperature range of 5°–15° C (t'Hart, 1960).

Grass tetany almost always occurs on grasses that grow well in cool weather with good moisture conditions (neither too dry nor too wet). In summer or in warmer weather, grass tetany disappears, but it may appear again when cool weather returns.

The most common forages inducing grass tetany are ryegrass (*Lolium perenne*), crested wheatgrass (*Agropyron desertorum* and *Agropyron cristatum*), orchardgrass (*Dactylis glomerata*), soft chess (*Bromus mollis*), mouse barley (*Hordeum leporinum*), wheat (*Triticum aestivum*), rye (*Secale cereale*), and oats (*Avena sativa*).

The disease is most common in early lactation, generally in late winter and early spring, although autumn tetany is known in Britain. It may follow efforts to increase animal production by application of fertilizers, especially those rich in nitrogen and potassium, or by the seeding of new species of grasses (such as crested wheatgrass in the western United States).

It would appear that the minimum level of magnesium in forage required to prevent tetany is 0.20 percent (Van der Molen and Padmos, 1965), with the proviso that high levels of nitrogen and potassium in the forage may increase the need for a higher level.

Thus, the presence of high levels of potassium or nitrogen in forage may be a major environmental factor in grass tetany. Workers in California have been able to induce grass tetany experimentally by administering a solution of potassium chloride and transaconitic acid or citric acid. Under these conditions, magnesium in the blood plasma was reduced in 24 h; tetany did not occur, however, unless both the acid and the potassium were administered (Bohman *et al.*, 1969).

As the level of magnesium in plants is a critical factor in grass tetany, it is interesting that Embleton (1966) suggests that if magnesium is below 10 percent of the cation-exchange capacity of a soil, plants generally respond to magnesium additions, suggesting that the plants may be deficient. This observation, tied to the finding that cool weather apparently results in a lower magnesium uptake by the plants, suggests that environment can influence the occurrence of grass tetany.

In Argentina, in the winter of 1963, an estimated 30,000 head of cattle died of grass tetany in Buenos Aires province, including not only dairy cattle but steers and beef breeding animals as well. In Santa Fe province, the dairy province of Argentina, grass tetany is annually a major hazard. One of the more common practices in Argentina is to provide a source of magnesium in a mineral supplement and to leave a crop of corn in the field as a reserve of dry matter when the animals are turned on to the winter pastures.

Although the distribution of grass tetany seems to correlate with low-magnesium soils (Barta, 1973), tetany is not necessarily a simple deficiency disease. It has been described as both a conditioned deficiency and a metabolic disease. The condition seen in early calving cows (December to March in the United States) fed grass hay containing 0.1 percent, or less, magnesium on a dry-matter basis, appears to be simple deficiency, where the availability of the magnesium may be less than 30 percent (Rook *et al.*, 1958). When grazing commences, the availability of magnesium may drop to less than 10 percent (Kemp *et al.*, 1961), largely because of interferences of the following kinds:

- High levels of NH_4^+ associated with high protein in spring herbages (Head and Rook, 1955).
- Depression of magnesium levels in herbage by high soil potassium, particularly if soil nitrogen is in excess (Smyth *et al.*, 1958; Kemp, 1960).
- Suppression of magnesium availability to the animal by the effects of high forage potassium (Kunkel *et al.*, 1953; Suttle and Field, 1969; House and Van Campen, 1971).
- Suppression of magnesium availability by lipids, which increase with nitrogen fertilization (Kemp *et al.*, 1966).
- Suppression of ionized magnesium in the blood by organic acids such as citric or transaconitic acid in browsed species such as crested wheatgrass (Burau and Stout, 1965; Bohman *et al.*, 1969).
- Endocrine involvement (Care, 1967).

Common preventive measures for grass tetany center on oral supplementation during lactation (Burns and Allcroft, 1967). Significant reductions of grass tetany in lactating ewes (classified as severely hypomagnesemic) have also been achieved by calcining residual clay loam soils in West Virginia with dolomitic limestone (Horvath, 1973, personal communication). This is unexpected in the light of studies in Britain (Todd, 1965) in which responses to top-dressed magnesium were favorable on acid sandy soils but limited on soils having higher cation-exchange capacity.

DEFICIENCY IN HUMANS

For many years, the existence of a pure magnesium deficiency state in humans was doubted. Yet, the existence of the state has been established (Flink *et al.*, 1954; Flink, 1956; Wacker *et al.*, 1962) and is characterized by spasmophilia (Durlach *et al.*, 1967), gross muscular tremor, choreiform movements, ataxia, tetany, predisposition to epileptiform convulsions (Hanna *et al.*, 1960), hallucinations, agitation, confusion, tremulousness, delirium, depression, vertigo, and muscular weakness. It may also show up as a low serum magnesium concentration associated with a normal serum calcium concentration and a normal blood pH, a low-voltage T-wave in the electrocardiogram (Caddell, 1967), a positive Chvostek and Trousseau sign, and prompt relief of the tetany when the serum magnesium concentration is restored to normal (Wacker *et al.*, 1962). Durlach (1967) proposes the presence of other manifestations of clinical magnesium deficiency, such as phlebothrombosis, constitutional thrombasthenia and hemolytic anemia, an allergic or osseous form of the deficiency, and oxalate lithiasis.

Because magnesium homeostasis involves more than vascular space, simple determination of serum magnesium does not always give a true indication of the total body magnesium stores. Indeed, hypomagnesemia can occur when the cellular content of magnesium is normal (L'Estrange and Axford, 1964; Richardson and Welt, 1965), and cellular depletion may exist without a concomitant lowering of serum values (Montgomery, 1960; MacIntyre *et al.*, 1961; Fourman and Morgan, 1962). Nevertheless, measurement of serum magnesium is a quick, simple, and effective approach to the evaluation of magnesium deficiency. If the clinical situation is suggestive and serum values are normal, erythrocyte (red blood cell) content and 24-h urinary excretion should be measured. If the magnesium content of erythrocytes or the urinary excretion of magnesium is normal, magnesium deficiency is very unlikely.

Production of a Pure Magnesium Deficiency in Normal Individuals

It is difficult to achieve a significant magnesium depletion by dietary restriction, because of the exceedingly efficient renal and gastrointestinal mechanism for conservation, although Dunn and Walser (1966) induced deficits approaching 10 percent of the total body magnesium by infusing sodium sulfate and adding calcium supplements to the magnesium-deficient diet. In general, urinary magnesium falls to trivial amounts within 4 to 6 days of magnesium restriction (Fitzgerald and Fourman, 1956; Barnes *et al.*, 1960). Randall *et al.* (1959) reported that severe body depletion of magnesium may result in psychiatric and neuromuscular symptoms.

The best study to date of magnesium deficiency in man is that of Shils (1964, 1969a, 1969b). Seven subjects were placed on a magnesium-deficient diet containing 8.5 mg of magnesium per day. Within 7–10 days, urinary and fecal magnesium decreased markedly, and at the height of the deficiency plasma magnesium fell to a range of 10–30 percent of control values. The red-cell magnesium declined more slowly and to a lesser degree. All male subjects developed hypocalcemia; the one female patient did not. Marked and persistent symptoms developed only in the presence of hypocalcemia. The serum potassium concentration decreased, and, in four of the five subjects in whom the measurement was made, the ^{42}K pool (that part of the organism in which ^{42}K diffuses) was decreased. The serum sodium concentration was not altered significantly. Three of the four subjects with the most severe symptoms also had metabolic alkalosis.

The most common neurologic sign was a positive Trousseau sign. Electromyographic changes, characterized by the development of myopathic potentials, and anorexia, nausea, and vomiting also occurred. When magnesium was added to the experimental diet, all clinical and biochemical abnormalities were corrected.

Studies in young rats show that sensitivity to stimuli-producing convulsions is correlated with a decrease in serum, cerebrospinal fluid (CSF) and brain magnesium concentrations (Chutkow and Grabow, 1972). Administration of magnesium to deficient animals produced a rapid restoration of CSF magnesium while correction of cellular defects requires several hours (Chutkow, 1974).

Clinical Conditions Associated with Depletion of Magnesium

Fasting Prolonged fasting is associated with a continued renal excretion of magnesium (Drenick *et al.*, 1969). After 2 months of fasting with adequate intake of water and vitamins, the deficit in some subjects may amount to 20 percent of the total body content of magnesium. The excess acid load presented to the kidney and the absence of carbohydrate intake might contribute to the persistent loss of magnesium. The greater the acidosis, the greater the excretion of magnesium. The ingestion of glucose decreases the urinary loss of magnesium.

Surgery Surgery is followed by a negative magnesium balance of approximately 3 days' duration, similar to changes observed after dietary restriction (Heaton, 1964). However, the loss usually does not result in symptomatic magnesium deficiency (Macbeth and Mabbott, 1964; Monsaingeon *et al.*, 1966; King *et al.*, 1973).

Gastrointestinal Disease Magnesium deficiency has been identified most often in patients with diseases of the gastrointestinal tract. It occurs frequently in patients with nontropical sprue (Balint and Hirschowitz, 1961; Goldman *et al.*, 1962; Fishman, 1965), other small-bowel disorders associated with steatorrhea (Booth *et al.*, 1963), and also as a result of abdominal irradiation (Vallee *et al.*, 1960) or after resection of a large portion of the small bowel (Fletcher *et al.*, 1960; Wacker *et al.*, 1962; Peterson, 1963; Opie *et al.*, 1964). It is relatively common in the postoperative period in association with prolonged nasogastric suction combined with the administration of large amounts of magnesium-free parenteral fluids (Flink, 1956; Flink *et al.*, 1957; Wacker and Vallee, 1958; Vallee *et al.*, 1960; Kellaway and Ewen, 1962; Thoren, 1962; Burhol *et al.*, 1966; Matko, 1966) or in association with intestinal and biliary fistulas (Fishman, 1965). Hypomagnesemia has also been reported in acute pancreatitis (Edmondson *et al.*, 1952) and alcoholic cirrhosis (Stutzman and Amatuzio, 1952; Flink *et al.*, 1955; Sullivan *et al.*, 1963). Celiac disease may result in striking magnesium loss in the stool, as high as four times the dietary intake (Goldman *et al.*, 1962). This remarkable loss can be reversed by a gluten-free diet.

Steatorrhea itself may be an important cause of magnesium loss, presumably owing to excretion of large amounts of magnesium soaps following excess fat intake.

Excessive fecal loss of magnesium may also occur following large amounts of oral calcium supplementation. In animals, magnesium absorption from the intestine varies inversely with calcium intake, possibly because of competition for a common transport system (Alcock and MacIntyre, 1962; Care and Van't Klooster, 1965). Data

from humans with steatorrhea are in accord with animal observations (Hanna *et al.*, 1960; Booth *et al.*, 1963). In one case during recovery from prolonged profuse diarrhea, an infant had convulsions associated with a low plasma magnesium concentration. Larger amounts of cow's milk ingested during recovery may well have activated asymptomatic, diarrhea-related magnesium depletion (Savage and McAdam, 1967).

In recent years, magnesium deficiency has been shown to play a prominent part in the so-called protein–calorie malnutrition syndromes. The magnesium content of muscle is decreased (Metcoff *et al.*, 1960; Montgomery, 1960; Caddell and Goddard, 1967), whereas serum magnesium concentrations may or may not be decreased even in patients with reduced tissue concentrations (Montgomery, 1960; Linder *et al.*, 1963) in whom markedly lowered urinary excretion is almost universally present. On institution of magnesium therapy, positive balance of magnesium, as well as calcium, phosphorus, and nitrogen, ensues (Montgomery, 1961; Linder *et al.*, 1963).

In a group of extremely ill Nigerian children with protein–calorie malnutrition, the symptoms of neuromuscular irritability (including tremors, tetany, and convulsions) and electrocardiographic changes (a short PR interval and flattened to inverted lateral precordial T-waves) suggested magnesium depletion. Parenteral administration of magnesium sulfate resulted in a statistically significant response (Caddell, 1967). Although the population at risk had a relatively poor magnesium intake, the clinical picture appeared only after prolonged diarrhea or severe vomiting, reflecting a "conditioned deficiency."

Magnesium deficiency has been observed in patients with chronic bowel disease being treated by intravenous hyperalimentation (Grand and Colodny, 1972) and in acute bowel disease including asiatic cholera (Kobayaski *et al.*, 1972; Lim and Jacob, 1972). It also occurs as a serious complication of jejunoileal intestinal bypass surgery (Swenson *et al.*, 1974).

Acute Alcoholism Patients exhibiting alcohol withdrawal signs and symptoms (Mendelson *et al.*, 1969) have low serum and cerebrospinal fluid levels of magnesium, low exchangeable magnesium levels (Martin and Bauer, 1962; Nielsen, 1963; Mendelson *et al.*, 1965), a lowered muscle content of magnesium (Jones *et al.*, 1969), and conservation of magnesium following intravenous loading (McCollister *et al.*, 1960). A transient decrease in serum magnesium may occur during the withdrawal state, even though prewithdrawal levels are normal. An ethanol-induced increase of magnesium in the urine occurs only when the blood-alcohol level is rising. It does not persist once the subject has established high blood-alcohol levels. Magnesium in red cells and plasma can be abnormally low (Smith and Hammarsten, 1959). Intracellular fluid levels of magnesium, as reflected in the erythrocyte, correlate better with clinical symptoms of alcoholism than do extracellular fluid levels. Such magnesium depletion in acute alcoholism most likely reflects an inadequate intake of magnesium, but could also reflect increased excretion of magnesium in the urine and feces (McCollister *et al.*, 1963; Dick *et al.*, 1969).

An abrupt and significant fall in serum magnesium levels may occur following cessation of drinking. This fall is associated with a transient decrease in concentration of other serum electrolytes and with respiratory alkalosis (Wolfe and Victor, 1969; Victor, 1973) and coincides with the onset of neuromuscular hyperexcitability that characterizes the withdrawal state (Mendelson *et al.*, 1969). A kinetic analysis of radiomagnesium turnover was performed in a group of alcoholic subjects partly repleted with respect to magnesium. Despite the continued presence of hypomagnesemia, there was little evidence of continued depletion of magnesium in the extracellular space or in the tissue pools (Wallach and Dimich, 1969).

Magnesium in the liver is decreased in cirrhosis (Wilke and Spilemann, 1968) and appears to reflect mainly the substitution of connective tissue of low magnesium content for high-magnesium parenchymal tissue.

Kidney Disease   Decreased serum concentrations of magnesium occur in patients with early renal disease, including glomerulonephritis, hydronephrosis, pyelonephritis, nephrosclerosis, and renal tubular acidosis (Smith and Hammarsten, 1958; Wacker and Vallee, 1958; Hanna, 1961). The reduction is presumably due to defective reabsorptive mechanisms. Hypermagnesemia associated with azotemia was first reported by Salveson and Linder (1923) and has been confirmed repeatedly (Wacker and Vallee, 1958; Robinson *et al.*, 1959; Randall *et al.*, 1964). In general, such hypermagnesemia correlates positively with nitrogen retention. Hypermagnesemia also occurs in acute renal failure (Wacker and Vallee, 1958).

The use of magnesium compounds in uremic patients for their antacid or purgative properties is dangerous and may produce toxic hypermagnesemia (Hoff *et al.*, 1940). There is at present little justification for magnesium medication, unless hypomagnesemia is documented and adequately monitored.

Magnesium Therapy for Renal Stones   In the early 1960's, it was reported that the administration of magnesium oxide by mouth to patients with idiopathic recurrent calcium oxalate stones reduced the frequency of stone formation. Since then, more extensive studies have been carried out on fairly large series of patients with recurrent urolithiasis and have confirmed the effectiveness of magnesium alone or in conjunction with pyridoxine in preventing stone formation. Neither the biochemical alteration leading to stone formation nor the mechanism by which magnesium (and pyridoxine) prevents stone formation is known (Anonymous, 1976). Martens and Harriss (1970) discuss the effects of magnesium on the precipitation kinetics of calcium phosphate (apatite) minerals, which are found in renal stones.

Diuretic Therapy   Depletion of magnesium is likely to occur as a result of diuretic therapy. The use of am-

monium chloride or mercurial diuretics can increase renal excretion of magnesium (Jabir *et al.*, 1957; Martin and Jones, 1961; Smith *et al.*, 1962). Symptomatic hypomagnesemia, induced by mercurial therapy, has been controlled by parenteral magnesium administration (Smith *et al.*, 1962). Increased urinary losses of magnesium have been documented repeatedly with the benzothiadizine diuretics, chlorothiazide, and hydrochlorothiazide (Wener *et al.*, 1959; Hanze, 1960; Wacker, 1961; Glaubitt and Rausch-Stroomann, 1962; Mininni and Zonno, 1962).

These considerations also appear pertinent to the use of digitalis preparations along with these diuretic agents. Although potassium supplements are usually prescribed for patients receiving oral diuretics (particularly if they are simultaneously receiving cardiac glycosides), other mineral supplementation is rarely given. An increased potential for arrhythmias occurs when cardiac glycosides are administered to magnesium-deficient animals (Vitale *et al.*, 1961, 1963).

Hypomagnesemia is prevalent in patients with digitalis toxicity and is associated with a high incidence of diuretic use (Bellar *et al.*, 1971, 1974). Patients with digitalis-induced arrhythmias and hypomagnesemia respond promptly when the magnesium deficiency is corrected.

Excessive Lactation   Magnesium deficiency, manifested as the tetany syndrome, may occur in women with excessive lactation (Greenwald *et al.*, 1963).

Exercise and Heat   A decrease in serum magnesium concentration, resulting from vigorous exercise, has been demonstrated by Bellar *et al.* (1972); the serum magnesium fell significantly from 24.18 ± 0.49 mg/l at the start to 23.3 ± 0.36 mg/l at the end of a 90-min experiment at room temperature. Eight other healthy subjects were exercised at an ambient temperature of 49° C (27° C wet bulb), conditions simulating a dry, desert environment. They showed a decrease in serum magnesium content from 22.72 ± 0.73 mg/l at the start to 21.99 ± 0.85 mg/l at 45 min and 20.90 ± 0.97 mg/l at 90 min.

The magnesium content of sweat collected during exercise ranged from 1.58 to 5.7 mg/l. This increased water loss by sweating may well account for the observed fall in serum magnesium concentrations. The magnesium content of sweat does not decrease during heat acclimatization (Consolazio *et al.*, 1963). In contrast, serum potassium concentration increases during prolonged exercise (Rose *et al.*, 1970).

EXCESS IN HUMANS

The symptoms of hypermagnesemia in man were first noted as a result of pharmacologic studies of the Mg^{2+} ion as a potential anticonvulsant and anesthetic agent (Peck and Meltzer, 1916). Infusions into animals and man led to impairment of neuromuscular transmission (Somjen *et al.*, 1966).

Cardiac conduction is affected at magnesium serum concentrations of 60.8–121.5 mg/l, resulting in increases of the PR wave interval and QRS wave duration and increased height of the T-wave in the electrocardiogram. Deep tendon reflexes are lost when the serum magnesium concentration approaches 121.5 mg/l and respiratory paralysis and general anesthesia may occur near 182.3 mg/l. Extremely high serum concentrations (in excess of 303 mg/l) cause cardiac arrest in diastole (Wacker and Vallee, 1958; Wacker and Parisi, 1968).

RECOMMENDATIONS FOR RESEARCH

1. Additional biological standards for magnesium should be defined by the National Bureau of Standards.

2. The mineral solubility relations that control the availability of magnesium in soils should be identified.

3. By the use of currently accepted analytical methods, the total magnesium content of various animal and human foods should be determined.

4. The relative importance of the various conditioning factors that govern the availability of magnesium to animals and humans should be determined.

5. An estimate should be made for the U.S. population as a whole to evaluate whether the recommended daily requirement of magnesium is being received.

6. A random sample of humans should be studied to determine whether excess magnesium intake (up to 1 g/day) has any beneficial effect.

7. A conference on magnesium metabolism is needed. This conference should include soil scientists, plant nutritionists, animal nutritionists, veterinarians, human nutritionists, and physicians and should serve to update knowledge and promote the transfer of information among these diverse, but necessarily related, disciplines. The symposium could be modeled after the Symposium on Magnesium and Agriculture, held at West Virginia University, in Morgantown, in 1959 (Beeson, 1959).

REFERENCES

Abrams, A., and C. Baren. 1967. The isolation of subunit structure of streptococcal membrane adenosine triphosphatase. Biochemistry 6:225–229.

Abrams, A., and C. Baren. 1968. Reversible attachment of adenosine triphosphatase to streptococcal membranes and the effect of magnesium ions. Biochemistry 7:501–507.

Aikawa, J. K. 1959. Gastrointestinal absorption of Mg^{28} in rabbits. Proc. Soc. Exp. Biol. Med. 100:293–295.

Aikawa, J. K., E. L. Rhoades, and G. S. Gordon. 1958. Urinary and fecal excretion of orally administered Mg^{28}. Proc. Soc. Exp. Biol. Med. 98:29–31.

Aikawa, J. K., G. S. Gordon, and E. L. Rhoades. 1960. Magnesium metabolism in human beings. J. Appl. Physiol. 15:503–507.

Albers, R. W. 1967. Biochemical aspects of active transport. Annu. Rev. Biochem. 36:727–756.

Alcock, N., and I. MacIntyre. 1962. Interrelations of calcium and magnesium absorption. Clin. Sci. 22:185–193.

Alfredson, K. S., and M. Walser. 1970. Magnesium excretion in the rat. [Is Mg secreted by the rat renal tubule?] Nephron 7:241–247.

Allcroft, R., and K. M. Burns 1968. Hypomagnesemia in cattle. N.Z. Vet. J. 16:109–128.

Allende, J. E., C. C. Allende, M. Gatica, and M. Matamala. 1964. Isolation of threonyl adenylate–enzyme complex. Biochem. Biophys. Res. Commun. 16:342–346.

Amos, R. L., G. J. Crissman, R. F. Keefer, and D. J. Horvath. 1975. Serum magnesium levels of ewes grazing orchardgrass topdressed with dolomite or calcite. J. Anim. Sci. 41 (1):198–202.

Anonymous. 1976. Magnesium oxide—Pyridoxine therapy for recurrent urolithiasis. Nutr. Rev. 34:18–20.

Averill, C. M., and F. W. Heaton. 1966. The renal handling of magnesium. Clin. Sci. 31:353–360.

Avioli, L. V., and M. Berman. 1966. Mg^{28} kinetics in man. J. Appl. Physiol. 21:1688–1694.

Balint, J. A., and B. Hirschowitz. 1961. Hypomagnesemia with tetany in non-tropical sprue. N. Engl. J. Med. 265:631–633.

Barnes, B. A., O. Cope, and T. Harrison. 1958. Magnesium conservation in human beings on low magnesium diet. J. Clin. Invest. 37:430–440.

Barnes, B. A., O. Cope, and E. B. Gordon. 1960. Magnesium requirements and deficits: Evaluation of two surgical patients. Am. Surg. 152:518–533.

Barta, A. 1973. Grass tetany in Ohio cattle. Ohio Rep., March–April, pp. 35–36.

Beck, G., G. Aubel-Sadron, and J. P. Ebel. 1967. Ribosomol RNA of magnesium-deficient yeast. Bull. Soc. Chim. Biol. 49(4):349–360. (In French)

Beeson, K. C. 1959. Magnesium in soils—Sources, availability, and zonal distribution. *In* Symposium on Magnesium and Agriculture, West Virginia University, Morgantown. 217 pp.

Bellar, G. A., W. B. Hood, T. W. Smith, W. H. Abelmann, and W. E. C. Wacker. 1971. Effect of magnesium on digitalis intoxication. Proc. 1st Int. Conf. Hum. Magnesium Metab., Vittel, France. Summaries of communications. p. 91.

Bellar, G. A., W. B. Hood, Jr., T. W. Smith, W. H. Ablemann, and W. E. C. Wacker. 1974. Correlation of serum magnesium levels and cardiac digitalis intoxication. Am. J. Cardiol. 33:225–228.

Bellar, G. A., J. T. Maher, I. H. Hartley, D. E. Bass, and W. E. C. Wacker. 1972. Serum magnesium and potassium concentrations during exercise in thermoneutral and hot conditions. The Physiologist 15:84.

Berglund, F., and R. P. Forster. 1958. Renal tubular transport of inorganic divalent ions by the aglomerular marine teleost, *Lophius americanus*. J. Gen. Physiol. 41:429–440.

Bohman, V. R., A. L. Lesperance, G. D. Harding, and D. L. Grunes. 1969. Induction of experimental tetany in cattle. J. Anim. Sci. 29:99–102.

Bolkens, E. T., B. H. Boyer, and R. B. Ritter. 1958. Microsomal particles in protein synthesis. R. B. Roberts. [ed.] Pergamon Press, New York. 168 pp.

Booth, C. C., N. Babouris, S. Hanna, and I. MacIntyre. 1963. Incidence of hypomagnesemia in intestinal malabsorption. Br. Med. J. 2:141–144.

Bowen, H. J. M. 1966. Trace elements in biochemistry. Academic Press. New York. pp. 61–101.

Breibart, S., J. S. Lee, L. A. McCord, and G. Forbes. 1960. Relation of age to radiomagnesium exchange in bone. Proc. Soc. Exp. Biol. Med. 105:361–363.

Brenner, S., F. Jacob, and M. Meselson. 1961. An unstable intermediate carrying information from genes to ribosomes for protein synthesis. Nature 190:576–581.

Bright, H. S. 1967. Divalent metal activation of β methyl aspartase. The importance of ionic radius. Biochemistry 6: 1191–1203.

Briscoe, A. M., and C. Ragan. 1966. Diurnal variations in calcium and magnesium excretion in man. Metabolism 15: 1002–1010.

Buckle, R. M., A. D. Care, C. W. Cooper, and H. J. Gitelman. 1968. The influence of plasma magnesium concentration on parathyroid hormone secretion. J. Endocrinol. 42:529–534.

Burau, R., and P. R. Stout. 1965. Trans-aconitic acid in range grasses in early spring. Science 150:766–767.

Burhol, P. G., J. Myren, and O. P. Foss. 1966. Electrolytes in gastric juice of man. II. A statistical analysis of magnesium before and after subcutaneous injection of large doses of histamine. J. Clin. Lab. Invest. 18:325–330.

Burns, H. N., and R. Allcroft. 1967. Hypomagnesemia in cattle. Br. Vet. J. 123:340–388.

Butler, E. J. 1963. The mineral element content of spring pasture in relation to the occurrence of grass tetany and hypomagnesaemia in dairy cows. J. Agric. Sci. 60:329–340.

Bygrave, F. L. 1967. The ionic environment and metabolic control. Nature 214:667–671.

Caddell, J. L. 1967. Studies in protein–calorie malnutrition. II. Doubleblind clinical trial to assess magnesium therapy. N. Engl. J. Med. 276:535–540.

Caddell, J. L., and D. R. Goddard. 1967. Studies in protein–calorie malnutrition. I. Chemical evidence for magnesium deficiency. N. Engl. J. Med. 276:533–535.

Care, A. D. 1967. Magnesium homeostasis in ruminants. World Rev. Nutr. Diet. 8:127–142.

Care, A. D., and A. T. Van't Klooster. 1965. In vivo transport of magnesium and other cations across wall of gastro-intestinal tract of sheep. J. Physiol. 177:174–191.

Chai, Y. S., and C. W. Carr. 1967. Ion-binding studies of RNA and Escherichia coli ribosomes. J. Mol. Biol. 25:331–345.

Chao, F. C. 1957. Dissociation of macromolecular ribonuclear protein of yeast. Arch. Biochem. Biophys. 70:426–431.

Chapman, H. D. [ed.] 1966. Diagnostic criteria for plants and soils. Division of Agricultural Sciences, University of California, Riverside. 793 pp.

Chesley, L. C., and I. Tepper. 1958. Some effects of magnesium loading upon renal excretion of magnesium and certain other electrolytes. J. Clin. Invest. 37:1362–1372.

Chutkow, J. G. 1964. Sites of magnesium absorption and excretion in the intestinal tract of the rat. J. Lab. Clin. Med. 63:71–79.

Chutkow, J. G. 1974. Metabolism of magnesium in central nervous system. Neurology 23:780–787.

Chutkow, J. G., and J. D. Grabow. 1972. Clinical and chemical correlations in magnesium deprivation encephalopathy of young rats. Am. J. Physiol. 223:1407–1414.

Clark, F. W. 1924. The data of geochemistry, 5th ed. U.S. Geol. Surv. Bull. 770. U.S. Government Printing Office, Washington, D.C. 841 pp.

Cohn, M. 1963. Magnetic resonance studies of metal activation of enzymic reactions of nucleotides and other phosphate sulphates. Biochemistry 2:623–629.

Cohn, M., and J. S. Leigh. 1962. Magnetic resonance investigations of ternary complexes enzyme–metal–substrate. Nature 193:1037–1040.

Consolazio, C. F., L. O. Matoush, R. A. Nelson, R. S. Harding, and J. E. Canham. 1963. Excretion of sodium, potassium, magnesium and iron in human sweat and the relation of each to balance and requirements. J. Nutr. 79:407–415.

Coussons, H. 1969. Magnesium metabolism in infants and children. Postgrad. Med. 46:135–139.

Cox, R. P., P. Gilbert, and M. J. Griffin. 1967. Alkaline inorganic phosphatase activity of mammalian cell alkaline phosphatase. Biochem. J. 105:155–161.

Das, H. K., A. Goldstein, and L. I. Lowney. 1967. Attachment of ribosomes to nascent messenger RNA in Escherichia coli. J. Mol. Biol. 24:231–245.

Dick, M., R. A. Evans, and L. Watson. 1969. Effect of ethanol on magnesium excretion. J. Clin. Pathol. 22:152–153.

Drenick, E. J., S. F. Hunt, and M. E. Swendseid. 1969. Magnesium depletion during prolonged fasting of obese males. J. Clin. Endocrinol. 29:1341–1348.

Dunn, M. J., and M. Walser. 1966. Magnesium depletion in normal man. Metabolism 15:884–895.

Durlach, J. 1967. Le magnésium en pathologie humaine. Gaz. Med. Fr. 74:3303–3320.

Durlach, J., F. Gremy, and S. Metral. 1967. La spasmophilie-forme clinique neuro-musculaire du déficit magnésien primitif. Rev. Neurol. 117:177–189.

Eaton, R. H., and D. W. Moss. 1967a. Inhibition of ortho-phosphatase and pyrophosphatase activities of human alkaline phosphatase. Biochem. J. 102:917–921.

Eaton, R. H., and D. W. Moss. 1967b. Organic pyrophosphatase substrates for human alkaline phosphatase. Biochem. J. 105:1307–1312.

Eaton, R. H., and D. W. Moss. 1967c. Alkaline orthophosphatase and inorganic pyrophosphatase activities in human serum. Nature 214:842–843.

Edmondson, H. A., C. J. Berne, R. G. Homann, and M. Wertman. 1952. Calcium, potassium, magnesium and amylase disturbances in acute pancreatitis. Am. J. Med. 12:34–42.

Embleton, T. W. 1966. Magnesium. Chapter 18 in Diagnostic criteria for plants and soils. H. D. Chapman [ed.]. Division of Agricultural Sciences, University of California, Berkeley. pp. 225–263.

Erdtman, H. 1927. Glycerophosphatspaltung durch Nierenphosphatase und ihre aktivierung. Z. Physiol. Chem. 172: 182–198.

Fernley, H. N., and P. G. Walker. 1967. Studies on alkaline phosphatase. Biochem. J. 104:1011–1018.

Field, A. C. 1961. Magnesium in ruminant nutrition. III. Distribution of Mg^{28} in the gastrointestinal tract and tissues of sheep. Br. J. Nutr. 15:349–359.

Fishman, R. A. 1965. Neurological aspects of magnesium metabolism. Arch. Neurol. 12:562–569.

Fitzgerald, M. G., and P. Fourman. 1956. Experimental study of magnesium deficiency in man. Clin. Sci. 15:635–647.

Fletcher, R. F., A. A. Henly, H. G. Sammons, and J. R. Squire. 1960. Case of magnesium deficiency following massive intestinal resection. Lancet 1:522–525.

Flink, E. B. 1956. Magnesium deficiency syndrome in man. J. Am. Med. Assoc. 160:1406–1409.

Flink, E. B., F. L. Stutzman, A. R. Anderson, T. Konig, and R. Fraser. 1954. Magnesium deficiency after prolonged parenteral fluid administration and chronic alcoholism complicated by delirium tremens. J. Lab. Clin. Med. 43:169–183.

Flink, E. B., T. J. Konig, and J. L. Brown. 1955. Association of low serum magnesium concentrations and deleterious effects of ammonia in hepatic cirrhosis and diabetes mellitus. J. Lab. Clin. Med. 46:814.

Flink, E. B., R. McCollister, A. S. Prasad, J. C. Melby, and R. P. Doe. 1957. Evidence for clinical magnesium metabolism. Ann. Intern. Med. 47:956–968.

Forster, R. P., and F. Berglund. 1956. Osmotic diuresis and its effect on total electrolyte distribution in plasma and urine of the aglomerular teleost, *Lophius americanus*. J. Gen. Physiol. 39:349–359.

Fourman, P., and D. B. Morgan. 1962. Chronic magnesium deficiency. Proc. Nutr. Soc. 21:34–41.

Gibbs, M. 1967. Photosynthesis. Ann. Rev. Biochem. 36:757–784.

Gibson, J., C. D. Upper, and I. C. Gunsalus. 1967. Succinyl coenzyme A synthetase from *Escherichia coli*. J. Biol. Chem. 242:2474–2477.

Gill, J. R. J., N. H. Bell, and F. C. Bartler. 1967. Effect of parathyroid extract on magnesium excretion in man. J. Appl. Physiol. 22:136–138.

Ginn, H. E., W. O. Smith, J. R. Hammarsten, and D. Snyder. 1959. Renal tubular secretion of magnesium in dogs. Proc. Soc. Exp. Biol. Med. 101:691–692.

Gitelman, H. J., S. Kukolj, and L. G. Welt. 1968a. Influence of parathyroid glands on hypercalcemia of experimental magnesium depletion in rat. J. Clin. Invest. 47:118–126.

Gitelman, H. J., S. Kukolj, and L. G. Welt. 1968b. Inhibition of parathyroid gland activity by hypermagnesemia. Am. J. Physiol. 215:483–485.

Glaubitt, D., and J. G. Rausch-Stroomann. 1962. Magnesium–Calcium und Phosphor-Bilanzen bei essentieller Hypertonie in Herzinsuffizierz unter der behandlung mit Hydro-Chlorothiazid. Klin. Wochenschr. 40:143–149.

Goldberg, A. 1966. Magnesium binding by *Escherichia coli* ribosomes. J. Mol. Biol. 15:663–673.

Goldman, A. S., D. D. VanFossan, and E. C. Baird. 1962. Magnesium deficiency in celiac disease. Pediatrics 29:948–952.

Gordon, J., and F. Lipmann. 1967. Role of divalent ions in poly U-directed phenylalanine polymerization. J. Mol. Biol. 23:23–33.

Graham, L. A., J. J. Caesar, and A. S. V. Burgen. 1960. Gastrointestinal absorption and excretion of Mg[28] in man. Metabolism 9:646–659.

Grand, R. J., and A. H. Colodny. 1972. Increased requirement for magnesium during parenteral therapy for granulomatous colitis. J. Pediatr. 81:788–790.

Greenwald, J. H., A. Dubin, and L. Cardon. 1963. Hypomagnesemic tetany due to excessive lactation. Am. J. Med. 35:854–860.

Gros, F., H. Hiatt, W. Gilbert, S. G. Kurland, R. W. Riseborough, and J. D. Watson. 1961. Molecular and biological characterization of messenger RNA. Cold Spring Harbor Symp. Quant. Biol. 26:111.

Grunes, D. L., P. R. Stout, and J. R. Brownell. 1970. Grass tetany in ruminants. Adv. Agron. 22:331–374.

Guidotti, G. 1976. The structure of membrane transport systems. *In* Trends Biochem. Sci. 1(1):11–13.

Hall, B. D., and P. Doty. 1959. The preparation and physical chemical properties of RNA from microsomes. J. Mol. Biol. 1:111–126.

Hanna, S. 1961. Plasma magnesium in health and disease. J. Clin. Pathol. 14:410–414.

Hanna, S., M. Harrison, I. McIntyre, and R. Fraser. 1960. Syndrome of magnesium deficiency in man. Lancet 2:172–175.

Hanze, S. 1960. Untersuchungen zur Wirkung verschiedener Diuretica auf die renale Magnesium—und Calcium—Ausscheidung. Klin. Wochenschr. 38:1168.

Head, M. J., and J. A. F. Rook. 1955. Hypomagnesemia in dairy cattle and its possible relationship to ruminal ammonia production. Nature 176:262–263.

Heaton, F. W. 1964. Magnesium metabolism in surgical patients. Clin. Chim. Acta. 9:327–333.

Heaton, F. W. 1965. The parathyroid glands and magnesium metabolism in the rat. Clin. Sci. 28:543–553.

Heaton, F. W. 1969. The kidney and magnesium homeostasis. Ann. N.Y. Acad. Sci. 162:775–785.

Hendrix, J. Z., N. W. Alcock, and R. M. Archibald. 1963. Competition between calcium, strontium and magnesium for absorption in the isolated rat intestine. Clin. Chem. 9:734–744.

Herd, R. P. 1966. Grass tetany in sheep. Aust. Vet. J. 42:160–164.

Hirschfelder, A. D., and V. G. Haury. 1934. Clinical manifestations of high and low plasma magnesium. Danger of epsom salt purgation in nephritis. J. Am. Med. Assoc. 102:1138–1141.

Hjerpe, C. A. 1964. Grass tetany in California cattle. J. Am. Vet. Med. Assoc. 144:1406.

Hoff, H. E., P. K. Smith, and A. W. Winkler. 1940. Effects of magnesium on the nervous system in relation to its concentration in serum. Am. J. Physiol. 130:292–297.

Horvath, D. J. 1959. So-called "grass tetany" in West Virginia. Survey and laboratory findings. *In* Symposium on Magnesium and Agriculture, D. J. Horvath [ed.]. West Virginia University, Morgantown. pp. 197–215.

House, W. A., and D. R. Van Campen. 1971. Magnesium metabolism of sheep fed different levels of potassium and citric acid. J. Nutr. 101:1483–1492.

Hughes, M. N. 1972. The inorganic chemistry of biological processes. John Wiley & Sons, London. 804 pp.

Jabir, F. K., B. D. Roberts, and R. A. Wormersly. 1957. Studies on renal excretion of magnesium. Clin. Sci. 16:119–124.

Jones, J. E., S. R. Shane, W. H. Jacobs, and E. B. Flink. 1969. Magnesium. Balance studies in chronic alcoholism. Ann. N.Y. Acad. Sci. 162:934–946.

Kellaway, G., and K. Ewen. 1962. Magnesium deficiency complicating prolonged gastric suction. N.Z. Med. J. 61:137–142.

Kemp, A. 1960. Hypomagnesaemia in milking cows: The response of serum magnesium to alterations in herbage composition resulting from potash and nitrogen dressings on pasture. Neth. J. Agric. Sci. 8:281–304.

Kemp, A., W. B. Deijs, O. J. Hernke, and A. J. H. Van Es. 1961. Hypomagnesaemia in milking cows: Intake and utilization of magnesium from herbage by lactating cows. Neth. J. Agric. Sci. 9:134–149.

Kemp, A., W. B. Deijs, and E. Kluvers. 1966. Influence of higher fatty acids on the availability of magnesium in milking cows. Neth. J. Agric. Sci. 14:290–295.

Kennell, D., and A. Kotoulas. 1967. Magnesium starvation of *Aerobacter aerogenes*. J. Bacteriol. 93:345–356.

Kerson, L. A., D. Garfinkle, and A. S. Mildvan. 1967. Computer simulation studies of mammalian pyruvate kinase. J. Biol. Chem. 242:2124–2133.

King, L. R., H. C. Knowles, Jr., and R. L. McLaurin. 1973. Calcium, phosphorus and magnesium metabolism following head injury. Ann. Surg. 177:126–131.

Kobayaski, A., K. Yabota, Y. Ohbe, K. Kobari, O. Kitamoto, and C. Vylaneo. 1972. Concentration of magnesium in plasma and feces of patients with cholera. J. Infect. Dis. 123:615–659.

Kuby, S. A., T. A. Mahowald, and A. E. Noltmann. 1962. Studies on ATP transphosphorylases. Biochemistry 1:748–762.

Kunkel, H. O., K. H. Burns, and B. J. Camp. 1953. A study of sheep fed high levels of potassium bicarbonate with particular reference to induced hypomagnesemia. J. Anim. Sci. 12:451–458.

Lagerkvist, U., and J. Waldenstrom. 1965. Some properties of purified valyl and lysyl ribonucleic acid synthetases from yeast. J. Biol. Chem. 240:2264–2265.

Lengemann, F. W. 1959. The metabolism of magnesium and calcium by the rat. Arch. Biochem. 84:278–285.

L'Estrange, J. L., and R. F. E. Axford. 1964. Study of magnesium and calcium metabolism in lactating ewes fed semi-purified diet low in magnesium. J. Agric. Sci. 62:353–368.

Lim, P., and E. Jacob. 1972. Tissue magnesium level in chronic diarrhea. J. Lab. Clin. Med. 80:313–321.

Linder, G. C., J. D. Hansen, and C. D. Karabus. 1963. Metabolism of magnesium and other inorganic cations and of nitrogen in acute kwashiorkor. Pediatrics 31:552–568.

Macbeth, R. A. L., and J. D. Mabbott. 1964. Magnesium balance in the postoperative patient. Surg. Gynec. Obstet. 118:748–760.

MacIntyre, I. 1960. Discussion on magnesium metabolism in man and animals. Proc. R. Soc. Med. 53:1037–1039.

MacIntyre, I., S. Hanna, C. C. Booth, and A. E. Read. 1961. Intracellular magnesium deficiency in man. Clin. Sci. 20:297–305.

MacIntyre, I., S. Ross, and V. A. Troughton. 1963. Parathyroid hormone and magnesium homeostasis. Nature 198:1058–1060.

Makman, M. H., and G. L. Cantoni. 1965. Isolation of seryl and phenylalanyl RNA synthetases from baker's yeast. Biochemistry 4:1434–1442.

Malstrom, B. G. 1956. Mechanism of metal-ion activation of enzymes: Studies on enolase. Almquist and Wiksells, Stockholm.

Marchesi, S. L., and D. Kennell. 1967. Magnesium starvation of Aerobacter aerogenes. J. Bacteriol. 93:357–366.

Marchesi, G. T., and G. E. Palade. 1967. The localization of magnesium–sodium–potassium activated ATPase on red cell ghost membranes. J. Cell. Biol. 35:385–404.

Marshall, R. G., C. J. Caskey, and M. Nirenberg. 1967. Fine structure of RNA code words recognized by bacterial, amphibian and mammalian transfer RNA. Science 155:820–826.

Martens, C., and R. Harriss. 1970. Inhibition of apatite precipitation in the marine environment by magnesium ions. Geochim. Cosmochim. Acta 34:621–625.

Martin, H. E., and F. K. Bauer. 1962. Mg^{28} studies in the cirrhotic and alcoholic. Proc. R. Soc. Med. 55:912–914.

Martin, H. E., and R. Jones. 1961. Effect of ammonium chloride and sodium bicarbonate on urinary excretion of magnesium, calcium and phosphate. Am. Heart J. 62:206–210.

Martindale, L., and F. W. Heaton. 1965. The relation between skeletal and extra-cellular fluid in vitro. Biochem. J. 97:440–443.

Massry, S. G., and J. W. Coburn. 1973. The hormonal and non-hormonal control of renal excretion of calcium and magnesium. Nephron 10:66–112.

Massry, S. G., J. W. Coburn, and C. R. Kleeman. 1969. Renal handling of magnesium in the dog. Am. J. Physiol. 216:1460–1467.

Matko, M. 1966. Magnesium deficiency in ulcerative colitis. J. Fla. Med. Assoc. 53:1063–1064.

McAleese, D. M., M. C. Bell, and R. M. Forbes. 1961. Mg^{28} studies in lambs. J. Nutr. 74:505–514.

McCarthy, B. J. 1962. The effect of magnesium starvation on the ribosome content of Escherichia coli. Biochem. Biophys. Acta 55:880–888.

McCollister, R. J., E. B. Flink, and R. P. Doe. 1960. Magnesium balance studies in chronic alcoholism. J. Lab. Clin. Med. 55:98–104.

McCollister, R. J., E. B. Flink, and M. D. Lewis. 1963. Urinary excretion of magnesium in man following the ingestion of ethanol. Am. J. Clin. Nutr. 12:415–420.

Mendel, L. B., and S. R. Benedict. 1909. The paths of excretion for inorganic compounds. IV. The excretion of magnesium. Am. J. Physiol. 25:1–33.

Mendelson, J. H., B. Barnes, C. Mayman, and M. Victor. 1965. The determination of exchangeable magnesium in alcoholic patients. Metabolism 14:88–98.

Mendelson, J. H., H. Ogata, and N. K. Mello. 1969. Effects of alcohol ingestion and withdrawal on magnesium states of alcoholics. Ann. N.Y. Acad. Sci. 162:918–933.

Metcoff, J., S. Frenk, I. Antanowicz, G. Gordillo, and E. Lopez. 1960. Relations of intracellular ions to metabolic sequences in muscle in kwashiorkor. Pediatrics 26:960–972.

Mildvan, A. S. 1970. Metals in enzyme catalysis. Chapter 9 In The Enzymes, P. D. Boyer [ed.]. Vol. II, 3d ed. (9 volumes) New York, Academic Press.

Mildvan, A. S., and M. Cohn. 1965. Kinetic and magnetic resonance studies of the pyruvate kinase reaction. J. Biol. Chem. 240:238–246.

Mildvan, A. S., and M. Cohn. 1966. Kinetic and magnetic resonance studies of the pyruvate kinase reaction. J. Biol. Chem. 241:1178–1193.

Mildvan, A. S., J. S. Leigh, and M. Cohn. 1967. Kinetic and magnetic resonance studies of pyruvate kinase. Biochemistry 6:1805–1818.

Mininni, G., and L. Zonno. 1962. L'influenza della idroclorotiazide sul magnesio serico. Minerva Pediatr. 14:1021–1024.

Monsaingeon, A., J. Thomas, Y. Nocquet, J. Savel, and J. Clostre. 1966. Sur le rôle du magnésium en pathologie chirurgicale. J. Chir. 91:437–454.

Montgomery, R. D. 1960. Magnesium metabolism in infantile protein malnutrition. Lancet 2:74–75.

Montgomery, R. D. 1961. Magnesium balance studies in marasmic kwashiorkor. J. Pediatr. 59:119–123.

Moore, P. B. 1966. Polynucleotide attachment to ribosomes. J. Mol. Biol. 18:8–20.

Morrison, J. F., and W. J. O'Sullivan. 1965. Kinetic studies of the reverse reaction catalyzed by ATP-creatine phosphotransferase. Biochem. J. 94:221–235.

Murdaugh, H. V., and R. R. Robinson. 1960. Magnesium excretion in the dog studied by stop–flow analysis. Am. J. Physiol. 198:571–574.

Nelson, N., A. Nelson, and E. Racker. 1972. Partial resolution of the enzymes catalyzing phosphorylation. Magnesium ATPase properties of heat activated coupling factors from chloroplasts. J. Biol. Chem. 247:6506.

Nielsen, J. 1963. Magnesium metabolism in acute alcoholics. Dan. Med. Bull. 10:225–233.

Novelli, G. D. 1967. Amino acid activation for protein synthesis. Ann. Rev. Biochem. 36:449–484.

Opie, L. H., B. G. Hunt, and J. M. Finlay. 1964. Massive small bowel resection with malabsorption and negative magnesium balance. Gastroenterology 47:415–420.

O'Sullivan, W. J., and M. Cohn. 1966. Magnetic resonance investigations of the metal complexes formed in the manganese activated creatine kinase reaction. J. Biol. Chem. 241:3104–3115.

Peck, C. H., and S. J. Meltzer. 1916. Anesthesia in human beings by the intravenous injection of magnesium sulfate. J. Am. Med. Assoc. 67:1131–1133.

Penefsky, H. S., M. E. Pulman, A. Datta, and E. Racker. 1960.

Partial resolution of the enzymes catalyzing oxidative phosphorylation. J. Biol. Chem. 235:3330–3336.

Petermann, M. L. 1960. Ribonucleoprotein from a rat tumor. The Jensen sarcoma. J. Biol. Chem. 235:1998–2003.

Peterson, V. P. 1963. Metabolic studies in clinical magnesium deficiency. Acta Med. Scand. 173:285–298.

Postma, P. W., and K. Van Dam. 1976. The ATPase complex from energy transducing membranes. Trends Biochem. Res., January, 1(1):16–17.

Pulman, M. E., and G. Schatz. 1967. Mitochondrial oxidations and energy coupling. Ann. Rev. Biochem. 36:539–610.

Pulman, M. E., H. S. Penefsky, A. Datta, and E. Racker. 1960. Partial resolution of the enzymes catalyzing oxidative phosphorylation. J. Biol. Chem. 235:3322–3329.

Randall, R. E., E. C. Rossmeisl, and K. H. Bleifer. 1959. Magnesium depletion in man. Ann. Intern. Med. 50:257–287.

Randall, R. E., M. D. Cohen, C. C. Spray, and E. C. Rossmeisl. 1964. Hypermagnesemia in acute renal failure: Etiology and toxic manifestations. Ann. Intern. Med. 61:73–88.

Ravel, J. M., S. Wang, C. Heinemyer, and W. Shive. 1965. Glutamyl and glutaminyl RNA synthetase of *Escherichia coli* W. J. Biol. Chem. 240:432–438.

Raynaud, C. 1962. Renal excretion of magnesium in the rabbit. Am. J. Physiol. 203:649–654.

Richardson, J. A., and L. Welt. 1965. Hypomagnesemia of vitamin D administration. Proc. Soc. Exp. Biol. Med. 118:512–514.

Robinson, R. R., H. V. Murdaugh, and G. Peschel. 1959. Renal factors responsible for hypermagnesemia of renal disease. J. Lab. Clin. Med. 53:572–576.

Rook, J. A. F., C. C. Balch, and C. Line. 1958. Magnesium metabolism in the dairy cow. J. Agric. Sci. 51:189–198.

Rose, L. I., D. R. Carroll, S. L. Lowe, E. W. Peterson, and K. H. Cooper. 1970. Serum electrolyte changes after marathon running. J. Appl. Physiol. 29:449–451.

Rubin, H. 1975. Central role for magnesium in coordinate control of metabolism and growth in animal cells. Proc. Natl. Acad. Sci. USA 72:3551.

Salas, M., J. J. Miller, A. J. Wahba, and S. Ochoa. 1967. Transfer of the genetic message. V. Effect of magnesium and formylation of methionine in protein synthesis. Proc. Natl. Acad. Sci. U.S.A. 57:1865–1869.

Salveson, A. A., and G. G. Linder. 1923. Observations on the inorganic bases and phosphates in relation to protein of blood and other body fluids in Bright's disease and in heart failure. J. Biol. Chem. 58:617–634.

Savage, D. C. L., and W. A. McAdam. 1967. Convulsions due to hypomagnesemia in an infant recovering from diarrhea. Lancet 2:234–236.

Schatz, G., H. S. Penefsky, and E. Racker. 1967. Partial resolution of the enzymes catalyzing oxidative phosphorylation. J. Biol. Chem. 242:2552–2560.

Schroeder, H. A., A. P. Nason, and I. H. Tipton. 1969. Essential metals in man. Magnesium. J. Chron. Dis. 21:815–841.

Schweet, R., and R. Heintz. 1966. Protein synthesis. Ann. Rev. Biochem. 35:723–758.

Selwyn, M. J. 1967. Preparation and general properties of a soluble ATPase from mitochondria. Biochem. J. 105:279–288.

Shacklette, H. T., J. C. Hamilton, J. G. Boerngen, and J. M. Bowles. 1971. Elemental composition of surficial materials in the conterminous United States. *In* Statistical studies in field geochemistry. U.S. Geol. Surv. Prof. Pap. 574-D. U.S. Government Printing Office, Washington, D.C. 71 pp.

Sherwood, L. M., I. Herrman, and C. A. Bassett. 1970. Parathyroid hormone secretion *in vitro*: Regulation by calcium and magnesium ions. Nature 225:1056–1058.

Shils, M. E. 1964. Experimental human magnesium depletion. I. Clinical observations and blood chemistry alterations. Am. J. Clin. Nutr. 15:133–143.

Shils, M. E. 1969a. Experimental production of magnesium deficiency in man. Ann. N.Y. Acad. Sci. 162:847–855.

Shils, M. E. 1969b. Experimental human magnesium depletion. Medicine 48:61–85.

Silver, L., J. S. Robertson, and L. K. Dahl. 1960. Magnesium turnover in the human studied with Mg^{28}. J. Clin. Invest. 39:420–425.

Smith, R. H., and A. B. McAllan. 1966. Binding of magnesium and calcium in the contents of the small intestine of the calf. Br. J. Nutr. 20:703–718.

Smith, R. K., A. W. Winkler, and B. M. Schwartz. 1939. The distribution of magnesium following the parenteral administration of magnesium sulfate. J. Biol. Chem. 129:51–56.

Smith, W. O., and J. F. Hammarsten. 1958. Serum magnesium in renal diseases. Arch. Intern. Med. 102:5–9.

Smith, W. O., and J. F. Hammarsten. 1959. Intracellular magnesium in delirium tremens and uremia. Am. J. Med. Sci. 237:413–417.

Smith, W. O., A. A. Kyriakopoulos, and J. F. Hammarsten. 1962. Magnesium depletion induced by various diuretics. J. Okla. Med. Assoc. 55:248–250.

Smyth, P. J., A. Conway, and M. J. Walsh. 1958. The influence of different fertilizer treatments on the hypomagnesaemia proneness of a rye grass sward. Vet. Rec. 70:846–848.

So, A. G., and E. W. Davie. 1964. The effects of organic solvents on protein biosynthesis and their influence on the amino acid code. Biochemistry 3:1165–1169.

Somjen, G., M. Helmy, and C. R. Stephen. 1966. Failure to anesthetize human subjects by intravenous administration of magnesium sulfate. J. Pharmacol. Exp. Ther. 154:652–659.

Storry, J. E., and J. A. F. Rook. 1962. The magnesium nutrition of the dairy cow in relation to the development of hypomagnesemia in the grazing animal. J. Sci. Food Agric. 13:621–627.

Stutzman, F. L., and D. S. Amatuzio. 1952. A study of serum and spinal fluid magnesium in normal humans. Arch. Biochem. 39:271–275.

Sullivan, J. F., H. G. Langford, M. J. Schwartz, and C. Farrel. 1963. Magnesium metabolism in alcoholism. Am. J. Clin. Nutr. 13:297–303.

Suttle, N. F., and A. C. Field. 1969. Studies on magnesium metabolism in ruminant nutrition. Br. J. Nutr. 23:81–90.

Swenson, S. A., Jr., J. W. Lewis, and K. R. Selby. 1974. Magnesium metabolism in man with special reference to jejunoileal bypass from obesity. Am. J. Surg. 127:250–255.

Szer, W., and S. Ochoa. 1964. Complexing ability and coding properties of synthetic polyribonucleotides. J. Mol. Biol. 8:823–837.

Takanami, M., and T. Okamoto. 1963. Interactions of ribosomes and synthetic polyribonucleotides. J. Mol. Biol. 7:323–333.

t'Hart, M. L. 1960. The influence of meteorological conditions and fertilizer treatment on pasture in relation to hypomagnesaemia. Br. Vet. Assoc. Conference on Hypomagnesaemia, Victoria Hall, Southampton Row, London. pp. 88–95.

Thoren, L. 1962. Magnesium deficiency: Studied in two cases of acute fulminant ulcerative colitis treated by colectomy. Acta Chir. Scand. 124:134–143.

Tibbetts, D. M., and J. C. Aub. 1937. Magnesium metabolism in health and disease. J. Clin. Invest. 16:491–501.

Tissières, A., and J. D. Watson. 1958. Ribonucleoprotein particles from *Escherichia coli*. Nature 182:778–780.

Tissières, A., J. D. Watson, D. Schlessinger, and R. B. Hollingworth. 1959. Ribonucleoprotein particles from *Escherichia coli.* J. Mol. Biol. 1:221–233.

Todd, J. R. 1965. The influence of soil type on the effectiveness of single dressings of magnesia in raising pasture magnesium content and in controlling hypomagnesemia. Br. Vet. J. 121:371–380.

Todd, J. R., and D. J. Horvath. 1970. Magnesium and neuro-muscular irritability in calves, with particular reference to hypomagnesaemic tetany. Br. Vet. J. 126:333–346.

Tourtelot, H. A. 1971. Chemical compositions of rock types as factors in our environment. *In* Environmental geochemistry in health and disease, H. L. Cannon and H. C. Hopps [eds.]. Memoir 123. Geological Society of America, Boulder, Colorado. pp. 13–29.

Tso, P. O. P., J. Bonner, and J. Vinograd. 1958. Studies and properties of microsomal ribonucleoprotein particles from pea seedlings. Biochem. Biophys. Acta 30:570–582.

Underwood, E. J. 1966. The mineral nutrition of livestock. FAO/CAB Publ. Central Press, Aberdeen. 237 pp.

Vallee, B. L., W. E. C. Wacker, and D. D. Ulmer. 1960. Magnesium deficiency tetany syndrome in man. N. Engl. J. Med. 262:155–161.

Van der Molen, H. 1964. Hypomagnesaemia grass fertilization in the Netherlands. Outlook Agric. 4:55–63.

Van der Molen, H., and L. Padmos. 1965. Plant and animal production on the nitrogen experimental farms during 1961–1964. Stikstof 9:4–19.

Vernon, W. B., and W. E. C. Wacker. 1977. Magnesium metabolism. Adv. Chem. Pathol. (In press).

Victor, M. 1973. The role of hypomagnesemic and respiratory alkalosis in the genesis of alcohol-withdrawal symptoms. Ann. N.Y. Acad. Sci. 215:235–248.

Vitale, J. J., E. E. Hellerstein, M. Nakamura, and B. Lown. 1961. Effect of magnesium deficient diet on puppies. Circ. Res. 9:387–394.

Vitale, J. J., H. Velez, C. Guzman, and P. Correa. 1963. Magnesium deficiency in cebus monkey. Circ. Res. 12:642–650.

Wacker, W. E. C. 1961. Effect of hydrochlorothiazide on magnesium excretion. J. Clin. Invest. 40:1086–1087.

Wacker, W. E. C. 1969. Biochemistry of magnesium. Ann. N.Y. Acad. Sci. 162:717–726.

Wacker, W. E. C., and A. F. Parisi. 1968. Magnesium metabolism. N. Engl. J. Med. 287:658–663, 712–717, 772–776.

Wacker, W. E. C., and B. L. Vallee. 1958. Magnesium metabolism. N. Engl. J. Med. 259:431–438.

Wacker, W. E. C., and R. J. P. Williams. 1968. Magnesium/calcium balances and steady states of biological systems. J. Theor. Biol. 20:65–78.

Wacker, W. E. C., F. D. Moore, D. D. Ulmer, and B. L. Vallee. 1962. Normocalcemic magnesium deficiency tetany. J. Am. Med. Assoc. 180:161–163.

Wallach, S., and A. Dimich. 1969. Radiomagnesium turnover studies in hypomagnesemic states. Ann. N.Y. Acad. Sci. 162:963–972.

Wallach, S., J. E. Rizek, A. Dimich, N. Prasad, and W. Siler. 1966. Magnesium transport in normal and uremic patients. J. Clin. Endocrinol. 26:1069–1080.

Warburg, O., and W. Christian. 1941. Isolierung und Kristallisation das garungsferments Enolase. Biochem. Z. 310:385–421.

Webster, L. T., Jr. 1965. Studies of the acetyl coenzyme A synthetase reaction. J. Biol. Chem. 240:4164–4169.

Webster, L. T., Jr. 1967. Studies of the acetyl coenzyme A synthetase reaction. J. Biol. Chem. 242:1232–1240.

Wener, J., R. Friedman, A. Mayman, and R. Schucher. 1959. Hydrochlorothiazide in management of cardiac oedema. Can. Med. Assoc. J. 81:221–227.

Wilke, H., and H. Spilemann. 1968. Untersuchungen uber den Magnesium-Gehalt der leben ben der Cirrhose. Klin. Wochenchr. 46:1162–1164.

Williams, R. J. P., and W. E. C. Wacker. 1967. Cation balance in biological systems. J. Am. Med. Assoc. 201:18–22.

Wilson, A. A. 1960. Magnesium homeostasis and hypomagnesemia in ruminants. Vet. Rev. 6:39–52.

Wold, F., and C. E. Ballou. 1957a. Studies on the enzyme enolase. I. Equilibrium studies. J. Biol. Chem. 227:301–312.

Wold, F., and C. E. Ballou. 1957b. Studies on the enzyme enolase. II. Kinetic studies. J. Biol. Chem. 227:313–328.

Wolfe, S. M., and M. Victor. 1969. The relationship of hypomagnesemia and alkalosis to alcohol withdrawal symptoms. Ann. N.Y. Acad. Sci. 162:973–984.

IV

Manganese

GENNARD MATRONE, *Chairman*

Everett A. Jenne, Joe Kubota, Ismael Mena, Paul M. Newberne

This chapter summarizes and highlights the significant facts known about manganese in the rock–soil–plant–animal chain and includes a discussion of its metabolism in plants, animals, and man in terms of both deficiencies and excesses. Attention to excesses of manganese becomes more relevant as man alters the manganese level in the environment by his present industrial activities. The potential substitution of manganese for lead in gasoline, and its potential consequences, will also be discussed.

GEOCHEMISTRY

The manganese concentration in the major rock types of the world varies widely. A summary of manganese in rocks and sediments is presented in Table 1. The range of total manganese content in soils and other regoliths of the conterminous United States is from <1 to 7,000 ppm with a geometric mean of 340 (Shacklette *et al.*, 1971). Many soils of the United States are formed from unconsolidated deposits of mixed rock origin. Among these soils, the sandy soils of the Southeastern Coastal Plain have the least amount (<100 ppm) of total manganese (Marbut, 1935; Shacklette *et al.*, 1971). Appreciably larger amounts of manganese occur in soils formed in unconsolidated deposits of the more arid regions; for example, a geometric mean of 600 ppm of manganese occurs in surface horizons in a wide range of California soils and somewhat lower amounts (470 ppm) in subsoil horizons (Bradford *et al.*, 1967). The regional patterns of total manganese in

soils reflect the influence of rock sources on soil parent materials and the nature of unconsolidated deposits on which soils are formed, as well as manganese losses through soil weathering.

Stream sediments and alluvial soils on floodplains generally reflect the manganese concentration of soils of the surrounding uplands. Consequently, differences in the manganese concentrations of fine-textured sediments may be found. For example, the clayey and silty sediments of the Rhine and Ems rivers and their estuaries in Europe have from 1,800 to 3,300 ppm of manganese (De Groot *et al.*, 1971).

The more prominent manganese minerals identified in sediments and soils are: birnessite [MnO_2 to (Na, Ca) Mn_7O_{14}]; Δ-MnO_x (where $x = 1.5$ to 1.99); todorokite [(Mn^{2+}, Mg, Ca) $Mn^{4+}O_2 \cdot 2H_2O$]; lithiophorite [$Li_2Mn_2^{2+}Al_8Mn_{10}^{4+}O_{35} \cdot 14H_2O$]; hollandite [$Ba(Mn^{2+}, Fe^{2+}) Mn_7^{4+}O_{16}$]; psilomelane [$BaMn^{2+}Mn_6^{4+}O_{16}(OH)_4$]; cryptomelane [$\alpha$-$MnO_2$ to $K(Mn^{2+}, Mn^{4+})_8O_{16}$]; nsutite [$\gamma$-$MnO_2$]; rhodochrosite [$MnCO_3$]; alabandite [$MnS$]; pyrolusite [$\beta$-$MnO_2$]; and hausmannite [$Mn_3O_4$] (Tiller, 1963; Taylor *et al.*, 1964; B. J. Anderson, unpublished data).

The available manganese status of soils has largely been estimated by determining the amount of the secondary forms of manganese released with various soil extractants. A consideration in the evaluation of the extractable forms of manganese is the fact that the amounts of manganese available to plants under natural conditions strongly reflect changes in the oxidation-reduction status of soils, especially due to changes in soil wetness and in soil pH. The primary minerals in soils are less important as sources of manganese for plant growth than are the secondary forms of manganese. The secondary forms, in general, are amorphic and represent the bulk of the active manganese fraction in soils. These mineral sources, and soil organic matter, appear to be the important sources of manganese for plants.

Manganese occurs primarily as the divalent ion in waters and soil solutions ($Mn^{2+} >> Mn^{4+} > Mn^{3+}$). The concentrations of Mn^{4+} are very low because of the low solubility of manganese dioxide minerals. The concentrations of Mn^{3+} are low because they readily reduce to Mn^{2+} and because of the disproportion reaction:

$$2Mn^{3+} + 2H_2O = Mn^{2+} + MnO_2 + 4H^+.$$

Dissolved manganese occurs as the free hydrated divalent ion and forms several inorganic complexes. In both fresh and marine waters, the predominant complexes are $MnHCO_3^+$, $MnSO_4^\circ$, $MnCl^+$, MnF^+, and $MnOH^+$ (Morgan, 1967; Hem, 1970; B. F. Jones, oral communication, May 1973). Manganese shows less tendency to form complexes with natural organics than do other first-transition-series elements.

Because of the unique susceptibility of manganese oxides to dissolution with decreasing redox potentials, the concentration of dissolved manganese in natural waters and in soil solutions usually fluctuates more than does the concentration of other first-transition-series elements. In a recent study in Missouri, the manganese content (geometric mean) of groundwaters in seven geohydrologic units ranged from <0.01 to 0.87 ppm (U.S. Geological Survey, 1972). Marine waters range from 0.00004 to 0.016 ppm but generally fall between 0.0002 and 0.005 ppm (Slowey, 1966). Because of a presumed redox reaction between Mn^{4+} and Fe^{2+}, Fe^{2+} does not generally begin to dissolve in significant amounts with a decrease in redox potentials until nearly all of the Mn^{4+} has been reduced to Mn^{2+} (Jenne, 1968). Oxygenation of acid anoxic waters, such as acid mine drainage, results in the rapid precipitation of amorphic $Fe(OH)_3$, but the precipi-

tation of manganese oxides occurs more slowly. In contrast to the less abundant first-transition-series metals, whose solute concentrations in waters are generally limited by sorption and coprecipitation processes, solute concentrations of manganese in waters and soils are probably limited by the solubility of the minerals listed above. In oxygenated waters, the solubility of manganese oxides probably controls the concentration of dissolved manganese. In mildly reducing systems, $MnCO_3$ is most likely to be the predominant control, whereas in anoxic systems both MnS and $MnCO_3$ serve as controls.

The 1962 Public Health Service drinking water standards list 0.05 ppm as the maximum permissible concentration of manganese, although this is presumably for esthetic and economic reasons. A concentration of 0.1 ppm is sufficient to cause undesirable discoloration when used for household purposes and may deposit in plumbing (Myers, 1961).

No information has been found that would indicate the relative availability of solute and particulate forms of manganese to fresh- or marine-water biota. For other than detritus feeders, all indications are that the uptake of trace elements is related to the solute concentration rather than the quantity in the particulate fraction, although the latter may be the greater by orders of magnitude. The various oxides, and the carbonate, sulfide, and natural organic compounds, are then the principal trace element sources for aquatic detritus feeders. No information is available regarding the relative availability to biota of manganese in these forms.

Chelation Chemistry of Divalent Manganese

Manganese, as noted above, is located in the first-transition-element group. In biological systems, Mn^{2+} almost always forms octahedral coordination compounds (Orgel, 1958). Divalent manganese forms primarily high spin complexes because it is a d^5 ion, requiring the CN^- ligand to bring about a low spin complex (Orgel, 1963). The ligand preferences for Mn^{2+} are similar to those of magnesium (i.e., those containing oxygen ligand atoms) except that Mn^{2+} has a greater preference for histidine (at

TABLE 1 Crustal Abundance of Manganese and Content in Rocks and Sediments

Crustal Abundance and Rock Type	Mean Concentration, ppm	Source
Crustal abundance	950	Riley and Chester, 1971
Shale, average	850	Krauskopf, 1967
Carbonates, average	1,100	Krauskopf, 1967
Carbonates, deep sea	1,000	Riley and Chester, 1971
Clay, deep sea	6,700	Riley and Chester, 1971
Near-shore sediments	8,500	Riley and Chester 1971
Granitic	500	Wedepohl, 1969
Intermediates	1,060	Wedepohl, 1969
Gabbroic–basaltic	1,430	Wedepohl, 1969
Peridotitic–anorthositic	1,210	Wedepohl, 1969
Alkalic	1,439	Wedepohl, 1969

the N_3 position), NH_2, and SH groups (Orgel, 1958). The formation constants or stability constants of Mn^{2+} with oxyanions such as organic acids and phosphates are more or less in the same order of magnitude as those for Mg^{2+} (Mahler, 1961). Since the coordination number of Mg^{2+} in biological systems is also almost always six, one might predict, and does find, that these two ions can be interchanged in the *in vitro* activation of many enzyme systems.

The basis for the antagonistic effects between iron and manganese reported for plants (Somers and Shive, 1942) and animals (Matrone *et al.*, 1959) also might be predicted on the similarity of the chemical parameter enumerated above, i.e., Mn^{2+} and Fe^{3+} are both d^5 ions and both preferentially form octahedral complexes (Orgel, 1958; Matrone, 1969).

DEFICIENCY AND EXCESS OF MANGANESE IN PLANTS

Characteristic symptoms of manganese deficiencies and toxicities are well documented for a wide range of plants, and changes in manganese concentrations with onset of deficiency or toxicity have been established (Labanauskas, 1966).

Deficiency

Specialized crops, fruit trees, and ornamental plants are especially sensitive to a deficiency of manganese (Labanauskas, 1966). Levels of 20 ppm of manganese in leaves appear to be an indicator of a deficiency of manganese for plant growth, independent of possible associated deficiencies due either to zinc or iron. Foliar applications, fertilization, and control of soil pH are used to increase the manganese supply for common crop plants.

While various aspects of manganese behavior in plants are known, recent reviews (Epstein, 1972; Price *et al.*, 1972) indicate that there is an absence of general agreement on the specific role of manganese in the growth of plants. A notable observation is the absence of *in vitro* specificity of manganese in enzymatic systems of plants. A deficiency of manganese results in a disorganization of plant chloroplasts, but marked changes in lamellar membranes of chloroplasts have not been observed (Anderson and Pyliotis, 1969). A possible requirement of manganese for the activation of the Hill reaction in photosystem II of photosynthesis has also been suggested (Cheniae, 1970). It has been observed that evolution of O_2 is associated with the loosely bound fraction of the manganese of chloroplasts (two-thirds of total manganese), and not with the firmly bound fraction (one-third manganese).

Excess

The existence of both a passive (nonmetabolic) and a metabolically dependent pathway of manganese uptake by plants has also been recognized (Moore, 1972). In the presence of an excessive supply of manganese, the uptake of manganese continues with a consequent buildup of manganese in the vegetative parts of plants. Most crop plants appear to tolerate as much as 200 ppm of manganese in their tissues without showing adverse effects. Under field conditions, excesses of manganese are most often observed in plants grown on acid soils. Rice appears to have a wide range of tolerance for manganese, ranging from 20 ppm in the deficient range to as much as 2,500 ppm in the toxic range (Tanaka and Yoshida, 1970). Whether the manganese concentrations noted in the leaves of rice are reflected in grain remains undetermined.

Toxic levels of manganese determined for common crop plants, however, are not applicable to native shrubs and trees. In humid areas, these plants grown on acid soils have 1,000 ppm or more of manganese (Smith *et al.*, 1956; Gerloff *et al.*, 1964; Stone, 1968; and Kubota *et al.*, 1970). Emergent and submerged aquatic plants also have large amounts of manganese. Of 18 species, 10 were found to have from 1,650 to 5,130 ppm of manganese (Boyd, 1968).

Soil Patterns of Deficiences and Excesses for Plants

Broad patterns of low and high areas of manganese in plants are difficult to depict geographically in relation to soil or geochemical environment. An absence of such patterns reflects the fact that deficiences or excesses of manganese are often species-dependent and vary widely with kinds of crop plants and with soil and crop management practices. In the United States, for example, a deficiency of manganese is most often associated with naturally wet areas that have been drained and cropped to soybeans (Kubota and Allaway, 1972). The leaching losses of available soil manganese and the presence of calcareous subsoils appear to be the dominant factors in the distribution of the manganese deficiencies observed. In areas of calcareous soils where soil manganese availability to plants is low, manganese deficiency often is observed in relatively manganese-sensitive plants.

DEFICIENCY AND EXCESS OF MANGANESE IN ANIMALS

Deficiency

Kemmerer *et al.* (1931) and Orent and McCollum (1931) were the first to demonstrate the essentiality of manganese for mammals. These investigations reported abnormal estrus cycles, failure to properly nourish the young, and decreased growth rate in mice; many of the young were born dead or weak, and, when analyzed, their tissues contained less than half the concentration of manganese found in normal, manganese-supplemented controls.

Shils and McCollum (1943) observed that a deficiency

of manganese in rats was associated with altered estrus cycles and a decrease in the functional capacity of mammary tissue. These investigators distinguished three degrees of deficiency, in order of severity: stage 1—viable young are produced but have incoordination or paralysis; stage 2—nonviable young are born, but die soon after birth; stage 3—estrus irregular, depressed, or absent; delay in opening of vaginal orifice, and failure to mate. Although reproduction in rats was deranged, their ovaries did not reveal any marked alterations as a result of the deficiency.

Manganese deficiency in cattle also is associated with slow, irregular estrus cycles and problems in conception; calves born to manganese-deficient cattle are weak and frail. Cows with adequate manganese have about 2 ppm manganese in the ovaries, while deficient cows average <1 ppm (Bentley and Phillips, 1951).

Plumlee et al. (1956) reported that female weanling pigs fed a diet low in manganese through one complete reproductive cycle (generation) had irregular estrus cycles, fetal resorption, and gave birth to small, weak pigs with reduced skeletal growth.

Thus, in many species of animals, manganese is a demonstrated essential element for normal reproduction, including conception, fetal growth and development, and normal growth during the postnatal period.

In addition to its importance in reproduction, manganese is essential to normal skeletal development (Caskey and Norris, 1938; Gallup and Norris, 1939; Hurley and Everson, 1963; Leach, 1971). In manganese-deficient animals of several species, bones are generally short, light, and fragile. The defect in formation of bone has recently been associated with a derangement in the chemical composition of the organic matrix of cartilage and bone (Leach, 1967). Manganese-deficient chicks have decreased amounts of tissue mucopolysaccharide (hexuronic acid); this may, in turn, relate to a requirement for Mn^{2+} by glycosyl transferase enzymes, including those needed for the synthesis of chondroitin sulfate (Leach et al., 1969; Leach, 1971).

Ataxia (incoordination) was observed and described in the early studies on manganese deficiency. Norris and Caskey (1939) reported congenital ataxia in chicks deficient in manganese. The same symptoms have been observed in rats (Shils and McCollum, 1943; Hurley et al., 1958), swine (Plumlee et al., 1956), and guinea pigs (Everson et al., 1959). Mice deficient in manganese were unable to maintain orientation when submerged in water (Erway et al., 1966). It is clear, then, that a consistent set of clinical symptoms occurs in manganese deficiency, including incoordination and lack of equilibrium. Hurley and Everson (1959), Hurley et al. (1960), and Erway et al. (1966) have revealed that the primary defect is delayed development of the osseous labyrinth and an absence of utricular and saccular otoliths of the vestibular portion of the inner ear.

Despite the many studies of the essentiality of dietary manganese, actual dietary requirements of manganese for different species of animals, including man, are largely surmised rather than definitive estimates. It would appear that the manganese required for reproduction may be greater than for growth. The evidence seems clear, however, that birds require more manganese than do mammals (Underwood, 1971). The requirement for birds is approximately 50 ppm of the total diet. For other animals, the estimates run from approximately 5 to 40 ppm (Cotzias, 1962; Underwood, 1971).

The difficulties in obtaining uniform estimates of the manganese requirements of animals are due, in part at least, to the level of antagonistic factors in different diets. A number of investigators have reported that the dietary levels of calcium and phosphorus affect the availability of dietary manganese. Excessively high intakes of calcium and phosphorus have been reported to accentuate the dietary requirements for manganese of swine (Miller et al., 1940), poultry (Mitchell, 1947), and calves (Hawkins et al., 1955). Wachtel et al. (1943) found that high calcium : phosphorus ratios in the diet of rats aggravated manganese deficiency. Lassiter et al. (1972) report conflicting results; rats fed diets with low calcium : phosphorus ratios increased ^{54}Mn excretion via feces and urine, and their tissues retained much less manganese.

The involvement of manganese as a metal activator of many enzyme reactions has been recorded in several reviews (Cotzias, 1962; Mahler, 1961; Vallee, 1962). In in vitro experiments, Mn^{2+} appears to substitute for Mg^{2+} wherever ATP (adenosine triphosphate) is involved in the enzyme reaction. Specificity of manganese in enzyme reactions, however, has been difficult to establish. Definitive evidence has been obtained that pyruvate carboxylase isolated from chicken mitochondria is a manganese-containing metalloprotein (Scrutton et al., 1966). The evidence for liver arginase containing manganese (Bentley and Phillips, 1951; Bach and Whitehouse, 1954) is less certain.

Cotzias et al. (1972) have demonstrated in the pallid strain of mutant mice a link between the transportation of manganese and that of L-dopa (L-3,4-dihydroxyphenyalanine) into the brain. This mutant exhibits congenital ataxia that can be prevented in offspring by feeding large amounts of manganese to mothers during pregnancy; furthermore, it is insensitive to the cerebral effects of L-dopa. These findings are in contrast to those with black C57B1/6J mice (different by one gene), which show neither large requirements for manganese nor abnormal cerebral responses to L-dopa.

Although brain concentrations of manganese are diminished in Purina-fed pallid mice, the intraperitoneal injection of L-dopa, and of L-tryptophan, produces smaller brain increases of L-dopa, dopamine, and serotonin in pallid mice than in black mice. Thus, a single gene appears to influence the transport into the brain of manganese and of two amino acids in the same direction—if pallid and black mice indeed differ by only a single gene.

Finally, a large body of in vitro work shows manganese to be an antagonist of Ca^{2+} in functions pertaining to release of acetylcholine and norepinephrine from motor, splenic, or vascular terminals. Ca^{2+} increases the release of these neurotransmitters, while manganese blocks it (Molinoff and Axelrod, 1971).

Excess

Metabolic Effects Animals fed a diet adequate in respect to other essential nutrients can tolerate high concentrations of dietary manganese, such as 1,000 ppm or more. The primary target of excess manganese appears to be iron metabolism. Hartman *et al.* (1955) showed that 2,000 ppm of manganese in the diet interferes with hemoglobin regeneration in lambs; Matrone *et al.* (1959) reported that this is also true for baby pigs and rabbits. In the experiments reported by Matrone *et al.* (1959), 2,000 ppm of manganese also decreased the rate of weight gain in baby pigs. When the iron content of the basal diet was increased from 25 ppm to 400 ppm, the depressing effect of 2,000 ppm of dietary manganese on hemoglobin regeneration was relieved.

Susan D. Potter and Gennard Matrone (unpublished data, 1973) raised weanling female mice on diets containing several levels of manganese, including a highest level of 2,000 ppm; when they reached maturity, the mice were bred to determine the effect of high manganese on growth and reproduction. The only tested level of manganese that appeared to affect growth and/or reproduction was 2,000 ppm.

Hartman (1956) found that the rate at which orally administered radioactive ^{59}Fe was incorporated into the red blood cells of phlebotomized anemic rabbits was slower for those animals receiving 2,000 ppm dietary manganese, as compared to control animals, whereas ^{59}Fe administered intravenously was incorporated in red blood cells at an equal rate for both experimental groups. The results suggested to these authors that manganese interfered with iron absorption. Other reports, however, indicate that the effect of excess dietary manganese in iron metabolism is more complicated. Diez-Ewald *et al.* (1968) reported that rats fed high doses of manganese had decreased stores of iron in their livers and an increased gastrointestinal absorption of iron. However, these investigators recognized that the etiology of the iron deficiency they observed might have been related to the demonstrated increase in gastrointestinal blood loss, which occurred under the conditions of their experiment.

Incorporation of manganese into the porphyrin molecule of red cells and increased gastrointestinal absorption of manganese have been reported in iron deficiency (Borg and Cotzias, 1958; Pollack *et al.*, 1965). Matrone *et al.* (1959) estimate the minimum level of dietary manganese capable of affecting hemoglobin formation is only 45 ppm for anemic lambs and lies between 50 and 125 ppm for anemic rabbits and baby pigs. It is apparent from these reports that the interrelationships of the metabolism of iron and manganese remain to be elucidated. Nonetheless, it also seems clear that rather high levels of dietary manganese are necessary, when adequate iron is present in the diet, before an adverse effect on iron metabolism occurs. The reason for the

toleration of large concentrations of dietary manganese may involve the homeostatic mechanism for maintaining body manganese. As elucidated by Britton and Cotzias (1966), this is due to controlled excretion rather than to regulated absorption.

Neurological Effects Experimental studies of manganese intoxication in animals are few and are most concerned with neuronal changes. In experimentally induced chronic manganese intoxication in dogs, Makarcenko (1956) reported that the first disorder observed is an alteration of the "higher nervous activity," which suggests a decrease in the ability to form conditional reflexes. Wasserman and Mihail (1964) suggest that the modification of the electrical excitability of the neuromuscular apparatus in manganic patients indicates a lesion of the cerebral functional centers, particularly in the cerebral cortex. In the one patient tested, intelligence, as measured by the Wechsler Adult Intelligence Scale, was normal (Rosenstock *et al.*, 1971).

The first experimental studies outlining the neurobehavioral effects of manganese intoxication in nonhuman primates were conducted by Mella (1924) using the common rhesus monkey (*Macaca mulatta*). Four monkeys were given $MnCl_2$ every other day for 18 months, the dose increasing from 5 mg per day to 15 or 25 mg. Mella reported: "The monkey at first develops movements which are choreic or choreo-athetoid in type, later passing into a state of rigidity accompanied by disturbances of motility; then fine tremors of the hands appear, and finally contracture of the hands with the terminal phalanges extended."

Direct exposure of a monkey (*M. mulatta*) to manganese ore dust containing 50 percent MnO_2 was tried by van Bogaert (1943) for 1 h daily for 100 days. Another monkey was given $MnSO_4$ (about 10–15 mg) daily for 300 days in its food. The animal exposed to the dust developed neurological symptoms of ataxia, wide-based gait, and intensive tremor; paralysis of the hind limbs appeared later. The other animal did not develop symptoms. Six months after termination of the treatment, chemical analysis detected no manganese in the organs of the animal poisoned through the respiratory tract; however, cerebellar atrophy, particularly in the Purkinje cells and granule cells, was observed, but there was no cytologic damage to the globus pallidus. Damage was instead scattered throughout the central nervous system, including the spinal cord.

Pentschew *et al.* (1963) gave two weekly intramuscular injections (50 mg each) of manganese dioxide to each of five mature rhesus monkeys; they observed symptoms of intoxication about 9 months after the first injection, at which time the animal became very excitable, especially when approached or confronted by people. In jumping, the animal often fell on its heels or buttocks instead of on its feet. In addition, there was general clumsiness, and in moving about there was a resemblance to moderate inebriation. No cogwheel phenomenon, tremor, or involuntary movements were observed.

Wasserman and Mihail (1964), in reviewing the experimental literature, concluded that all experimental studies undertaken between 1921 and the date of their writing "have always succeeded in reproducing in the animal the morphological lesions of the central nervous system by the administration of manganese compounds if the doses and exposure times were sufficient."

However, real differences probably exist in the sensitivity of various species and in the manner in which they express the intoxication. Even within the primate order, there are wide differences among the species, both in the outward manifestation of manganese intoxication and in the histopathologic picture. The generality of the behavioral and neurological changes is difficult to assess quantitatively, because some investigators have doubtlessly overlooked alterations of potential significance. It is also likely that in some species the symptoms may be so subtle that they cannot be noticed without special procedures. In addition, the correlation between symptom and pathology is still obscure.

In some species, behavioral changes are evident even though histological alterations cannot be found. Squirrel monkeys (*Saimiri sciureus*) injected subcutaneously either two or three times with 200 mg MnO_2 at monthly intervals began showing signs of muscular rigidity, flexor posturing of the extremities, or fine rapid tremors of the distal extremities 2 months after the first injection (Neff *et al.*, 1969). Two of the monkeys were unable to climb about the cage when prodded; one exhibited an exaggerated startle reaction, hitting its head on the cage ceiling when approached, a symptom noted also by Pentschew *et al.* (1963). Some monkeys exhibited "obstinate progression," which reached near somersault proportions. Despite marked reductions in the dopamine and serotonin content of the caudate nucleus, there were no histological changes in the cerebral cortex, caudate nucleus, thalamus, hypothalamus, subthalamic nucleus, substantia nigra, cerebellum, or other areas of the lower brain stem, whether in neurons, glial elements, or vascular supply. This is further evidence that clinical and biochemical abnormalities may precede the histological changes and suggests that some of the clinical manifestations are primarily biochemical, not anatomical, in nature.

Behavioral changes following intoxication have scarcely been reported, and dose–response relationships are unknown; nevertheless, of all experimental animals, the mimicry of human symptoms is greatest in chimpanzees. Single injections in rhesus monkeys are without apparent effect, and repeated intramuscular injections of MnO_2 at levels of 500 mg/kg of body weight produces deranged milk production and secretion (Riopelle, 1969). The effects observed in chimpanzees were considerably more elaborate, although the testing performed was inadequate and informal. Approximately 3 months after injection of MnO_2 at 500 mg/kg in multiple sites of two young chimpanzees (about 4 and 8 yr old), gross activity decreased. The animals held unusual positions and postures for up to 30 s. In general, their deportment was similar to that before the injections, but the time course of any action was slowed somewhat, and vigorous activity was absent. A first sign of effects was difficulty in climbing a fence or in returning to the ground. The lower lip, which often droops in chimpanzees, drooped more than normal in manganese intoxication. The subsequent course of the illness in the chimpanzees was rapidly downhill. A masklike motionless expression became obvious, periods of somnolence and possibly unconsciousness were exhibited, and locomotion became virtually impossible. Cogwheel rigidity was absent. Athetoid extension and dystonic posturing of their fingers and arms was evident.

Gibbons (*Hylobates lar*) injected in a similar manner exhibited little change other than a tendency to hang on the fence and to walk bipedally slower than they did before injection.

METABOLISM OF MANGANESE IN MAN

Homeostatic Mechanism

Ingested manganese and most of the manganese inhaled are eventually absorbed in the intestine, i.e., brought up through muco/ciliary action and swallowed (Mena *et al.*, 1969). A small fraction (3 percent of a measured dose) is absorbed (Mena *et al.*, 1969), and concentrates rapidly (Mena *et al.*, 1967) in organs rich in mitochondria, such as the liver, pituitary, or pancreas (Cotzias *et al.*, 1971). Homeostasis of manganese is efficiently controlled by the excretory route (Cotzias *et al.*, 1968; Mahoney and Small, 1968). The liver excretes manganese to the bile, and the rate of clearance is a function of body burden (Cotzias *et al.*, 1968). Normal individuals excrete ^{54}Mn from the total body with $t_{1/2}$ (half-life) of 37 ± 7 days, while miners who have been overexposed (air concentrations of stable $^{55}Mn > 5$ mg/m^3) excrete ^{54}Mn with $t_{1/2}$ of 15 ± 2 days (Cotzias *et al.*, 1968). Whole blood concentrations in these two populations are 25 ± 1.6 and 11 ± 1.1 μg/l, respectively (P < 0.001). High blood concentrations of manganese are indicators of exposure to the metal, but not necessarily of manganese poisoning. The miners have apparently developed the capacity to excrete manganese more rapidly than nonexposed individuals.

Possible individual susceptibility has been related to variations in the intestinal absorption of manganese; these are rather rare cases, however. Anemic individuals with increased iron absorption also have increased manganese absorption (Mena *et al.*, 1969). Absorption of ^{54}Mn in a normal man is 3 ± 0.5 percent and in anemic patients is 7.5 ± 2 percent, while ^{54}Fe absorption was 11 ± 10 percent and 64 ± 22 percent, respectively (Mena *et al.*, 1969). Information on absorption of manganese in infants is not available. These factors must be considered in studying chronic overexposure to small amounts of manganese and in analyzing susceptibility to massive overexposure to manganese. The latter leads to manganese poisoning.

HEALTH EFFECTS OF MANGANESE EXCESS

Chronic manganese poisoning is an environmental disease of manganese miners and of workers in manganese mills. It occurs most frequently in the manganese mining villages of Russia, India, North Africa, Yugoslavia, Cuba, and Chile (Mena *et al.*, 1967). Its estimated incidence is as high as 25 percent of the exposed population in some areas and as low as 2–4 percent in Chile. It occurs after variable periods of exposure to ore dust by inhalation.

Symptoms

Chronic manganese poisoning has both psychiatric and neurological manifestations, the most crippling of which have been related to the extrapyramidal system of the brain. The psychiatric symptoms are transient, but the neurological damage may be irreversible, although some patients experience partial regression of their symptomatology after early removal from exposure (Mena *et al.*, 1967).

Consistently appearing in all cases in Chile, but notably absent in reports from steel foundries and ore-crushing plants in the United States (Greenhouse, 1971), has been the occurrence of a psychotic period at the onset of the disease, defined by the villagers as "manganic madness." It is characterized by hallucinations, delusions, and compulsions; in most cases, the patients are aware of the abnormal nature of these phenomena. The psychosis lasts from 1 to 3 months, whether or not the patients are immediately removed from the mines (Mena *et al.*, 1967). Toward the end of the psychotic period, or immediately after it, neurological symptoms characteristic of extrapyramidal involvement emerge. In the Chilean patients, these included loss of facial expression, rigidity, slowness of movements, diminution of postural reflexes, and impairment of speech. A few patients developed a dystonia similar to the spontaneously occurring dystonia, *musculorum deformans*. In one study of U.S. workers in a crushing plant, rigidity was notably absent, while the slowness (hypokinesia) and impairment of balance were predominant (Rosenstock *et al.*, 1971).

Background for Treatment

The successful treatment of Wilson's disease with metal-binding agents seemed to provide a precedent for treating chronic manganese poisoning, since the two diseases present some clinical similarities. This notion was weakened when excesses of manganese were found only in the tissues of healthy, exposed manganese miners, whereas crippled ex-miners who were no longer exposed had cleared these loads. Their brain damage seemed to have been caused by flooding with manganese, but their symptoms persisted after such flooding had been terminated. Even if some parts of their brains still contained an excess of metal, this must have been in a tightly sequestered state; it thus appeared that the brain had suffered a structural injury due to manganese.

Similarity to Parkinsonism

Manganese poisoning has many features in common with Parkinson's disease, in which the structural damage to the brain consists of depigmentation of the substantia nigra and the metabolic changes consist of diminished melanin in the substantia nigra and diminished catecholamines and serotonin in the corpus striatum (Neff *et al.*, 1969). The function of melanin is still unknown, but the biogenic amines are neurotransmitters. Upon systemic administration, these amines are bound or inactivated in the periphery and are prevented from entering the brain; therefore, inactive precursors must be administered from which they can be synthesized by the brain. A common precursor of both melanin and catecholamines is the amino acid 3,4-dihydroxyphenylalanine (dopa). The administration of L-dopa to Parkinsonian patients was found to improve them significantly, regardless of the cause of the disease (Committee on Biologic Effects of Atmospheric Pollutants, 1973); this suggests that the metabolic sequelae were related to the localization, not to the nature, of the brain damage.

Although the pathology of chronic manganese poisoning has not yet been sufficiently studied, it was speculated that at least some of the symptoms common to the two diseases might be due to similar metabolic sequelae within surviving neurons. In Parkinson's disease, slowly increasing doses of L-dopa have produced marked improvement of rigidity and hypokinesia, and high doses have decreased or stopped tremor. During treatment, some previously hypokinetic patients have developed involuntary movements; other side effects have been the emergence of mental aberrations and intermittent loss of the therapeutic action of L-dopa (Cotzias *et al.*, 1971).

Response to L-Dopa

In manganic patients, the response to L-dopa has been a function of the neurological pattern of symptoms. Rigid, hypokinetic patients with loss of postural reflexes and impairment of gait have responded to doses greater than 3 g per day with marked to total reduction of rigidity, improvement of postural reflexes and gait, and correction of hypokinesia, but no improvement of speech (Mena *et al.*, 1970). The therapeutic effects lasted while L-dopa was given (for periods of up to 4 yr), but the symptoms reemerged after 7 to 10 days on placebo therapy. Notably absent in these patients have been side effects such as involuntary movements, mental aberrations, or intermittent loss of therapeutic effect. Several of these patients returned to minor menial jobs.

In a second pattern of symptoms, dystonic manganic patients were given doses of 4 to 5 g of L-dopa per day with improvement of the dystonia and diminution of the passive muscular tonus. However, physical strain and emotional stress were able to trigger the appearance of dystonic crisis. After periods of 3 to 4 months, L-dopa lost its effectiveness and dystonia reemerged with greater intensity than the pretreatment level. Placebo administration for 10 to 30 days caused this abnormality to regress, and L-dopa therapy was reinstituted with the same therapeutic effects as before.

A third pattern was represented by one patient without rigidity but rather with muscular hypotonus, tremor, slowness, and impaired postural reflexes. Treatment with 1.2 g of L-dopa per day caused a marked aggravation of hypotonia, impairment of postural reflexes, and further impairment of gait; 3 g per day also caused worsening of tremor. Placebo administration restored pretreatment levels after 48 h.

In the United States, Rosenstock et al. (1971) have reported that in a patient working in a steel foundry, L-dopa improved the masklike face, markedly improved rigidity, slightly improved slowness, but did not improve dystonia. Greenhouse (1971) has reported, in four patients from a manganese ore-crushing plant, a clinical pattern of impairment of postural reflexes and slowness of movements without major extrapyramidal symptoms such as rigidity and tremor; these patients did not respond to doses of 5 g of L-dopa per day. These investigators have not reported major side effects with L-dopa treatment.

In summary, rigid, hypokinetic Chilean miners have responded in a sustained way to treatment with L-dopa; an American industrial worker with rigidity responded less well; and American industrial workers with impairment of stability as the major finding did not respond at all.

Our knowledge of the effect of overexposure of humans to manganese is limited to observations made on miners and industrial workers. Thus, maximum permissible amounts in air have been established only for these male workers: for an 8-h work day, the levels are 5 mg/m³ in the United States and 0.3 mg/m³ in the USSR (Committee on Biologic Effects of Atmospheric Pollutants, 1973). What should be the maximum permissible dose of manganese in air for 24-h/day exposure of women, pregnant women, infants, and children is not known. Yet this information will be necessary if lead in gasoline and fuel oil is replaced by methyl manganese tricarbonyl (MMT) compounds. These compounds have been used in gasoline since 1974. Archipova et al. (1963, 1965) have reported on toxicity of manganese cyclopentadienyltricarbonyl; however, because MMT has a different chemical structure, these results do not bear on its toxicity. Additional information will be needed concerning the toxicity of the oxidative inorganic compounds generated (e.g., MnO_2, Mn_3O_4).

A crude estimate can be made of the increase of manganese in the air should it be used to replace lead in gasolines. One gram of manganese per gallon of gasoline would increase the levels of manganese (in the atmosphere) two to eight times from 0.10 μg/m³ to 0.2–0.8 μg/m³ (Committee on Biologic Effects of Atmospheric Pollutants, 1973). Thus, the current average inhalation of 3 μg/day by an adult man would be increased to 6–24 μg/day (assuming an inhalation of 30 m³ of air per day in an adult man). Lesser amounts of manganese added per gallon of gasoline would reduce the amount of Mn/m³ proportionately. Although the ultimate fate of inhaled manganese has not been fully determined, probably most of it will eventually be swallowed and absorbed through the intestine.

Under normal conditions, the dietary manganese intake of man is approximately 3 mg/day. Of this 3 mg, 3 percent is absorbed; and after a rapid reexcretion phase, a net amount of 1 percent is retained. This amounts to a net retention of approximately 30 μg of Mn/day. Departure from the normal adult pattern has been shown in adults with iron deficiency; increased iron absorption is coupled with 200–300 percent increase of manganese absorption (Mena et al., 1969). Other circumstances under which one might expect a departure from the adult pattern are: (1) rate of intestinal absorption of manganese in infants and newborns, and (2) transfer of manganese across the blood/brain barrier in very early life when the blood/brain barrier is not fully developed.

RECOMMENDATIONS FOR RESEARCH

Areas for further research on manganese include:

1. Interrelationships of manganese metabolism with iron metabolism. The basis of the interaction of manganese and iron has been documented in this report on chemical and biological grounds. Although the effects of high levels of dietary manganese in the presence of adequate dietary levels of iron have been investigated rather extensively, only a few studies have been reported where the effect of excess manganese was observed under dietary conditions of iron deficiency or in anemic experimental subjects. The limited data available indicate that under conditions of anemia, levels of manganese as low as 50 ppm may inhibit hemoglobin synthesis. This phenomenon should be investigated in depth, and, if confirmed, the mechanism of the interference of manganese with iron metabolism should be elucidated.

2. Interrelationship between manganese metabolism and calcium and phosphorus levels in the diet.

3. Susceptibility to manganese poisoning of infants, older persons, and women of childbearing age.

4. Drug interference with manganese metabolism.

5. Fate and toxicity of inhaled MnO_2 and other products from combustion of methyl manganese tricarbonyl compounds (in gasoline); effects of particle size.

6. Intestinal absorption of manganese in newborn infants and in young children.

7. Blood/brain barrier to manganese in prenatal life; toxicity of manganese in prenatal life; relation of manganese and catecholamines metabolism in prenatal life and adult life.

8. Studies on selective manganese chelation therapy in animals and chronically overexposed people.

9. Studies of the relative biological availability and

toxicity of manganese incorporated into food as compared to the data available when the manganese is admixed with diet or added to the animal's milk or water.

REFERENCES

Anderson, J. M., and N. A. Pyliotis. 1969. Studies with manganese-deficient spinach chloroplasts. Biochim. Biophys. Acta 189:280–293.

Archipova, O. G., M. S. Tolgskaya, and T. A. Kochetkova. 1963. Toxic properties of manganese cyclopentadienyltricarbonyl antiknock substance. Hyg. Sanit. 28:29.

Archipova, O. G., M. S. Tolgskaya, and T. A. Kochetkova. 1965. Toxicity within a factory of the vapor of new antiknock compound cyclopentadienyltricarbonyl. Hyg. Sanit. 30:40–44.

Bach, S. J., and D. B. Whitehouse. 1954. Purification and properties of arginase. *In* Proceedings of the Biochemical Society. Biochem. J. 57:xxxi. (Abstract)

Bentley, O. G., and P. H. Phillips. 1951. The effect of low manganese rations upon dairy cattle. J. Dairy Sci. 34:396–403.

Borg, D. C., and G. C. Cotzias. 1958. Incorporation of manganese into erythrocytes as evidence for a manganese porphyrin in man. Nature 182:1677.

Boyd, C. E. 1968. Some aspects of aquatic plant ecology. Reservoir Fishery Resources Symposium, Athens, Georgia, 1967.

Bradford, G. R., R. J. Arkley, P. F. Pratt, and F. L. Bair. 1967. Total content of nine mineral elements in fifty selected benchmark soil profiles of California. Hilgardia 38(14):541–556.

Britton, A. A., and G. C. Cotzias. 1966. Dependence of manganese turnover on intake. Am. J. Physiol. 211:203–206.

Caskey, C. D., and L. C. Norris. 1938. Further studies on the role of manganese in poultry nutrition. Poult. Sci. 17:433.

Cheniae, G. M. 1970. Photosystem II and O_2 evolution. Ann. Rev. Plant Physiol. 21:467–498.

Committee on Biologic Effects of Atmospheric Pollutants. 1973. Manganese. Medical and Biologic Effects of Environmental Pollutants Series. National Academy of Sciences, Washington, D.C. 191 pp.

Cotzias, G. C. 1962. Manganese. *In* Mineral metabolism: An advanced treatise. C. L. Comar and F. Bronner [eds.], Vol. 2, The elements, Part B. Academic Press, New York, pp. 403–442.

Cotzias, G. C., K. Horiuchi, S. Fuenzalida, and I. Mena. 1968. Chronic manganese poisoning: Clearance of tissue manganese concentrations with persistence of the neurological picture. Neurology 18(4):376–382.

Cotzias, G. C., P. S. Papavasiliou, J. Ginos, A. Steck, and S. Duby. 1971. Metabolic modification of Parkinson's disease and of chronic manganese poisoning. Ann. Rev. Med. 22:305–326.

Cotzias, G. C., L. C. Tang, S. T. Miller, D. Sladic-Simic, and L. S. Hurley. 1972. A mutation influencing the transportation of manganese, L-dopa and L-tryptophan. Science 176:410–412.

De Groot, A. J., J. J. M. Goeji, and C. Zegers. 1971. Contents and behaviour of mercury as compared with other heavy metals in sediments from the rivers Rhine and Ems. Geol. Mijnbouw 50(3):393–398.

Diez-Ewald, M., L. R. Weintraub, and W. H. Crosby. 1968. Interrelationship of iron and manganese metabolism. Proc. Soc. Exp. Biol. Med. 129:448.

Epstein, E. 1972. Mineral nutrition of plants: Principles and perspectives. John Wiley & Sons, New York. 412 pp.

Erway, L., L. S. Hurley, and A. Fraser. 1966. Neurological defect: Manganese in phenocopy and prevention of a genetic abnormality of inner ear. Science 152:1766–1768.

Everson, G. J., L. S. Hurley, and J. F. Geiger. 1959. Manganese deficiency in the guinea pig. J. Nutr. 68:49–56.

Gallup, W. D., and L. C. Norris. 1939. The amount of manganese required to prevent perosis in the chick. Poult. Sci. 18:76–82.

Gerloff, G. C., D. G. Moore, and J. T. Curtis. 1964. Mineral content of native plants of Wisconsin. Wis. Agric. Exp. Stn. Res. Rep. 14.

Greenhouse, A. H. 1971. Manganese intoxication in the United States. Trans. Am. Neurol. Assoc. 96:248–249.

Hartman, R. H. 1956. A study of the mode of action of manganese in the utilization of iron by rabbits and lambs. Ph.D. Thesis (Animal Industry), North Carolina State University.

Hartman, R. H., G. Matrone, and G. H. Wise. 1955. Effect of high dietary manganese on hemoglobin formation. J. Nutr. 55:429–439.

Hawkins, G. E., Jr., G. H. Wise, G. Matrone, R. K. Waugh, and W. L. Lott. 1955. Manganese in the nutrition of young dairy cattle fed different levels of calcium and phosphorus. J. Dairy Sci. 38:536–547.

Hem, J. D. 1970. Study and interpretation of the chemical characteristics of natural water. U.S. Geol. Surv. Water-Supply Pap. 1473. U.S. Government Printing Office, Washington, D.C. 363 pp.

Hurley, L. S., and G. J. Everson. 1959. Delayed development of righting reflexes in offspring of manganese-deficient rats. Proc. Soc. Exp. Biol. Med. 102:360–362.

Hurley, L. S., and G. J. Everson. 1963. Influence of timing of short-term supplementation during gestation on congenital abnormalities of manganese-deficient rats. J. Nutr. 79:23–27.

Hurley, L. S., G. J. Everson, and J. E. Geiger. 1958. Manganese deficiency in rats: congenital nature of ataxia. J. Nutr. 66:309–319.

Hurley, L. S., E. Wootten, G. J. Everson, and C. W. Asling. 1960. Anomalous development of ossification in the inner ear of offspring of manganese-deficient rats. J. Nutr. 71:15–19.

Jenne, E. A. 1968. Controls on Mn, Fe, Co, Ni, Cu, and Zn concentrations in soils and water: The significant role of hydrous Mn and Fe oxides. Adv. Chem. 73:337–387.

Kemmerer, A. R., C. A. Elvehjem, and E. B. Hart. 1931. Studies on the relations of manganese to the nutrition of the mouse. J. Biol. Chem. 92:623–630.

Krauskopf, K. B. 1967. Introduction to geochemistry. McGraw-Hill, New York. 721 pp.

Kubota, J., and W. H. Allaway. 1972. Geographic distribution of trace element problems in micronutrients in agriculture. Soil Science Society of America, Madison, Wisconsin. pp. 525–554.

Kubota, J., S. Rieger, and V. A. Lazar. 1970. Mineral composition of herbage browsed by moose in Alaska. J. Wildl. Manage. 34:565–569.

Labanauskas, C. K. 1966. Manganese. *In* Diagnostic criteria for plants and soils, H. D. Chapman [ed.]. University of California, Division of Agricultural Sciences, Riverside. pp. 264–285.

Lassiter, J. W., W. J. Miller, F. M. Pate, and A. P. Gentry. 1972. Effect of dietary calcium and phosphorus on ^{54}Mn metabolism following single tracer intraperitoneal and oral doses in rats. Proc. Soc. Exp. Biol. Med. 139:345–348.

Leach, R. M., Jr. 1967. Role of manganese in the synthesis of mucopolysaccharides. Fed. Proc. 26:118–120.

Leach, R. M., Jr. 1971. Role of manganese in mucopolysaccharide metabolism. Fed. Proc. 30:991.

Leach, R. M., Jr., A. M. Muenster, and E. Wien. 1969. Studies on the role of manganese in bone formation. II. Effect upon chondroitin sulfate synthesis in chick epiphyseal cartilage. Arch. Biochem. Biophys. 133:22–28.

Mahler, H. R. 1961. Mineral metabolism, C. L. Comar and F.

Bronner [eds.], Vol. 1, Part B. Academic Press, New York. pp. 773–801.

Mahoney, J. P., and W. J. Small. 1968. Studies on manganese. III. The biological half-life of radiomanganese in man and factors which affect this half-life. J. Clin. Invest. 47:643–653.

Makarcenko, A. F. 1956. Izmenenia nervnoi sistemi pri intoxicatii margantsem [Changes in the nervous system during intoxication with manganese]. Kiev.

Marbut, C. F. 1935. Soils of the United States. USDA atlas of American agriculture, Pt. 3. U.S. Government Printing Office, Washington, D.C. 98 pp.

Matrone, G. 1969. Biochemistry and mechanisms of action of trace elements. Proc. 8th Int. Congr. Nutr. 213. Prague, Czechoslovakia. pp. 171–175.

Matrone, G., R. H. Hartman, and A. J. Clawson. 1959. Studies of a manganese–iron antagonism in the nutrition of rabbits and baby pigs. J. Nutr. 67:309–317.

Mella, H. 1924. The experimental production of basal ganglion symptomatology in Macacus rhesus. Arch. Neurol. Psychiatry 11:406–417.

Mena, I., O. Marin, S. Fuenzalida, and G. C. Cotzias. 1967. Chronic manganese poisoning: Clinical pictures and manganese turnover. Neurology 17:128.

Mena, I., K. Horiuchi, K. Burke, and G. C. Cotzias. 1969. Chronic manganese poisoning: Individual susceptibility and absorption of iron. Neurology 19:1000–1006.

Mena, I., J. Court, S. Fuenzalida, P. S. Papavasiliou, and G. C. Cotzias. 1970. Modification of chronic manganese poisoning: Treatment with L--Dopa or 5-OH tryptophane. N. Engl. J. Med. 282: 5–10.

Miller, R. C., T. B. Keith, M. A. McCarty, and W. T. S. Thorp. 1940. Manganese as a possible factor influencing the occurrence of lameness in pigs. Proc. Soc. Exp. Biol. Med. 45:50–51.

Mitchell, H. H. 1947. The mineral requirements of farm animals. J. Anim. Sci. 6:365.

Molinoff, P. B., and J. Axelrod. 1971. Biochemistry of catecholamines. Ann. Rev. Biochem. 40:465–500.

Moore, D. P. 1972. Mechanism of micronutrient uptake by plants. In Micronutrients in agriculture. Soil Science Society of America, Madison, Wisconsin. pp. 171–198.

Morgan, J. J. 1967. Chemical equilibria and kinetic properties of manganese in natural waters. In Principles and applications of water chemistry, S. D. Faust and J. V. Hunter [eds.]. Fourth Rudolfs Conference. John Wiley & Sons, New York. pp. 561–624.

Myers, H. C. 1961. Manganese deposits in western reservoirs and distribution systems. J. Am. Water Works Assoc. 53: 579–588.

Neff, N. H., R. E. Barrett, and E. Costa. 1969. Selective depletion of caudate nucleus dopamine and serotonin during chronic manganese dioxide administration to squirrel monkeys. Experientia 25:1140–1141.

Norris, L. C., and C. D. Caskey. 1939. A chronic congenital ataxia and osteodystrophy in chicks due to manganese deficiency. J. Nutr. 17:16–17. (Abstract)

Orent, E. R., and E. V. McCollum. 1931. Effects of deprivation of manganese in the rat. J. Biol. Chem. 92:651.

Orgel, L. E. 1958. Enzyme–metal–substrate complexes as coordination compounds. In Metals and enzyme activity. Biochem. Soc. Sym. 15. Cambridge University Press, Cambridge. p. 14.

Orgel, L. E. 1963. An introduction to transition-metal chemistry: Ligand-field theory. Butler and Tanner, Ltd., London. p. 42.

Pentschew, A., F. F. Ebner, and R. M. Kovatch. 1963. Experimental manganese encephalopathy in monkeys. J. Neuropath. Exp. Neurol. 22:488–499.

Plumlee, M. P., D. M. Thrasher, W. M. Beeson, F. N. Andrews, and H. E. Parker. 1956. The effects of a manganese deficiency upon the growth, development, and reproduction of swine. J. Anim. Sci. 15:352–367.

Pollack, S., J. N. George, R. C. Reba, R. M. Kaufman, and W. H. Crosby. 1965. The absorption of nonferrous metals in iron deficiency. J. Clin. Invest. 44:1470–1473.

Price, C. A., H. E. Clark, and E. A. Funkhouser. 1972. Functions of micronutrients in plants. In Micronutrients in agriculture. Soil Science Society of America, Madison, Wisconsin, pp. 231–239.

Riley, J. P., and R. Chester. 1971. Introduction to marine chemistry. Academic Press, New York. 465 pp.

Riopelle, A. J. 1969. Manganese effects in chimpanzees. In G. E. Crane and R. Gardner, Jr. [eds.]. Psychotropic drugs and dysfunctions of the basal ganglia: A multidisciplinary workshop. Public Health Service Publ. No. 1938. U.S. Government Printing Office, Washington, D.C. pp. 47–52.

Rosenstock, H. A., D. G. Simons, and J. S. Meyer. 1971. Chronic manganism. Neurologic and laboratory studies during treatment with levodopa. J. Am. Med. Assoc. 217(10):1354–1358.

Scrutton, M. C., M. F. Utter, and A. S. Mildvan. 1966. Pyruvate carboxylase. IV. The presence of tightly bound manganese. J. Biol. Chem. 241:3480–3487.

Shacklette, H. T., J. C. Hamilton, J. G. Boerngen, and J. M. Bowles. 1971. Elemental composition of surficial materials in the conterminous United States. U.S. Geol. Surv. Prof. Pap. 574-D. 71 pp.

Shils, M. E., and E. V. McCollum. 1943. Further studies on symptoms of manganese deficiency in the rat and mouse. J. Nutr. 26:1–19.

Slowey, J. F., Jr. 1966. Studies on the distribution of copper, manganese, and zinc in the ocean using neutron activation analysis. In The chemistry and analysis of trace metals in sea water, D. W. Hood [ed.]. AEC Contract AT(40-1)-2799 Final Report, Texas A&M University, Proj. 276. pp. 1–105.

Smith, F. H., K. C. Beeson, and W. E. Price. 1956. Chemical composition of herbage browsed by deer in two wildlife management areas. J. Wildl. Manage. 20:359–367.

Somers, I. I., and J. M. Shive. 1942. The Fe–Mn relation in plant metabolism. Plant Physiol. 17:582–602.

Stone, E. L. 1968. Micronutrient nutrition of forest trees: A review in forest fertilization. Theory and practice. TVA, Muscle Shoals, Alabama. pp. 132–175.

Tanaka, A., and S. Yoshida. 1970. Nutritional disorders of the rice plant in Asia. Int. Rice Res. Inst. Tech. Bull. 10. p. 51.

Taylor, R. M., R. M. McKenzie, and K. Norrish. 1964. The mineralogy and chemistry of manganese in some Australian soils. Aust. J. Soil Res. 2: 235–248.

Tiller, K. G. 1963. Weathering and soil formation on dolerite in Tasmania with particular reference to several trace elements. Aust. J. Soil Res. 1(1):74–90.

Underwood, E. J. 1971. Trace elements in human and animal nutrition, 3d ed. Academic Press, New York. 543 pp.

U.S. Geological Survey. 1972. Environmental geochemistry: Geochemical survey of Missouri, plans and progress for fifth six-month period (July–December, 1971). U.S. Geol. Surv. Open-File Rep. No. 1706. Branch of Regional Geochemistry, USGS, Denver and Reston. 145 pp.

Vallee, B. L. 1962. Mineral metabolism, C. L. Comar and F. Bronner [eds.], Vol. 2, Part B. Academic Press, New York. p. 743.

van Bogaert, L. 1943. On the cortico-cerebellar atrophy observed in the monkey subjected to poisoning by aerosols. Bull. Acad. R. Med. Belg. 8:99.

Wachtel, L. W., C. A. Elvehjem, and E. B. Hart. 1943. Studies on the physiology of manganese in the rat. Am. J. Physiol. 140:72–82.

Wasserman, M., and G. Mihail. 1964. Significant indicators for early detection of manganism in miners. Acta Med. Legalis Soc. Liege 17:61–89.

Wedepohl, K. H. 1969. Composition and abundance of common igneous rocks. *In* Handbook of geochemistry, K. H. Wedepohl [ed.]. Vol. 1. Springer-Verlag, New York–Heidelberg–Berlin. pp. 227–271.

V

Nickel

FORREST H. NIELSEN, *Chairman*

Horace T. Reno, Lee O. Tiffin, Ross M. Welch

Nickel is ubiquitous in nature, occurring in air, water, rocks, and soils, and is incorporated into all biological materials (Schroeder *et al.,* 1962; Bowen, 1966; Underwood, 1971). Factors important to the translocation of nickel in the biosphere and the food chain are: (1) geochemical and biogeochemical processes that release nickel from geochemical reserves in forms available to living organisms; (2) biological processes involved in the accumulation and metabolism of nickel in plant and animal tissue; and (3) man-related processes that release nickel or its compounds to the environment for accumulation in the biosphere, geosphere, atmosphere, and hydrosphere.

Nickel enters man's food in two ways: (1) by movement along the food chain from naturally occurring sources in rocks and minerals to soils, to plants, and then to animals or man, and (2) through by-products of man's activities (e.g., nickel residues in food resulting from the use of nickel-containing fungicides, fertilizers, soil amendments, and sewage sludges in the production of agronomically important crops; the use of nickel alloys in food-processing equipment; and contamination of food and feed with industrial and municipal solid, liquid, and aerosol wastes containing nickel residues). In general, the former route provides natural barriers against high levels of nickel in foods because of the high phytotoxicity of nickel to edible plants (Vanselow, 1966; Allaway, 1968). The latter route may result in the accumulation of nickel in foods to levels much higher than would occur naturally in food and feed crops.

The biological essentiality of nickel for chicks and rats has recently been reported (Sunderman *et al.,* 1972a; Nielsen *et al.,* 1975a, 1975b), but its essentiality for plants and man has yet to be proven.

Nickel is relatively nontoxic orally, and nickel contamination when ingested probably does not present a serious health hazard. Some forms of nickel (e.g., metallic nickel dust and nickel carbonyl) are carcinogenic, however, when inhaled. Some forms of nickel may cause contact dermatitis.

GEOCHEMISTRY

Rocks, Minerals, and Soil

Nickel comprises about 0.008 percent of the earth's crust. By far the largest part of this is in the igneous rocks, of which nickel comprises approximately 0.01 percent. Nickel concentrations of the common rocks found in the geological units of the upper part of the earth's crust are listed in Table 2. The chemical composition of soils and waters depends, to a large extent, on the composition of closely associated rocks.

Among igneous rocks in the lithosphere, the ultramafic rocks are the richest in nickel, ranging from 140 ppm in gabbro to an average of 2,000 ppm in peridotite. Diorite contains 40 ppm, and granitic rocks contain about 8 ppm. Among sedimentary rocks, shale and carbonate rocks contain an average of approximately 50 ppm nickel, whereas

TABLE 2 Nickel Content (ppm) of Rocks, Soils, and Minerals

Sample Type	Range Usually Reported	Average
Ultramafic igneous	270–3,600	2,000
Basalts and gabbros	45–410	140
Granitic rocks	2–20	8
Shales and clays	20–250	68
Limestones	–	20
Sandstones	–	2
Soils	5–500	40
Phosphorites	10–1,000	100
Coals	3–100	20

SOURCES: Turekian and Wedepohl (1961), Rankama and Sahama (1950), Abernethy *et al.* (1969).

sandstone contains about 2 ppm nickel. As shown in Figure 2, rock types low in silica are high in nickel (except for carbonate), and those high in silica are relatively low in nickel. Soils contain between 5 and 500 ppm nickel (Table 2). Those soils carrying less than 5 ppm are too acidic to support normal plant growth. Based on the available data, it is estimated that the average farm soil in the United States contains more than 40 ppm nickel.

In general, the total nickel content of a soil is a poor measure of the availability of the nickel to plants. A better measure of nickel availability to plants from soils is the extractable (soluble plus exchangeable) nickel content of the soils. Various aqueous solutions used as extractants to measure the availability of nickel to plants from soils include 1 *N* (Normal, 1 g equivalent of a solute per liter) potassium chloride, 1 *N* ammonium acetate, acetic acid adjusted to various pH levels, 2.5 percent acetic acid (pH 2.5), and 0.05 *M* ethylenediaminetetraacetic acid (EDTA) (pH 7). The nickel content of plants appears to be closely correlated to the extractable nickel of soils as determined by any of these extractants (Mitchell *et al.*, 1957; Swaine and Mitchell, 1960; Mitchell, 1964, 1971; Halstead, 1968; Halstead *et al.*, 1969; Sillanpää and Lakanen, 1969; Roth *et al.*, 1971).

The content of extractable nickel in different soil types varies a great deal depending on a number of complex factors (Swaine and Mitchell, 1960; Ng and Bloomfield, 1962; Mitchell, 1964, 1971; Jenne, 1968). Among these are soil parent material, soil physical factors (e.g., soil texture, structure, aeration, temperature, water content, and degree of development), soil chemical factors (e.g., pH, organic matter content, lime level, and oxidation-reduction potential), and soil biological factors (e.g., microbial activity, plant-root excretions, and depletions of mineral elements by plant roots and microorganisms). As a result of the dynamic nature of soil processes, the extractable nickel content of a soil is in a constant state of flux. Extractable nickel in most soils, except for those derived from serpentine parent material, range from less than 0.01 ppm to 2.6 ppm (Mitchell *et al.*, 1957; Swaine and Mitchell, 1960; Mitchell, 1964, 1971; Vanselow, 1966; Halstead, 1968; Halstead *et al.*, 1969; Sillanpää and Lakanen, 1969; Roth *et al.*, 1971). The values for extract-

able nickel in phytotoxic soils (e.g., serpentine soils) derived from ultrabasic rocks are much higher and range from 3 to 70 ppm (Vanselow, 1966). The toxicity of these soils to plants should not be attributed entirely to high levels of extractable nickel. Other causal factors of toxicity are involved, such as high concentrations of manganese (Williams, 1967), inhibition of iron metabolism by high nickel and manganese levels in the plants (Williams, 1967), unfavorable magnesium : calcium exchange ratios, and possibly toxic chromium concentrations (Vanselow, 1966; Lyon *et al.*, 1968; Peterson, 1971).

Water

Nickel concentrations in seawater range from 0.0001 to 0.0005 ppm. In most groundwaters, nickel has not been detected, and, where found, it is probably in colloidal form (U.S. Geological Survey, 1965).

During rock weathering, nickel tends to form an insoluble hydrolysate; therefore, any nickel is likely to occur in small amounts in surface or groundwaters, unless it is present as a result of industrial pollution (Kopp and Kroner, 1968).

Kopp and Kroner (1968) reported that nickel was found in United States surface waters with a frequency of 16 percent and at an overall mean concentration of 19 ppb, and it was observed in at least one sample in every major drainage basin of the United States. The highest value found was 130 ppb in the Cuyahoga River, near heavily industrialized Cleveland, Ohio.

The nickel content of selected samples taken in 1962 of public water supplies of the 100 largest cities in the

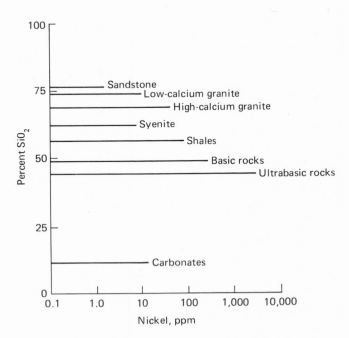

FIGURE 2 Average nickel content in various sedimentary and igneous rocks (Turekian and Wedepohl, 1961).

United States was reported by the U.S. Geological Survey (Durfor and Becker, 1964) to range from 4 to 56 μg/l. The samples taken were of water at its source, in storage, and in various stages of treatment. These data strengthen the conclusion that most of the nickel in surface and groundwaters above a concentration of about 10 μg/l probably reflects man's activities. Nickel in natural waters apparently would present a health hazard only in most unusual circumstances.

Atmosphere

Nickel in the atmosphere ranges from less than 0.0016 μg/m^3 to 0.052 μg/m^3 according to the air quality data reported by the National Air Surveillance Networks. Extensive sampling has shown that it is most concentrated in urban areas in the vicinity of heavily traveled highways (National Air Pollution Control Administration, 1968a) and is probably derived from asphalt and automobile tires. The existence of large areas of the earth with less than 0.0016 μg/m^3 nickel in the atmosphere suggests that man's activities are responsible for most atmospheric nickel.

Coal

Nickel, which occurs in virtually all coals, along with a number of other trace elements, is retained largely in the ash when coal is burned. Abernethy *et al.* (1969) have shown that nickel content of coal varies with geographic origin.

Of 600 analyses of coals taken from eight eastern states, the average ash content of the coals was 9.3 percent, and the average nickel content of ash was 0.0209 percent. This compares with an average ash content of 10.5 percent containing 0.0262 percent nickel for 123 analyses of coals from seven midwestern states, and an average 9.8 percent ash with only 0.0054 percent nickel content for 104 analyses of coals from eight western states. It follows, therefore, that the nickel content of United States coals varies from an average of about 0.06 lb/ton in the midwestern states to 0.04 lb/ton in the eastern states to 0.01 lb/ton in the western states.

Petroleum

The nickel content of domestic crude oils has been reported to range from 1.4 to 64 ppm and that of imported crude oils from 0.3 to 29.5 ppm. The median content of the domestic crude was 4.3 ppm, and the average of the analyses reported was 14.2 ppm; the median of the imported crude was 6 ppm, and the average was 10 ppm (National Air Pollution Control Administration, 1968b).

According to C. W. Kelly (personal communication, 1973), petroleum is refined in a closed system, and there is little chance that nickel-bearing materials can escape to the atmosphere during refining processes. Exhaustive tests have indicated that nickel invariably remains with the heavier and higher boiling fractions of the crude-oil

processes. It is therefore eventually concentrated in residual fuel oils and in asphalt.

The nickel content of typical commercial residual fuel oils, like those burned in electrical power generating plants, reported in the Petroleum Products Handbook, ranges from nil to 20 ppm as follows (Guthrie, 1960):

Type	No. 4 Fuel Oil			No. 5 Fuel Oil			No. 6 Fuel Oil				
	A	B	C	A	B	E	F	B	C	G	H
Ni (ppm)	0.2	—	8	—	10	10	20	10	20	20	12	20

Complete analyses of stack gases, fly ash, and residual material at power generating plants are not available. In the past, nickel has been used as an additive in certain petroleum fuels, but apparently is not now so used.

Ore Deposits

Nickel ore deposits are formed as veins, stringers, or fissure fillings containing pentlandite (FeNiS$_2$), chalcopyrite (CuFeS$_2$), and pyrrhotite (Fe$_x$S$_{x+1}$). Other nickel ore deposits are formed by lateritic weathering of ultramafic ferromagnesium silicate rocks. Part of the nickel ore consists of hydrous nickel–magnesium silicates collectively referred to as "garnierite." Several nickel silicates, each an analog of a magnesium mineral (given in brackets), can occur as constituents of "garnierite" : nimite, (Ni, Mg, Fe, Al)$_3$ (Si,Al)$_2$O$_5$(OH)$_4$ [chlorite]; willemseite, (Ni, Mg)$_3$Si$_4$O$_{10}$(OH)$_4$ [talc]; pecoraite, Ni$_3$Si$_2$O$_5$(OH)$_4$ [clinochrysotile]; nepouite (Ni,Mg)$_3$Si$_2$O$_5$(OH)$_4$ [lizardite]; and pimelite, (Ni, Mg)$_3$Si$_4$O$_{10}$(OH)$_2 \cdot$4H$_2$O [stevensite] (Faust, 1966; Faust *et al.*, 1969; DeWaal, 1970).

UTILIZATION

Production

The principal nickel-producing areas of the world are shown in Figure 3. According to the U.S. Bureau of Mines (1975), the United States was eighth in nickel production in 1972 with 16,864 short tons of ore shipped. The seven leading nickel-producing countries are: Canada, USSR, New Caledonia, Australia, Cuba, Indonesia, and the Dominican Republic. All Canadian, Finnish, Rhodesian, and South African nickel is produced from sulfide ores. Part of that produced in the USSR and Australia is from sulfides; the remaining countries (Figure 3) produce from secondary ores.

Extraction

Nickel sulfide ores are mined chiefly underground; nickel minerals are concentrated by physical methods, and the concentrate in most instances is smelted pyrometallurgically. Secondary nickel ores are mined from open pits. Because the nickel minerals cannot be concentrated by physical means, the nickel is extracted

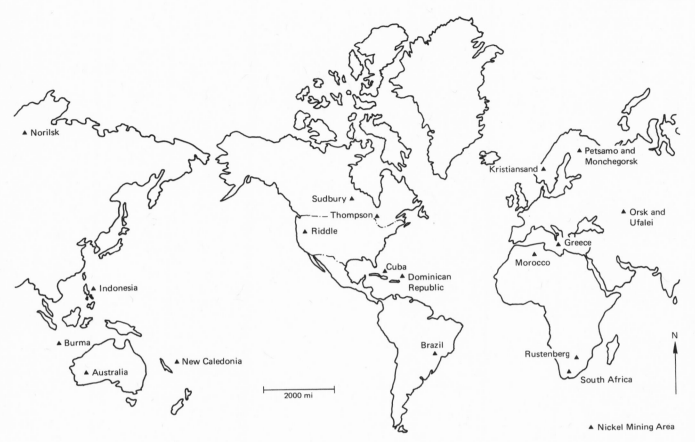

FIGURE 3 Principal nickel-producing areas of the world (Howard-White, 1963).

either in a chemical form by leaching or in the form of ferronickel by smelting.

Basically, the pyrometallurgical treatment includes five operations: concentrating, roasting, smelting, converting, and refining. There is practically no airborne effluent during concentration. The rejected portion of the ore is returned underground for mine fill or put in tailings ponds, which in time are stabilized by a cover of suitable vegetation.

Roasting generates a metallurgical smoke consisting of gases (carbon dioxide, water vapor, oxygen, and sulfur oxides), dust (partially reacted fine particles of concentrate, furnace lining, and fuel), and fume (solid material that has been volatilized and subsequently condensed). The dust composition depends on the type of material being roasted.

Effluents from the smelters and converters are similar to those emitted by the roasters. Normally, however, they are at higher temperatures than gases from the roasters and may not contain as large a percentage of the sulfur oxides if smelting and converting is preceded by roasting. In a well-operated nickel electrolytic refining operation, there is no measurable effluent to the atmosphere.

Nickel laterite ores are processed by smelting to pro-duce an iron–nickel matte, smelting to produce ferro-nickel, leaching with ammonia, or leaching with sulfuric acid. In contrast to underground nickel mines, dust is generated in loading, transporting, and in the blending and drying yards. Gases, dust, and fumes are emitted from the smelting furnaces as they are from the furnaces smelt-ing sulfide ores; however, in those furnaces that produce ferronickel, the sulfur oxide emissions are not as much of a problem. The ammonia and sulfuric-acid leaching sys-tems are in a closed circuit from which there is no emis-sion to the atmosphere. The precipitated nickel carbonate is roasted to remove carbon dioxide, which, however, carries some nickel oxide with it to the atmosphere.

Final Usage

Most nickel is used in metal alloys to make a wide variety of consumer hard goods. The nickel used to make stain-less steel reaches practically every household in the land, either in cooking utensils, flatware, or kitchen appliances. Other nickel alloys are strong, corrosion-resistant, en-gineering materials; they are used throughout industrial plants and municipal facilities, in some household water systems, or in place of stainless steel.

ANALYTICAL METHODS

There are adequate methods available for the analysis of nickel at the sensitivities required for most materials. At present the method of Nomoto and Sunderman (1970), using atomic-absorption spectrometry, is probably the best available for the determination of small quantities of nickel in biological tissues, although it is quite tedious. Methods are still needed for the determination of the chemical and physical forms of nickel in certain materials, such as fly ash, which can affect environmental quality (and thus human health).

NICKEL IN THE FOOD CHAIN

Soil Factors

In general, nickel availability to plants is associated with sorption and desorption of nickel ions on exchange sites in soil minerals. Most of these sites are present in organic matter or hydrous iron- and manganese-oxide coatings on soil particles. Sorption and desorption at these sites is controlled primarily by the solute concentration of nickel and other heavy metals, pH, oxidation-reduction potential, and the amount and strength of inorganic complex ion formers and organic chelates in the soil solution (Jenne, 1968). Impeded soil drainage, or anaerobic conditions, affect nickel mobilization and, hence, availability to plants (Mitchell, 1971).

Most investigations on the pedological factors affecting uptake by plants have been concerned with the poor growth of plants on serpentine soils high in extractable nickel. Visible symptoms of nickel toxicity resemble those of iron chlorosis in cereal crops, manganese deficiency in tomato plants, and zinc deficiency in citrus trees (Vanselow, 1966).

A high percentage of the nickel in peaty serpentine soils exists in organic complexes; its uptake by plants is reduced by raising the soil pH, generally by liming (Pratt et al., 1964; Vanselow, 1966; Roth et al., 1971). Halstead (1968) studied the effect of different soil amendments on the yield and nickel content of oat plants grown in serpentine soil. Additions of $Ca(OH)_2$ and/or organic matter to these soils increased the yield of oat tops by a factor of four and markedly reduced the nickel concentration in the tops. Halstead et al. (1969) reported that oats and alfalfa plants, grown in soils to which varying concentrations of nickel were added, reached nickel concentrations (dry weight) of 60 ppm in oat grain, 28 ppm in oat straw, and 44 ppm in alfalfa tops during toxicity. [For further references on the toxicity of nickel to plants, see Vanselow (1966) who has reviewed the extensive literature on this subject.]

Nickel Content

The average range of nickel concentrations in plant tops under normal conditions is 0.05–5 ppm dry weight (Schroeder et al., 1962; Vanselow, 1966; Sauchelli, 1969; Underwood, 1971). Tables 3 and 4 show the levels of nickel (both wet and dry weight) in plant materials commonly used as foods. Some of the variations seen in these values could be due to the dry versus wet weight basis, but other unaccounted-for variations may also be present; in addition, the validity of these older data are suspect because of the analytical methodology.

Accumulation and Essentiality in Plants

Nickel concentrations of more than 50 ppm (dry wt) in plants are usually toxic, but certain species found on serpentine soils may contain nickel levels far in excess of this. Vanselow (1966), Lyon et al. (1970), Wild (1970), and Peterson (1971) list most of these species. Table 5 shows nickel data for several nickel accumulator plants and the nickel content of the soils in which they grow. Although some plants do not accumulate large quantities of nickel, they may, nevertheless, be useful in locating nickel. Cannon (1960, 1971) and Timperley et al. (1970) list plants useful in mineral prospecting for nickel and other metal deposits.

TABLE 3   Nickel in Cereal Grains

Grain	Concentration, ppm
	Wet weight
Wheat, winter, seed	0.16
Wheat, Japanese	N.D.[a]
Wheat flour, Japanese	N.D.
Wheat flour, all-purpose	0.30–0.54
Wheat, crushed, Vermont	0.75
Bread, whole-wheat, stone-ground	1.33
Wheatena cereal	N.D.
Wheaties cereal	3.00
All-bran cereal	0.74
Grape Nuts cereal	0.13
Buckwheat, seed	6.45
Rye, seed	2.70
Oats, seed	1.71–2.60
Oats, precooked, quick	2.35
Corn meal, New Hampshire	N.D.
Corn oil	N.D.
Rice, Japanese, polished	0.50
Rice, Japanese, unpolished	1.80
Rice, Japanese, polished (204 samples)	0.65
Rice, American, polished	0.47
Rice, puffed	0.30[b]
	Dry weight
Corn, grain, mature	0.14
Oats, grain, mature	0.45
Rice, polished	0.02
Buckwheat, seed	1.34
Barley	4.00–6.00
Wheat, grain, mature	0.35–35

[a]Not detected.
[b]SOURCES: This and all previous data are from Schroeder et al. (1962). Ensuing data are from Vanselow (1966).

TABLE 4 Nickel in Vegetables and Fruits

Vegetable or Fruit	Concentration, ppm
	Wet weight
Potato, raw	0.56
Peas, fresh, frozen	0.30
Peas, canned	0.46
Peas, split, dried	1.66
Beans, string, frozen	0.65
Beans, string, canned	0.17
Beans, navy, dried	1.59
Beans, yellow-eye, dried	0.69
Beans, red kidney, dried	2.59
Spinach, fresh	0.35
Celery, fresh	0.37
Beet greens	1.94
Swiss chard, organic	0.71
Escarole, fresh	0.27
Chicory, fresh	0.55
Lettuce, garden, organic	1.14
Lettuce, head	0.14
Kale, organic	1.12
Kohlrabi, leaves, organic	0.47
Cabbage, white	0.14–0.32
Cabbage, red	0.24
Cauliflower leaves	0.19
Broccoli, fresh, frozen	0.33
Tomato, fresh	0.02
Tomato juice, canned	0.05
Apple, raw	N.D.[a]
Apple, raw	0.08
Banana	0.34
Pear	0.20[b]
	Dry weight
Spinach	2.40
Squash	4.60
Tomato	0.01–0.154
Cabbage	3.30
Carrot, root	0.30
Carrot, leaves	1.80
Cress, water, tops	0.50
Cress, water, leaves	0.13
Mushroom	3.50
Peas	2.00–2.25
Potato	0.08–0.37
Onion	0.16
Lettuce	1.51
Lentils	1.61
Haricot beans	0.59
Orange	0.16
Apricot	0.64
Plum	0.90
Pear	0.90
Fig	1.20

[a]Not detected.
[b]SOURCES: This and all previous data in table are from Schroeder *et al.* (1962). Ensuing data are from Vanselow (1966), except for lettuce and data following that, which are from Bertrand and Mokragnatz (1930).

Hodgson (1970) developed a generalized model relating various environmental factors and the trace element composition of plants:

$$P = f(R, S, A, p),$$

where P is the content of a given element in plant tissue, R is the elemental content of the soil parent (i.e., the rocks), S represents soil-forming factors, A is the available concentration of the element in the soil, and p is the interaction of the plant with its environment. Information is meager concerning the concentration of nickel in plants as affected by the plant response parameter p.

Tissue concentrations of nickel among plant species and varieties grown on the same soils vary widely (Fleming, 1963; Guha and Mitchell, 1966; Lyon *et al.*, 1968; Halstead *et al.*, 1969; Lyon *et al.*, 1970; Peterson, 1971; Ashton, 1972). The nickel concentration among the various plant parts of a single species also varies widely (Fleming, 1963; Guha and Mitchell, 1966; Halstead *et al.*, 1969; Sauchelli, 1969). Nickel tends to accumulate in the seeds of various plants to levels as much as three times those in the stems and leaves (Fleming, 1963; Halstead *et al.*, 1969; Patterson, 1971). Maturity and time of sampling (seasonal variations) also affect the nickel levels found in plants (Guha and Mitchell, 1966; Sauchelli, 1969). The nickel content of plants may also vary geographically (Zook *et al.*, 1970).

The essentiality of nickel for plant growth has not been proven, although there are reports of slightly beneficial effects of nickel on plant growth (*see* Vanselow, 1966, and references cited therein; Bertrand and de Wolf, 1965). Bertrand and de Wolf (1967) suggest that nickel may be an essential element for the growth of a green alga, *Chlorella vulgaris*; maximum growth rates appeared at nutrient solution concentrations of 3 ppb.

Nickel may interact with iron in plant nutrition. Crooke (1955) and Crooke and Knight (1955) showed that both uptake and toxicity symptoms of nickel in plants are reduced if the culture solution is high in iron. In other work, Halstead (1968) showed that the addition of $Ca(OH)_2$ and/or organic matter to an unproductive serpentine soil increased plant yields and decreased nickel concentrations in the plants. The effect of these treatments appeared to be primarily on the solubility of soil nickel rather than on plant physiology.

Little work has been done with respect to interactions of nickel with other elements in plant nutrition and how they may be related to nickel deficiency or toxicity in humans and animals. Dixon *et al.* (1975) suggest that nickel may be an essential trace element for jack beans, because their findings show that jack bean urease is a nickel metalloenzyme.

Translocation in Plants

Tiffin (1971) used the radioisotope [63]Ni to study uptake and translocation of nickel in several agricultural species, including tomatoes, carrots, corn, cucumbers, and

TABLE 5   Nickel in Soils and Accumulator Plants

Plant	Country	Ni in Soil, ppm	Ni in Plants, ppm	Source
		Dry weight		
Dicoma macrocephala	Rhodesia	7,375	1,401	Wild, 1970
Becium obovatum	Rhodesia	3,563	261	Wild, 1970
Vellozia equisetoides	Rhodesia	5,927	380	Wild, 1970
Blepharis bainessi	Rhodesia	6,250	625	Wild, 1970
Becium homblei	Rhodesia	5,622	171	Wild, 1970
		Ash weight		
Lichen (unknown species)	New Zealand	1,700	8,300	Lyon and Brooks, 1970
Pimelea suteri	New Zealand	4,600	5,500	Lyon and Brooks, 1970
Myosotis monroi	New Zealand	3,800	8,000	Lyon and Brooks, 1970
Leptospermum scoparium	New Zealand	5,000	4,900	Lyon and Brooks, 1970
Hebe odora	New Zealand	2,400	2,600	Lyon and Brooks, 1970
Gentiana corymbifera	New Zealand	2,500	1,300	Lyon and Brooks, 1970
Myrsine divaricata	New Zealand	1,700	2,000	Lyon and Brooks, 1970
Cassinia vauvilliersii var. *serpentina*	New Zealand	2,500	2,000	Lyon and Brooks, 1970
Alyssum bertolonii	Italy	—	100,000 (as NiO)	Gambi, 1967

peanuts. Table 6 gives his data for tomatoes. Two plants were placed in each liter of nutrient solution containing a nickel concentration as shown and 50 μCi (microcuries) [63]Ni. Plants were topped and exudate was collected for 10 h. Electrophoretic distribution of nickel in the exudates is shown in Figure 4. The exudate nickel concentrations were similar to the original nutrient concentrations, except on the lowest nutrient treatment; there, the exudate nickel was double that supplied in the nutrient. The highest nutrient nickel level depressed exudate volume. However, the proportion of the absorbed nickel that was released in the xylem exudate was about the same (5–6 percent) for the three nutrient concentrations.

Figure 4 shows the electrophoretic patterns for nickel in the three exudates listed in Table 6. All exudates show the anodic and cathodic migration of nickel, but the proportions of the metal in these fractions are very different for the three exudates. In exudate 1, 85 percent of the nickel ran as a negatively charged fraction; the remaining 15 percent ran cathodically. The distributions for exudate 2 were 35 and 65 percent, and for exudate 3 were 5 and 95

percent. These patterns demonstrate the limited binding capacity (< 175 ppb nickel equivalent) of the anodic carrier in these exudates.

The anodic bands of nickel (effluents 15 to 16, Figure 4) do not indicate the presence of more than one nickel complex, although in recent studies with peanut-root sap, Tiffin (unpublished data, 1973) has determined that at least two nickel carriers are involved in the anodic migration of nickel. This was demonstrated by running nickel-free root sap electrophoretically, then adding [63]Ni to the effluent fractions before rerunning them electrophoretically in pH 5.4 buffer. The results showed that an anionic component in effluent 16 bound nickel and migrated in the rerun to about the same anodic position. With or without nickel, this carrier had the same electrophoretic mobility. Another nickel-free compound was found in effluents near the origin, indicating that at pH 5.4 this compound was neutrally charged. Bound to nickel, however, this agent acquired negative charge and migrated at a rate nearly identical to that of the complex in tomato exudate mentioned above. The carrier agents in tomato

TABLE 6   Uptake and Translocation of Nickel in Tomato

Datum No.	Ni in Nutrient Solution, μg/l	Total Ni Absorbed (Ni_A), μg	Exudate Volume, ml	Ni in Exudate, μg/l	Total Ni Released (Ni_R), μg	Ni_R/Ni_A (%)
1	29	23.5	24	58.7	1.4	6.0
2	294	135	23	329	7.6	5.6
3	2,940	782	12	3,258	39.1	5.0

SOURCE: Tiffin, 1971.

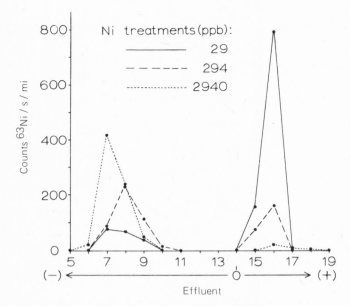

FIGURE 4 Continuous electrophoretic distribution of nickel in tomato xylem exudate. The curves show effluent nickel distribution for three exudates that are characterized in Table 6. Electrophoretic conditions: S&S No. 470-C paper, 20 μM sodium acetate at pH 5.4, 800 v, 4 ± 1° C (Tiffin, 1971).

and peanut have not been identified. Although there is general agreement that major cations are free of organic association in their transport from roots to leaves, the affinity of organic and amino acids for the micronutrients and related heavy metals, such as nickel, suggests their association with organic agents in transport. A limited capacity for nickel binding in xylem exudates is demonstrated in Figure 4. After saturation of the very stable anodic carrier, excess nickel ran cathodically to the position of the inorganic nickel control. At physiological levels of nickel (< 175 ppb in tomato xylem exudate, and perhaps higher levels in other plant exudates), there appears to be sufficient high-affinity carrier to bind the metal and transport it to the aerial portions of the plants (Tiffin, 1971).

Plant Tolerance and Availability to Animals and Man

In comparison with other heavy metals, nickel is very toxic to plants (Patterson, 1971). Where tolerances exist, they obviously involve the ability to prevent interference with plant metabolism and growth. Except for the work carried out by Tiffin (1971, 1972), however, little is known about the chemical form of nickel in plants, nor is anything known about the availability and toxicity to animals and man of natural plant forms of nickel. Vohra *et al.* (1965) have reported that nickel forms a stable complex with phytic acid. Food high in phytin may decrease and alter the bioavailability and biotoxicity of nickel to animals and man.

Lagerwerff (1967), Mitchell (1964), and Patterson (1971) have reviewed the nickel contamination of soils from various man-made sources. Several agricultural practices contaminate soils and plant foods with nickel. Mitchell (1964) has listed the range of trace element contents expected in the more commonly used fertilizers. Rock phosphate, superphosphate, basic slag, and potassium salts contain from 1 to 10 ppm nickel (dry wt). These amounts are seldom sufficient to change the total nickel content of soils appreciably, but they may change the available nickel level in soil.

One of the greatest potential sources of nickel contamination of agricultural soils results from the expanding application to them of sewage sludges from densely populated and industrialized areas. Patterson (1971) found that the total nickel content in sewage sludges from England and Wales ranged from 20 to 5,000 ppm (dry wt); nickel, soluble in 0.5 N acetic acid, ranged from 6.8 to 2,400 ppm; and the extractable percentage ranged from 14.5 to 92.7 percent. Mitchell (1964) reported that sewage sludges may contain nickel concentrations as high as 1,000 ppm. Kick *et al.* (1971) studied the availability of nickel to plants from two soils treated with sewage sludge, containing 73 ppm nickel (dry wt) mixed with inorganic salts. The approximate 1–4 ppm content of nickel (depending on harvest date) found in *Lolium peranne* grown with sewage sludge increased to 50–77 ppm when a high level of inorganic nickel (66 mg/kg of sludge, dry wt) was applied. Addition of the inorganic nickel without the sewage sludge to a third soil suppressed the growth of *L. peranne* completely. Thus, they concluded that sewage-sludge treatment of soils containing levels of nickel toxic to plants depresses that toxicity without changing the pH of the soils. The sludge treatments did not prevent the accumulation of high levels of nickel in the plant.

The fact that sewage sludges appear to protect plants against nickel toxicity, yet still allow them to accumulate nickel to high levels, shows that such use could eliminate an effective barrier (i.e., high phytotoxicity of nickel) against accumulation of high levels of nickel in natural foods. For this reason, studies on the bioavailability and biotoxicity of the forms of nickel in plants to animals and man are urgently needed.

Another potential source of man-made nickel residues in food crops from agricultural practices is the use of fungicidal sprays incorporating nickel (Stewart and Ross, 1969). However, at present, the use of nickel as a fungicidal agent has not been approved.

Lagerwerff and Specht (1970) have reported the contamination of grasses along roadsides with nickel. The nickel content of the grass ranged from 1.3 to 5 ppm (dry wt) and decreased with distance from the road. The nickel concentration also decreased with depth in the soil profile. The authors suggested that the use of nickel-bearing gasoline and abrasion of nickel-containing automobile parts (including tires) may explain the gradient of nickel concentrations in plants and soils along roadsides (Smith *et al.*, 1975).

NICKEL IN HUMANS

Metabolism

Schroeder *et al.* (1962) estimated that the usual oral intake of nickel by American adults ranges from 300 to 600 µg/day. Wide variations in nickel ingestion may occur. For example, a 2,300-calorie diet consisting of 100 g protein, 250 g carbohydrate, and 100 g fat, based on meat, milk, eggs, refined white bread, Wheatena, butter, and corn oil, would provide 3–10 µg/day; whereas a diet based on oysters, meat, milk, eggs, oats, whole wheat or rye bread, certain vegetables, potatoes, and legumes, with little added fat, would provide 700–900 µg/day.

Nodiya (1972) and Horak and Sunderman (1973) showed that nickel is poorly absorbed from ordinary diets and is mostly excreted in the feces. Nodiya performed nickel-balance studies on 10 Russian males, aged 17 yr, who ingested an average of 289 µg/day (SD ± 23; range = 251–309). Fecal excretion of nickel averaged 258 µg/day (SD ± 23; range = 219–278). Horak and Sunderman measured fecal nickel excretion in 10 healthy subjects in the United States (age 22–65; 4 males, 6 females), who ingested varied diets prepared in their own homes. They found that fecal excretion of nickel averaged 258 ± 126 µg/day (range = 80–540). Nickel has been found in the bile of rats, so some fecal nickel may arise from this source (Smith and Hackley, 1968).

Ten to 100 times less nickel is excreted in the urine than in the feces. Perry and Perry (1959) found that 24 healthy adults excreted an average of 30 µg/day of nickel in the urine, whereas Sunderman (1965) noted a mean daily urinary excretion of 19.8 µg in 17 normal subjects. In another study, Nomoto and Sunderman (1970) found that 26 healthy subjects had a mean daily urinary excretion of 2.4 µg/day. There is evidence that appreciable losses of nickel also occur in the sweat. Horak and Sunderman (1973) found a mean concentration of 49 µg/l in sweat collected in plastic bags that encased the arms of 5 healthy men during sauna bathing. Consolazio *et al.* (1964) determined that approximately 8.3 µg nickel is lost in the sweat daily.

Distribution in the Human Body

Schroeder and Nason (1971) suggested that the human body contains approximately 10 mg of nickel. Of this total, they suggested that 0.16 mg is in the blood, with 0.09 mg in the plasma and 0.07 mg in the red blood cells. Table 7 shows that values obtained for nickel in whole human blood varied between 4.8 to 327 ppb; in serum and plasma, between 2.6 and 62 ppb. It is difficult to explain this variability, but the later analyses with improved methods tend to be lower and are probably more valid. In human serum, nickel exists as ultrafiltrable nickel ($\simeq$ 40 percent of the total nickel in serum); albumin-bound nickel ($\simeq$ 34 percent); and a nickel–metalloprotein ($\simeq$ 26 percent) (Sunderman *et al.*, 1972b). This metalloprotein has been named "nickeloplasmin" and is a macroglobulin with an estimated molecular weight of 7×10^5. Purified nickeloplasmin migrates as a single protein band in the α_2-globulin region when subjected to disc gel and immunoelectrophoresis (Sunderman *et al.*, 1971a). Soestbergen and Sunderman (1972) suggested that ultrafiltrable nickel in serum exists primarily in five nickel complexes and that ultrafiltrable nickel receptors play an important physiological role in nickel homeostasis by serving as diffusible vehicles for the extracellular transport and renal excretion of nickel.

Studies of other human tissues indicate that the body does not readily retain nickel and does not accumulate this element with age in any organ examined so far, except perhaps in lens tissue. Swanson and Truesdale (1971) found 7.1–24 ppm nickel (dry basis) in various parts of the lens from adults in the 50–85 age-group, whereas < 0.004–0.089 ppm nickel (dry basis) were found in these areas in the 0–20 age-group. Schroeder and Nason (1969) found that the average nickel concentrations in hair from 79 men (0.97 ppm) were significantly lower than in hair samples from 25 women (3.96 ppm). In men,

TABLE 7   Nickel in Human Whole Blood, Serum, or Plasma

Nickel, ppb	Method[a]	Population	No. Subjects	Whole Blood (B), Serum (S), or Plasma (P)	Source
42	ES	—[b]	153	B	Imbus *et al.* (1963)
327	ES	California	47	B	Butt *et al.* (1964)
27	AA	Germany	63	B	Schaller *et al.* (1968)
4.8	AA	Connecticut	17	B	Nomoto and Sunderman (1970)
53–62[c]	ES	California	48	S	Butt *et al.* (1964)
22	S	Florida	23	S	Sunderman (1967)
7.8	ES	Germany	59[d]	S	Mertz *et al.* (1968)
2.6	AA	Connecticut	40	S	Nomoto and Sunderman (1970)
21	AA	Germany	26	P	Schaller *et al.* (1968)

[a]S = spectrophotometry; ES = emission spectrography; AA = atomic absorption.
[b]Industrial workers from Ohio, New York, Florida, Colorado, and Oregon.
[c]Range of mean values; nickel not detected in 18 sera.
[d]Includes 25 healthy subjects and 34 patients.

there was more nickel in red than in brown hair. They also did not observe any significant change in hair concentrations with advancing age. Nechay and Sunderman (1973) reported mean concentrations of nickel in hair from 20 subjects (13 men, 7 females) to be 0.22 ppm. They observed no significant difference in nickel concentrations in hair from men or women, nor was any correlation noted between nickel concentrations and the color of the hair samples. There was a slight but significant diminution in nickel concentrations with advancing age. These data are obviously disparate and clearly indicate the need for further work. Nusbaum *et al.* (1965) reported that human bone ash contains 96–148 ppm nickel, with the 0–20-yr age-group showing the highest concentrations. Perry *et al.* (1962) found that nickel is present in low concentrations (< 0.5–5.0 ppm in tissue ash) in liver, kidney, lung, heart, aorta, spleen, and brain. Schroeder and Nason (1971) have indicated that 18 percent of the body nickel is stored in skin.

EFFECTS ON HEALTH

Health Hazards

Nickel in minerals, rocks, and soils apparently does not present a hazard to animal and human health, and normal use of nickel and its alloys by humans does not usually present a health hazard. However, nickel-bearing surgical implants and prostheses may corrode and become toxic, and dermatitis may be caused by contact with nickel-bearing alloys and nickel-plated jewelry. Burning of fossil fuels (coal and petroleum) introduces nickel into the atmosphere in concentrations that may create a local health hazard. Mining, smelting, refining, and alloying nickel processes generate smoke, dust, and fumes that may, similarly, create a human health hazard.

Metabolism in Health and Disease

Serum nickel levels are significantly increased in patients with acute myocardial infarction (D'Alonzo and Pell, 1963; Nomoto and Sunderman, 1970; Sunderman *et al.*, 1970; McNeely *et al.*, 1971; Sunderman, *et al.*, 1971b; Sunderman *et al.*, 1972b). High concentrations of nickel were also found in the serum of patients with acute stroke and acute burns (> 25 percent body surface) by McNeely *et al.* (1971). This study also showed that low serum levels may occur in patients with hepatic cirrhosis and chronic uremia. In contrast to infarction, serum nickel levels did not increase in patients with myocardial ischemia.

Elevated blood-nickel concentrations have been found in women with toxemia of pregnancy (Leonov *et al.*, 1971) or with uterine cancer (Arsagova, 1971). Pathological alterations of nickel concentrations in blood and sera probably reflect changes in blood constituents such as albumin, nickeloplasmin, and ultrafiltrable nickel complexes in some diseases. Acute trauma with fractured bones, acute

delirium tremens, and muscular dystrophy had no effect on serum nickel (McNeely *et al.*, 1971).

Other studies have shown that tissue nickel levels change in certain diseases. An increased retention of nickel has been found in psoriatic skin (Gaul and Staud, 1934). Soroka *et al.* (1972) reported that in all forms of schizophrenia there is an increased concentration of nickel in the brain, liver, and spleen and a decreased concentration in the kidney, lungs, and adrenal glands. Swanson and Truesdale (1971) noted a substantial increase in nickel levels in cataractous human lens tissue. The apparent specificity of the changes in blood or tissue nickel levels suggests that this element plays a role in the etiology of some diseases.

Studies on the toxicity of nickel to man leave no doubt that this element, like zinc, manganese, and chromium, is relatively nontoxic orally and that nickel contamination of food or water need not necessarily present a serious health hazard. One population exposed to nickel contamination is that of Sudbury, Ontario, the principal nickel-producing area in the world. McNeely *et al.* (1972) compared water- and air-survey data from Sudbury and from Hartford, Connecticut. Sudbury air contained an average nickel concentration of 533 μg/1,000 m^3; municipal tap water, 200 ppb. Hartford air contained 36 μg/1,000 m^3; municipal water, 1.1 ppb. McNeely *et al.* (1972) found that healthy hospital employees in the Sudbury area have higher levels of nickel in urine (7.9 μg/day) and serum (4.6 ppb) than did a similar population living in Hartford (urine [2.5 μg/day] and serum [2.6 ppb]). This suggests that measurements of nickel in serum and urine may serve as biological indices of environmental exposure to nickel. Records of the International Nickel Co., which operates the mines (nickel sulfide ore) and related industry in the Sudbury area, indicate that the nickel contamination is apparently not harmful, as the employees' health is above the average of the Canadian population (H. T. Reno, personal communication, 1972). In a similar fashion, U.S. Bureau of Mines personnel found no ill effects to the health of workers or inhabitants of the Riddle, Oregon, area, the only nickel-mining operation (nickel oxide ore) in the United States in 1972 (H. T. Reno, personal communication, 1972).

The relatively nontoxic nature of ingested nickel is demonstrated by toxicologic studies with rats (Phatak and Patwardhan, 1950, 1952), chicks (Weber and Reid, 1968), mice (Weber and Reid, 1969), and dairy cattle (O'Dell *et al.*, 1970). Depending on species and age, toxic effects were not observed until 250–1,600 ppm of nickel were present in the diet. Signs of toxicity included depression in the activity of certain enzymes in mice and a reduction of nitrogen retention in chicks and dairy calves. Chicks, mice, and calves also exhibited reduced growth. Schroeder and Mitchener (1971), however, reported findings that indicate much lower levels of nickel can be toxic. Feeding 5 ppm nickel as a soluble salt in the drinking water to reproducing rats resulted in 9.1 percent young deaths and 30.6 percent runts in the first generation, 10.2 percent young deaths and 5.1 percent runts in the second, and 21 percent young deaths and 6.2 percent runts in the third. The size of the litters decreased somewhat with each generation, and few males were born in the third generation. Thus, nickel fed in doses that do not interfere with growth or survival of weanling or adult rats may be intolerable when fed to reproducing rats.

Evidence for a nickel deficiency that impairs human health has not been found. However, since certain people may have restricted intakes of nickel, or perhaps increased needs in certain pathological disorders, and in view of recent findings with experimental animals (Sunderman et al., 1972a; Nielsen et al., 1975a; Nielsen et al., 1975b), this possibility should not be dismissed. Through a series of experiments (Nielsen and Sauberlich, 1970; Nielsen and Higgs, 1971; Nielsen and Ollerich, 1974; Nielsen et al., 1975a, 1975b), a diet based on dried skim milk and ground corn containing 3–4 ppb nickel has been developed. Feeding this diet to chicks and rats in trace-metal-deficient controlled environmental systems resulted in impaired liver metabolism and morphology. The findings in chicks included a reduced ability to oxidize α-glycerophosphate, decreased phospholipids, and ultrastructural degeneration in the liver. The ultrastructural abnormalities included dilation of the cisternae of the rough endoplasmic reticulum, enlargement of the mitochondria with loss of density in the matrix, dilation of the perinuclear space, condensation of peripheral nuclear chromatin, and pyknotic nuclei. Sunderman et al. (1972a) demonstrated a dilation of the perimitochondrial rough endoplasmic reticulum in the hepatocytes of nickel-deprived chicks. Deficient rats exhibited slower growth, lower hematocrits and liver cholesterol, and ultrastructural changes in the liver with the most obvious difference in the amount and organization of the rough endoplasmic reticulum (Nielsen et al., 1975b; Schnegg and Kirchgessner, 1975). To date, a minimum dietary requirement for nickel (52 ppb) has only been established for chicks (Nielsen et al., 1975a). Assuming that humans require approximately the same, it may be possible that some people may not be consuming adequate nickel. The diet that Schroeder et al. (1962) calculated to supply 3–10 μg of nickel per day would contain only 6.6–22 ppb nickel. Thus, studies are needed to ascertain whether nickel is essential for humans and whether nickel deficiency in humans is a naturally occurring phenomenon.

Nickel and Cancer

Various forms of nickel are known to be carcinogenic to experimental animals (Sunderman, 1971). The carcinogenic potency of nickel compounds may be inversely related to their solubilities in water. For example, metallic nickel dust, nickel sulfide, nickel carbonate, nickel oxide, nickel carbonyl, and nickelocene are carcinogenic and only sparingly water soluble, while soluble salts of nickel (e.g., nickel chloride, nickel sulfate, and nickel ammonium sulfate) are not known to be carcinogenic.

Epidemiological evidence links nickel dusts at nickel smelters with nose and lung cancer in workers. The risk of cancer increases with increased dust levels. Furthermore, exposure to conversion, by roasting, of nickel sulfide (Ni_3S_2) to nickel oxide (NiO) is liable to cause lung cancer. According to recent epidemiological studies these occupational hazards have apparently been eliminated in nickel refineries or in other industries where workers are exposed to nickel (Mastromatteo, 1967; Sunderman, 1968).

Nickel may be linked to lung cancer and cigarette smoking. Cigarettes contain from 1.9 to 6.2 μg nickel, and 10–20 percent of the total nickel in tobacco is released into the mainstream smoke (Sunderman and Sunderman, 1961; Sunderman et al., 1968). Nickel carbonyl may be stable in tobacco smoke, and studies are under way to ascertain whether nickel carbonyl is a causal factor of lung cancer in cigarette smoking.

Polycyclic aromatic hydrocarbons apparently are synergistic with nickel compounds in carcinogenesis. Cigarette-smoking workers in nickel refineries may be subjected to a higher risk of lung cancer because of this synergism (Doll et al., 1970).

Treagan and Furst (1970) suggest that exposures to nickel could increase the rate of replication of tumor viruses.

RECOMMENDATIONS

So far, no geographic distribution of nickel that can be correlated to disease has become apparent, although nickel does affect human health in unique situations. The following recommendations for research are therefore made:

1. Develop analytical methods to determine the chemical and physical state of nickel in stack gases of nickel smelters and oil- and coal-burning plants, and improve existing methods for the determination of nickel in all kinds of materials.

2. Obtain new data on the content of nickel in all kinds of materials to replace the older suspect data.

3. Measure precisely the nickel content of emissions from selected nickel-processing industrial plants and from oil- and coal-burning plants.

4. Determine the exact form(s) of metallic nickel and nickel-bearing chemical compounds that cause(s) cancer.

5. Maintain detailed health records for the people who live in areas where nickel is mined and in areas that contain high concentrations of nickel in the rocks and soils (such as the Sudbury district in Ontario, Canada; the island of New Caledonia in the South Pacific; and Riddle, Oregon, in the United States).

6. Isolate and identify the biological forms of nickel in food sources, and determine the bioavailability and biotoxicity of these forms to animals and humans.

7. Continue studies on the effect of nickel in foods from the agricultural use of sewage sludges. Nickel in sludge from various sources should be closely monitored.

8. Determine the precise metabolic role of nickel in

animals, including man. Ascertain through further experiments the level of ingested nickel that can affect reproduction in animals. If relatively low levels of ingested nickel are found to be toxic for reproduction in animals, carry out surveys to see whether this is a problem in certain human populations.

9. Ascertain whether nickel deficiency occurs in humans.

10. Design studies to ascertain whether nickel is an essential element for higher plants.

REFERENCES

Abernethy, R. F., M. J. Peterson, and F. H. Gibson. 1969. Spectrochemical analyses of coal ash for trace elements. U.S. Bur. Mines Rep. Invest. 7281. U.S. Department of the Interior, Washington, D.C. 20 pp.

Allaway, W. H. 1968. Agronomic controls over the environmental cycling of trace elements. Adv. Agron. 20:235–274.

Arsagova, N. S. 1971. On the nickel and manganese content in patients with uterine cancer. Vopr. Onkol. 17:53–56. (In Russian)

Ashton, W. M. 1972. Nickel pollution. Nature 237:46–47.

Bertrand, D., and A. de Wolf. 1965. Le nickel comme engrais complémentaire. C. R. Acad. Sci. (Paris) 261:5195–5197.

Bertrand, D., and A. de Wolf. 1967. Nickel, a dynamic trace element for higher plants. C. R. Acad. Sci. (Paris) Ser. D 265:1053–1055.

Bertrand, G., and M. Mokragnatz. 1930. Réparition du nickel et du cobalt dans les plantes. C. R. Acad. Sci. (Paris) 190:21–25.

Bowen, H. J. M. 1966. Trace elements in biochemistry. Academic Press, New York. 241 pp.

Butt, E. M., R. E. Nusbaum, T. C. Gilmour, S. L. Didio, and Sister Mariano. 1964. Trace metal levels in human serum and blood. Arch. Environ. Health 8:52–57.

Cannon, H. L. 1960. Botanical prospecting for ore deposits. Science 132:591–598.

Cannon, H. L. 1971. The use of plant indicators in groundwater surveys, geologic mapping, and mineral prospecting. Taxon 20:227–256.

Consolazio, C. F., R. N. Nelson, L. O. Matoush, R. C. Hughes, and P. Urone. 1964. Trace mineral losses in sweat. U.S. Army Medical Research and Nutrition Laboratory Report 284. Fitzsimmons General Hospital, Denver, Colorado. 14 pp.

Crooke, W. M. 1955. Further aspects of the relationship between nickel toxicity and iron supply. Ann. Appl. Biol. 43:465–476.

Crooke, W. M., and A. H. Knight. 1955. The relationship between nickel toxicity symptoms and the absorption of iron and nickel. Ann. Appl. Biol. 43:454–464.

D'Alonzo, C. A., and S. Pell. 1963. A study of trace metals in myocardial infarction. Arch. Environ. Health 6:381–385.

DeWaal, S. A. 1970. Nickel minerals from Barberton, South Africa. II. Nimites—A nickel rich chlorite. Am. Min. 55:18–30.

Dixon, N. E., C. Gazzola, R. L. Blakeley, and B. Zerner. 1975. Jack bean urease (EC 3.5.1.5). A metalloenzyme. A simple biological role for nickel? J. Am. Chem. Soc. 97:4131–4133.

Doll, R., L. G. Morgan, and F. E. Speizer. 1970. Cancers of the lung and nasal sinuses in nickel workers. Br. J. Cancer 24:623–632.

Durfor, C. N., and E. Becker. 1964. Public water supplies of the 100 largest cities in the United States, 1962. U.S. Geol. Surv. Water Supply Pap. 1812. U.S. Government Printing Office, Washington, D.C. 364 pp.

Faust, G. T. 1966. Hydrous, nickel magnesium silicate—garnierite. Am. Min. 51:279–298.

Faust, G. T., J. J. Fahey, B. Mason, and E. J. Dwornik. 1969. Pecoraite—$Ni_6 Si_4 O_{10} (OH)_8$ nickel analog of clinochrysotile, formed in the Wolf Creek meteorite. Science 165:59–60.

Fleming, G. A. 1963. Distribution of major and trace elements in some common pasture species. J. Sci. Food Agric. 14:203–208.

Gambi, O. V. 1967. Prima dati sulla localizzazione istologica del nichel in *Alyssum bertolonii*. Desv. Giorn. Bot. Ital. 101:59–60.

Gaul, L. E., and A. H. Staud. 1934. Clinical spectroscopy: The quantitative retention of nickel in psoriasis; observations on forty-six cases. Arch. Dermatol. Syphilol. 30:697–703.

Guha, M. M., and R. L. Mitchell. 1966. The trace and major element composition of the leaves of some deciduous trees. II. Seasonal changes. Plant Soil 24(1):90–112.

Guthrie, V. B. [ed.] 1960. Petroleum products handbook, 1st ed. McGraw-Hill, New York. pp. 8–25.

Halstead, R. L. 1968. Effect of different amendments on yield and composition of oats grown on a soil derived from serpentine material. Can. J. Soil Sci. 48:301–305.

Halstead, R. L., B. J. Finn, and A. J. MacLean. 1969. Extractability of nickel added to soils and its concentration in plants. Can. J. Soil Sci. 49:335–342.

Hodgson, J. F. 1970. Chemistry of trace elements in soils with reference to trace element concentration in plants. Proc. 3d Annu. Conf. Trace Subst. Environ. Health, June 24–26, 1969, D. D. Hemphill [ed.] University of Missouri, Columbia. pp. 45–58.

Horak, E., and F. W. Sunderman, Jr. 1973. Fecal nickel excretion by healthy adults. Clin. Chem. 19:429–430.

Howard-White, F. B. 1963. Nickel. An historical review. D. Van Nostrand Co., New York. 350 pp.

Imbus, H. R., J. Cholak, L. H. Miller, and T. Sterling. 1963. Boron, cadmium, chromium, and nickel in blood and urine. Arch. Environ. Health 6:286–295.

Jenne, E. A. 1968. Controls on Mn, Fe, Co, Ni, Cu and Zn concentrations in soils and water: The significant role of hydrous Mn and Fe oxides. Adv. Chem. Ser. 73:337–387.

Kick, H., R. Nosbers, and J. Warnusz. 1971. The availability of Cr, Ni, Zn, Cd, Sn and Pd for plants. Int. Symp. Soil Fert. Evaln. (New Delhi) Proc. 1:1039–1045.

Kopp, J. F., and R. C. Kroner. 1968. Trace metals in waters of the United States: A five-year summary of trace metals in rivers and lakes of the United States (October 1, 1962–September 30, 1967). U.S. Department of Interior, Federal Water Pollution Control Administration, Cincinnati, Ohio. 32 pp. and 16 appendixes.

Lagerwerff, J. V. 1967. Heavy-metal contamination of soils. *In* Agriculture and the quality of our environment, H. C. Brady [ed.]. American Association for the Advancement of Science Publ. 85. pp. 343–364.

Lagerwerff, J. V., and A. W. Specht. 1970. Contamination of roadside soil and vegetation with cadmium, nickel, lead and zinc. Environ. Sci. Techol. 4:583–586.

Leonov, V. A., I. K. Gurskaya, V. I. Medvedeva, and M. V. Chichko. 1971. Disturbances of manganese, nickel, chromium, copper, and molybdenum exchange between mother and fetus in late pregnancy toxicoses. Dokl. Akad. Nauk. Beloruss. SSR 15:656–657. (In Russian)

Lyon, G. L., and R. R. Brooks. 1970. Trace elements in plants from serpentine soils. N.Z. J. Sci. 13:133–139.

Lyon, G. L., R. R. Brooks, P. J. Peterson, and G. W. Butler. 1968. Trace elements in a New Zealand serpentine flora. Plant Soil 29:225–240.

Mastromatteo, E. 1967. Nickel: A review of its occupational health aspects. J. Occup. Med. 9:127–136.

McNeely, M. D., F. W. Sunderman, Jr., M. W. Nechay, and H. Levine. 1971. Abnormal concentrations of nickel in serum in cases of myocardial infarction, stroke, burns, hepatic cirrhosis, and uremia. Clin. Chem. 17:1123–1128.

McNeely, M. D., M. W. Nechay, and F. W. Sunderman, Jr. 1972. Measurements of nickel in serum and urine as indices of environmental exposure to nickel. Clin. Chem. 18:992–995.

Mertz, D. P., R. Koschnick, G. Wilk, and K. Pfeilsticker. 1968. Metabolism of trace elements in men. I. Serum concentrations of cobalt, nickel, silver, cadmium, chromium, molybdenum, and manganese. Z. Klin. Chem. Klin. Biochem. 6:171–174. (In German)

Mitchell, R. L. 1964. Trace elements in soils. In Chemistry of the soil, F. E. Bear [ed.]. Reinhold, New York. pp. 320–368.

Mitchell, R. L. 1971. Trace elements in soils. In Trace elements in soils and crops. Tech. Bull. 21. Ministry of Agriculture, Fisheries, and Food, London. pp. 8–20.

Mitchell, R. L., J. W. S. Reith, and I. M. Johnston. 1957. Trace-element uptake in relation to soil content. J. Sci. Food Agric. 8:51–59.

National Air Pollution Control Administration. 1968a. Air quality data from the national air sampling network. 1966 ed. U.S. Department of Health, Education, and Welfare. U.S. Government Printing Office, Washington, D.C. pp. 109–111. (Available from NTIS, 5285 Port Royal Rd., Springfield, Va. 22151, as PB 195–155.)

National Air Pollution Control Administration. 1968b. National inventory of sources and emissions: Cadmium, nickel, and asbestos. Prepared by W. E. Davis and Associates as APTD-68, APTD-69, and APTD-70. (Published in 1973 by U.S. Government Printing Office and available from NTIS, 5285 Port Royal Rd., Springfield, Va. 22151, as PB 192-250, PB 192-251, and PB 192-252.)

Nechay, M. W., and F. W. Sunderman, Jr. 1973. Measurements of nickel in hair by atomic absorption spectrometry. Ann. Clin. Lab. Sci. 3:30–35.

Ng, Siew Kee, and C. Bloomfield. 1962. The effect of flooding and aeration on the mobility of certain trace elements in soils. Plant Soil 16:108–135.

Nielsen, F. H., and D. J. Higgs. 1971. Further studies involving a nickel deficiency in chicks. Proc. 4th Annu. Conf. Trace Subst. Environ. Health, June 23–25, 1970, D. D. Hemphill [ed.]. University of Missouri, Columbia. pp. 241–246.

Nielsen, F. H., and D. A. Ollerich. 1974. Nickel: A new essential trace element. Fed. Proc. 33(6):1767–1772.

Nielsen, F. H., and H. E. Sauberlich. 1970. Evidence of a possible requirement for nickel by the chick. Proc. Soc. Exp. Biol. Med. 134:845–849.

Nielsen, F. H., D. R. Myron, S. H. Givand, and D. A. Ollerich, 1975a. Nickel deficiency and nickel–rhodium interaction in chicks. J. Nutr. 105:1607–1619.

Nielsen, F. H., D. R. Myron, S. H. Givand, T. J. Zimmerman, and D. A. Ollerich, 1975b. Nickel deficiency in rats. J. Nutr. 105:1620–1630.

Nodiya, P. I. 1972. Cobalt and nickel balance in students of an occupational technical school. Gig. Sanit. 37:108–109. (In Russian)

Nomoto, S., and F. W. Sunderman, Jr. 1970. Atomic absorption spectrometry of nickel in serum, urine, and other biological materials. Clin. Chem. 16:477–485.

Nusbaum, R. E., E. M. Butt, T. C. Gilmour, and S. L. Didio. 1965. Relation of air pollutants to trace metals in bone. Arch. Environ. Health 10:227–232.

O'Dell, G. D., W. J. Miller, W. A. King, S. L. Moore, and D. M. Blackmon. 1970. Nickel toxicity in the young bovine. J. Nutr. 100:1447–1453.

Patterson, J. B. E. 1971. Metal toxicities arising from industry. In Trace elements in soils and crops. Tech. Bull. 21. Ministry of Agriculture, Fisheries, and Food, London. pp. 193–207.

Perry, H. M., Jr., and E. F. Perry. 1959. Normal concentrations of some trace metals in human urine: Changes produced by ethylenediaminetetraacetate. J. Clin. Invest. 38:1452–1463.

Perry, H. M., Jr., I. H. Tipton, H. A. Schroeder, and M. J. Cook. 1962. Variability in the metal content of human organs. J. Lab. Clin. Med. 60:245–253.

Peterson, P. J. 1971. Unusual accumulations of elements by plants and animals. Sci. Prog. (Oxford) 59:505–526.

Pettijohn, F. J. 1957. Sedimentary rocks. Harper & Brothers, New York. p. 8.

Phatak, S. S., and V. N. Patwardhan. 1950. Toxicity of nickel. J. Sci. Ind. Res. (India) 9:70–76.

Phatak, S. S., and V. N. Patwardhan. 1952. Toxicity of nickel—Accumulation of nickel in rats fed on nickel-containing diets and its elimination. J. Sci. Ind. Res. (India) 11:173–176.

Pratt, P. F., F. L. Blair, and G. W. MacLean. 1964. Nickel and copper chelation capacities of soil organic matter. Int. Congr. Soil Sci., Trans. 8th (Bucharest) III:243–248.

Rankama, K., and Th. G. Sahama. 1950. Geochemistry. Chicago University Press, Chicago. 911 pp.

Roth, J. A., E. F. Wallihan, and R. G. Sharpless. 1971. Uptake by oats and soybeans of copper and nickel added to a peat soil. Soil Sci. 112:338–342.

Sauchelli, V. 1969. Trace elements in agriculture. Van Nostrand Reinhold, New York. pp. 217–222.

Schaller, K. H., A Kühner, and G. Lehnert. 1968. Nickel as trace element in human blood. Blut 17:155–160. (In German)

Schnegg, A., and M. Kirchgessner. 1975. Veränderungen des Hämoglobingehaltes der Erythrozytenzahl und des Hämatokrits bei Nickelmangel. Nutr. Metab. 19:268–278.

Schroeder, H. A., and M. Mitchener. 1971. Toxic effects of trace elements on the reproduction of mice and rats. Arch. Environ. Health 23:102–106.

Schroeder, H. A., and A. P. Nason. 1969. Trace metals in human hair. J. Invest. Dermatol. 53:71–78.

Schroeder, H. A., and A. P. Nason. 1971. Trace-element analysis in clinical chemistry. Clin. Chem. 17:461–474.

Schroeder, H. A., J. J. Balassa, and I. H. Tipton. 1962. Abnormal trace metals in man—Nickel. J. Chron. Dis. 15:51–65.

Sillanpää, M., and E. Lakanen. 1969. Trace-element content of plants as a function of readily soluble soil trace elements. J. Sci. Agric. Soc. (Finland) 41:60–67.

Smith, I. C., T. L. Ferguson, and B. L. Carson. 1975. Metals in new and used petroleum products and by-products—Quantities and consequences, Chapter 7 In Role of trace metals in petroleum, T. F. Yen [ed.]. Ann Arbor Science Publishers, Ann Arbor, Michigan. pp. 123–149.

Smith, J. C., and B. Hackley. 1968. Distribution and excretion of nickel-63 administered intravenously to rats. J. Nutr. 95:541–546.

Soestbergen, M. V., and F. W. Sunderman, Jr. 1972. ^{63}Ni complexes in rabbit serum and urine after injection of ^{63}NiCl$_2$. Clin. Chem. 18:1478–1484.

Soroka, V. R., V. Ya. Arsenm'ev, and M. S. Mukhaev. 1972. Nickel metabolism during schizophrenia. Zh. Nevropatol. Psikhiatr. im. S. S. Korsakova 72:69–72. (In Russian)

Stewart, D. K. R., and R. G. Ross. 1969. Nickel residues in apple fruit and foliage following a foliar spray of nickel chloride. Can. J. Plant Sci. 49:375–377.

Sunderman, F. W., Jr. 1965. Measurements of nickel in biological materials by atomic absorption spectrometry. Am. J. Clin. Pathol. 44:182–188.

Sunderman, F. W., Jr. 1967. Spectrophotometric measurement of serum nickel. Clin. Chem. 13:115–125.

Sunderman, F. W., Jr. 1968. Nickel carcinogenesis. Epidemiology of respiratory cancer among nickel workers. Dis. Chest 54:527–534.

Sunderman, F. W., Jr. 1971. Metal carcinogenesis in experimental animals. Food Cosmet. Toxicol. 9:105–120.

Sunderman, F. W., and F. W. Sunderman, Jr. 1961. Nickel Poisoning. XI. Implication of nickel as a pulmonary carcinogen in tobacco smoke. Am. J. Clin. Pathol. 35:203–209.

Sunderman, F. W., Jr., N. O. Roszel, and R. J. Clark. 1968. Gas chromatography of nickel carbonyl in blood and breath. Arch. Environ. Health 16:836–843.

Sunderman, F. W., Jr., S. Nomoto, A. M. Pradhan, H. Levine, S. H. Bernstein, and R. Hirsch. 1970. Increased concentrations of serum nickel after acute myocardial infarction. N. Engl. J. Med. 283:896–899.

Sunderman, F. W., Jr., M. I. Decsy, S. Nomoto, and M. W. Nechay. 1971a. Isolation of a nickel—α_2 macroglobulin from human and rabbit serum. Fed. Proc. 30:1274. (Abstract)

Sunderman, F. W., Jr., S. Nomoto, and M. Nechay. 1971b. Nickel metabolism in myocardial infarction. II. Measurements of nickel in human tissues. Proc. 4th Annu. Conf. Trace Subst. Environ. Health, June 23–25, 1970, D. D. Hemphill [ed.]. University of Missouri, Columbia. pp. 352–356.

Sunderman, F. W., Jr., S. Nomoto, R. Morang, M. W. Nechay, C. N. Burke, and S. W. Nielsen. 1972a. Nickel deprivation in chicks. J. Nutr. 102:259–267.

Sunderman, F. W., Jr., M. I. Decsy, and M. D. McNeely. 1972b. Nickel metabolism in health and disease. Ann. N.Y. Acad. Sci. 199:300–312.

Swaine, D. J., and R. L. Mitchell. 1960. Trace-element distribution in soil profiles. J. Soil Sci. 11:347–368.

Swanson, A. A., and A. W. Truesdale. 1971. Elemental analysis in normal and cataractous human lens tissue. Biochem. Biophys. Res. Commun. 45:1488–1496.

Tiffin, L. O. 1971. Translocation of nickel in xylem exudate of plants. Plant Physiol. 48:273–277.

Tiffin, L. O. 1972. Translocation of micronutrients in plants. *In* Micronutrients in agriculture. J. J. Mortvedt, P. M. Giordano, and W. L. Lindsay [eds.]. Soil Science Society of America, Madison, Wisconsin. pp. 199–229.

Timperley, M. H., R. R. Brooks, and P. J. Petersen. 1970. Prospecting for copper and nickel in New Zealand by statistical analysis of biogeochemical data. Econ. Geol. 65:505–510.

Treagan, L., and A. Furst. 1970. Inhibition of interferon synthesis in mammalian cell cultures after nickel treatment. Res. Commun. Chem. Pathol. Pharmacol. 1:395–402.

Turekian, K. K., and K. H. Wedepohl. 1961. Distribution of the elements in some major units of the earth's crust. Geol. Soc. Am. Bull. 72:175–192.

Underwood, E. J. 1971. Trace elements in human and animal nutrition, 3d ed. Academic Press, New York. pp. 170–176.

U.S. Bureau of Mines. 1975. Mineral industry surveys: Nickel in October 1974. U.S. Department of the Interior, Washington, D.C. 12 pp.

U.S. Geological Survey. 1965. Quality of surface waters of the United States. U.S. Geol. Surv. Water Supply Pap. 1961. U.S. Government Printing Office, Washington, D.C. 779 pp.

Vanselow, A. P. 1966. Nickel. *In* Diagnostic criteria for plants and soils, H. D. Chapman [ed.]. Division of Agricultural Sciences, University of California, Riverside.

Vohra, P., G. A. Gray, and F. H. Kratzer. 1965. Phytic acid–metal complexes. Proc. Soc. Exp. Biol. Med. 120:447–449.

Weber, C. W., and B. L. Reid. 1968. Nickel toxicity of growing chicks. J. Nutr. 95:612–616.

Weber, C. W., and B. L. Reid. 1969. Nickel toxicity in young growing mice. J. Anim. Sci. 28:620–623.

Wild, H. 1970. Geobotanical anomalies in Rhodesia. 3. The vegetation of nickel bearing soils. Kirkia 7:1–62.

Williams, P. C. 1967. Nickel, iron and manganese in the metabolism of the oat plant. Nature 214:628.

Zook, E. G., F. E. Greene, and E. R. Morris, 1970. Nutrient composition of selected wheats and wheat products. VI. Distribution of manganese, copper, nickel, zinc, magnesium, lead, tin, cadmium, chromium, and selenium, as determined by atomic absorption spectroscopy and colorimetry. Cereal Chem. 47(6):720–731.

VI

Silicon

HOWARD C. HOPPS, *Chairman*

*Edith M. Carlisle, Justin A. McKeague,
Raymond Siever, Peter J. Van Soest*

Silicon, atomic number 14, has an atomic weight of 28.09 and three naturally occurring stable isotopes, 28, 29, and 30. There are, in addition, five radioactive isotopes. Silicon is second only to oxygen as the most common element in the earth's crust, and its inorganic compounds, the silicates, make up the great bulk of rocks and soils exposed at the surface. Despite its abundance, silicon has received relatively little attention as an element of biological importance, except for a long history of discussion about the essentiality of silicon for some plants and more than a century of concern about silicosis, a disease that was early related to inhalation of silica dust by miners and mill workers.

The oxides of silicon, however, have been exhaustively investigated by geologists and geochemists. Industrial chemists have put silica and sodium silicates to practical use in a variety of ways, beginning with the ancient use of water glass as glue. Since the 1950's, silica chemistry has had a firm foundation based on knowledge of the nature of its solids and solutions, their chemical reactions, and polymerization–depolymerization reactions. Much of the literature on industrial silicate chemistry is summarized by Iler (1955) and Sosman (1965); the latter emphasizes solid phases.

Silicon has been recognized as an essential element for certain lower organisms, notably diatoms, for many years. Evidence is accumulating that silicon may play an essential role in certain plants, e.g., *Equisetum*, rice, and sugarcane, and, since 1972, silicon has been shown to be essential for certain higher animals (chicken and rat), where it contributes to the formation of both bone and cartilage.

In terms of the relationship of silicon to human disease, several relatively new entities have been recognized, notably siliceous urolithiasis. Moreover, the pathogenesis of one of the "old" diseases—silicosis—has become much clearer. A comparatively newly recognized disease, asbestosis, is of great current concern because of the realization that exposure to asbestos is widespread (i.e., not simply an occupational hazard) and that asbestos is carcinogenic. The problem of exposure to asbestos fibers, and of the consequences of such exposure, has high priority among the several federal agencies concerned with occupational health and with environmental protection.

GEOCHEMISTRY AND OCCURRENCE

Silica abundance in natural waters and the general outline of the geochemical cycle of silicon has been sketched by Siever (1957, 1972), as shown in Figures 5 and 6. Silicon is released from silicate rocks by weathering and enters the natural water system at that point. Some dissolved silica in soil water enters the biological cycle through absorption by plant roots. The bulk of the dissolved silica migrates to the ocean via surface water and groundwater.

The removal of silica from ocean water has been a source of controversy; most scientists favor removal en-

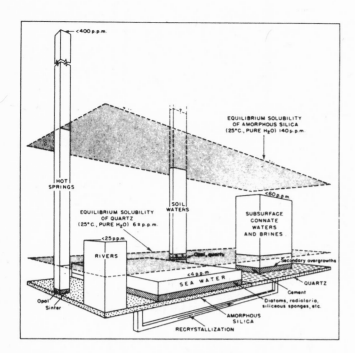

FIGURE 5 Silica abundance in natural waters. Volume of "buildings" roughly proportional to absolute amounts of silica in solution. Height of "buildings" roughly proportional to concentrations of silica in solution. Reproduced by permission from Raymond Siever and *The American Mineralogist* 42:836, Figure 1. Copyright by Mineralogical Society of America, 1957.

tirely by biological populations (mainly diatoms and radiolaria), but others consider removal by sorption on detrital clay minerals in the ocean important. This argument has been summarized by Siever and Woodford (1973).

Regardless of the precise proportion, it is generally agreed that the vast bulk of dissolved silicon in the ocean is removed steadily by silica-secreting organisms. Their activity depletes silicon in surface waters of the ocean, which contain levels of 0.1–1.0 ppm silicon in most areas. Concentration of silicon in bottom waters tends to be somewhat higher; some silicon is released by dissolution of the silica shells sedimented to the bottom after death of the organisms. In some of these diatom and radiolarian ooze sediments, silicon values are as high as 12–15 ppm. Additional silicon may come from back diffusion from interstitial water of recently deposited sediment.

The most important factors in controlling the concentration of silicon in groundwater and surface water are rock types, precipitation, and runoff rates. Potable waters contain 2–20 ppm silicon, which constitutes normally about 5–10 percent of the total dissolved solids. Certain alkaline waters contain from a few to several thousand ppm silicon. The higher silica values come from areas underlain by volcanic rocks such as basalts and andesites. The lower range of values, some as low as 2–3 ppm, comes from limestone terrains where little silicate is available for weathering. Laterite soils, occurring in deeply wea-

thered, leached terrains, are low in silicon. Thus, in some tropical rain forests, most of the available silicon is in the silicon-accumulator plants rather than in the soil. High runoff levels, of course, also decrease silicon concentrations by extreme dilution of groundwater and surface water. Davis (1964, 1969) has summarized the controls on silica concentrations in streams and groundwater.

Solid Phases

Silicon occurs largely in nature as the oxide silica (SiO_2). Silica exists in several polymorphic modifications at low to moderate pressures: quartz, cristobalite, tridymite, and amorphous or glassy forms.

At temperatures below 100° C, the two important phases are quartz and amorphous silica. Quartz is the stable phase; amorphous silica persists metastably for geologically significant times and is thus important in biological systems. Virtually all quartz is coated with a surface layer at least several hundred angstroms thick of amorphous silica, formed by disordering and hydration of the crystalline quartz surface. Amorphous silicas display a variety of states of aggregation, hydration, and density, depending on their modes of formation. These range from fused silica glass (the densest anhydrous form) to porous, extensively hydrated silica gels. Opal is a special variety of amorphous silica that has a lower range of hydration states and higher densities than most gels. Opal eventually assumes the short-range crystalline structure of cristobalite. Biogenic silica is either amorphous or some organically complexed material having similar physical

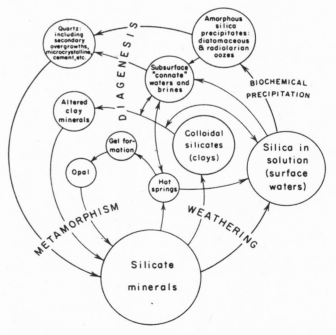

FIGURE 6 The silica cycle. Reproduced by permission from Raymond Siever and *The American Mineralogist* 42:837, Figure 2. Copyright by Mineralogical Society of America, 1957.

properties. No occurrence of crystalline quartz of biological origin is known.

Hundreds of silicate minerals have been described, all containing the fundamental silicon tetrahedron (SiO_4^{-4}) building block. A great many combinations of cations that have polymeric or monomeric crystalline structures are known. These include the clay minerals, and other hydrous aluminosilicates, that are prominent constituents of soils. Others that have been implicated in biological systems are the asbestos minerals—a serpentine group (chrysotile), and an amphibole group of hydrous magnesium silicates.

Organic Carbon Compounds of Silica

Organosilicon compounds (silicones), containing a silicon–carbon bond, are synthetically produced and remain a question mark in biological and geological systems. Reports on the occurrence of silicic esters in nature are ambiguous (Rochow, 1951; Siever and Scott, 1963). Any such compounds that might be present would be transient species, or so labile as to defy extraction and analysis (Van Soest, 1970). Some evidence suggests that organic-complexed silicon does occur in plants (Van Soest and Lovelace, 1969).

The most recent review of silica organic geochemistry is by Siever and Scott (1963). Iler (1955) summarizes the organic silicon chemistry literature as of 1954.

Silica Solubility and Rates of Solution

Siever (1972) has reviewed the general solution geochemistry of silica. Quartz solubility is very low, in the range of 3–5 ppm silicon at 25° C (All parts-per-million figures for Si compounds are given in terms of silicon, unless otherwise specified). Amorphous silica solubility is an order of magnitude higher, 60–70 ppm. Because it is metastable, it controls dissolved silica abundance over the short term in natural aqueous systems. The rate of solution of amorphous silica is strongly affected by pH; for example, amorphous silica at pH 6 takes approximately 10 times as long to achieve equivalent concentrations as it does at pH 8. Increasing ionic strength of the solvent does not affect solubility, but does seem to increase slightly the rate of solution. The dominant species in undersaturated or saturated solutions at pH values between 4 and 9 is H_4SiO_4, monomeric silicic acid. In supersaturated or alkaline solutions, it may polymerize to varying extents.

Both solubility and rate of solution increase significantly as temperature rises. The solubility of amorphous silica is approximately doubled between 25° and 75° C. The solubility of quartz rises even more steeply.

Gel formation involves polymerization of H_4SiO_4 units in a supersaturated solution. The rate of gel formation increases with degree of supersaturation, amount of impurities, and ionic strength. Gel rate is fastest in the pH region between 4 and 7, decreasing at higher and lower pH levels. Most geological and biological systems do not permit stabilization of supersaturated silica sols.

Silicate minerals vary in their solubilities and solubility rates. These can be characterized by solubility diagrams such as those given by Garrels and Christ (1965). Equilibrium solubilities of many aluminosilicates are low, ranging over levels from 1–10 ppm silicon; few are known accurately. Some forms, such as anorthite feldspar ($CaAl_2Si_2O_8$), have much higher solubilities. Dissolved silica released to solution by silicates is generally monomeric H_4SiO_4, but there is some evidence of polysilicic acid polymers as well (Weitz et al., 1950). Such polymers in undersaturated solution depolymerize to the monomer over hours or days. Polymeric reactions of this sort are not known to occur in biological fluids.

The rates at which silicic acid is released from silicate minerals depend to a certain extent on their solubilities; the more unstable the species in general, the more rapid the dissolution. Silica and silicates probably follow the same general kinetic paths of dissolution. Initial release of silica (also cations and, depending on pH, aluminum species) is rapid for the first few hours, followed by a period of several weeks (or months) during which the rate of release decreases parabolically. After some additional weeks (or months), the release of silica further decreases at a much lower rate. Within a year's time, the release has essentially stopped.

Silicon in Soil Solutions and Soil Extracts

Silicon in soil solutions may be thought of as the beginning point of the biological cycle of silicon. In general terms, when water enters the soil, silicon is released rapidly, and near-equilibrium concentrations of silicon in solution (usually H_4SiO_4) are reached within a few hours. Monosilicic acid is taken up with water by plants and deposited as amorphous silica (and probably organic-complex forms) in the plant tissue as water is transpired. Russell (1961) estimated that twice as much silicon is cycled annually by plants as is lost in drainage waters. Animals feeding on such plants ingest both soluble silicon in the sap and amorphous silica deposits. Most of the silicon ingested by animals is excreted in the feces; a minor portion in the urine. A very small amount is retained, however, and plays an essential role in some species (Carlisle, 1972a).

The silicon concentrations of most soil solutions probably fall within the range of 2–20 ppm; however, concentrations of silicon ranging from less than 1 ppm to nearly 200 ppm have been reported in aqueous extracts of soils (Acquaye and Tinsley, 1964; Jones and Handreck, 1967; McKeague and Cline, 1963c). Although much of this range reflects differences in the soils involved, part is probably the result of different procedures used in obtaining the extracts. Three general kinds of solutions were analyzed:

1. Extracts of soils at their natural water contents or at "field capacity" (1/3 bar). Concentrations of silicon in such extracts of some soils in Australia ranged from 14 to 19 ppm (Jones and Handreck, 1965b, 1967).

2. Extracts of soil–water suspensions of various ratios,

commonly 1:1 to 1:5 (McKeague and Cline, 1963a; Acquaye and Tinsley, 1964; Jones and Handreck, 1967; Weaver *et al.*, 1971; Elgawhary and Lindsay, 1972). Concentrations of silicon range from 1 to 25 ppm generally in such extracts. Extracts of water-saturated soils contained silicon at concentrations of 3–17 ppm (Raupach, 1957) and 4–10 ppm (McKeague and Cline, 1963a).

3. Drainage water from soils (McKeague and Cline, 1963c; Acquaye and Tinsley, 1964) commonly contains less than 10 ppm silicon. Unpublished data of F. J. Sowden (personal communication, 1973) for tile effluent in the Ottawa area shows silicon concentrations of 7–8 ppm.

Although the concentrations of silicon in these kinds of solution are not widely different, the values for those from soils below saturation are generally higher than those for drainage waters or for suspensions of low soil:water ratios.

The concentration of silicon in solution in a soil extract is determined by the following factors:

• *The degree of weathering* The concentration of silicon in soil solution is much greater in soils developed in young volcanic ash deposits than in old, highly weathered soils composed mainly of oxides of iron and aluminum. This seems consistent with results for drainage waters; for example, Kobayashi (1960) reported an average silicon concentration of 21 ppm in Japanese rivers draining areas of volcanic rocks and 4.8 ppm in rivers draining areas of sedimentary rocks.

• *The pH* Increasing the pH to about 9 decreases the silicon concentration. For example, changing the pH from 4.5 to 7.9 resulted in a change in silicon concentration from 6.4 to 0.9 ppm in a 1:4 soil:borate buffer suspension (McKeague and Cline, 1963a). For a soil at 1/3 bar water tension, silicon in solution decreased from 33 to 11 ppm when the pH was changed from 5.4 to 7.2 (Jones and Handreck, 1967). Increasing the pH beyond approximately 9.5 appears to result in sharply increased concentrations of silicon in solution, presumably resulting from the formation of $H_3SiO_4^-$. For example, Kelley and Brown (1939) reported a silicon concentration of 175 ppm in a 1:5 soil:water extract of a soil having a pH of 10.4. Much higher silicon concentrations, beyond 1,200 ppm, occur in some alkaline lakes (Jones *et al.*, 1967), but such concentrations are probably very rare in soil solutions. Furthermore, such soils would be of little significance in the production of plants. Although pH has a marked effect on the concentration of silicon in solution in a given soil, the concentrations of silicon in solution in soils generally cannot be predicted accurately from pH alone.

• *Iron and aluminum oxides of high surface area* These substances have a marked capacity to sorb H_4SiO_4 (Beckwith and Reeve, 1963; McKeague and Cline, 1963b; Jones and Handreck, 1967). Recent work throws some light on the mechanism of adsorption of dissolved silica by goethite (Hingston *et al.*, 1967, 1968) and soils (Obihara and Russell, 1972).

• *Temperature* Silicon in soil solution increases with temperature. For example, McKeague and Cline (1963a) reported almost double the concentration of silicon in extracts of 1:2 soil:water mixtures at 25° C compared with those maintained at 4° C. Similar temperature effects were reported by Fanning and Pilson (1971) for silicon in solution in marine sediments.

• *Treatment of soil and method of equilibration* Drying and grinding the soil before equilibration with water may result in either higher or lower concentrations of silicon in solution than those obtained with moist soil otherwise treated in the same way (Acquaye and Tinsley, 1964). Shaking soil–water suspensions may result in markedly higher concentrations of silicon in solution than those obtained without shaking, and equilibrium values may not be attained (McKeague and Cline, 1963a). Shaking quartz–water suspensions at 25° C yielded concentrations of silicon in solution of as much as 200 ppm (Morey *et al.*, 1962). Increasing the soil:water ratio results in increased concentrations of silicon in solution (McKeague and Cline, 1963a, 1963c; Jones and Handreck, 1967).

• *Degree of leaching of soil* Highly leached soils generally have lower concentrations of silicon in solution than do weakly leached soils of similar composition (McKeague and Cline, 1963c).

• *Freezing* Freezing and subsequent thawing of a soil has been reported to result in a decrease in the concentration of silicon in solution (Slavnyy and Vorob'yeva, 1962), but few data are available.

• *Reduction of soil as a result of flooding* Flooding may double the concentration of silicon in soil solution, particularly if appreciable organic matter is introduced (McKeague and Cline, 1963c; Ponnamperuma, 1964). This effect is thought to be the result of the reduction of ferric hydroxides, which have a high capacity to absorb H_4SiO_4 (Ponnamperuma, 1964).

Tiller (1967, 1968) has shown that clays sorb more heavy metal cations such as zinc, cobalt, and nickel in the presence of monosilicic acid in solution than in the absence of added silicon. He considered that the adsorption of H_4SiO_4 by the clays probably provided additional adsorption sites for the heavy metals.

ANALYTICAL METHODS

The analytical chemistry of silicon is well known. In high concentrations, such as 1,000 ppm or more, classical gravimetric analyses will suffice, but in the more typical concentration range of 0.1–10.0 ppm, colorimetric determination by silicomolybdate complexing is the preferred method. For concentrations above 5 ppm, the yellow (oxidized) form is appropriate for monomeric dissolved silica; for lower concentrations, the reduced blue silicomolybdate is used. Colorimetry is standardized by reference to pure quartz crystal, which can be obtained in a remarkably pure state. Analyses of solutions with phosphate or arsenate are more complicated because both form molybdate complexes. In such solutions, either conditions must be chosen to avoid interferences (e.g., adjustment of pH) or the interfering substances must be separated.

The wide range of silicon content reported in animal tissues by some of the early French and German investigators probably reflects interference by phosphorus in the analytical method used (earlier work reviewed by King and Belt, 1938). Improvements in analytical techniques have allowed more accurate determinations to be made (King, 1939; King *et al.*, 1955), but, even with improved techniques, precautions must be taken to avoid interference by phosphorus in biological tissues. Conversely, silicon interferes with the determination of phosphorus in drainage water from soils, as shown by Sowden (1972), among others. Unless precautions are taken to correct for or to eliminate the interference from silicon, phosphorus values may be in error.

In analyzing soils, different results may be obtained depending on whether the soil is moist or dry before equilibration with water (Acquaye and Tinsley, 1964). McKeague and Cline (1963a) also report markedly higher concentrations for shaken than for unshaken soil–water suspensions and when soil:water ratios are increased.

Because of the simplicity and ready availability of newer analytical methods, such as atomic absorption, the number of analyses of silicon in geological and biological materials has been steadily growing for the past decade. With certain complex biochemical mixtures (e.g., bone), however, the present methodology needs further development. Between 1969 and 1972, Carlisle (personal communication, 1973) tested the suitability for animal tissues of 23 methods for quantitative analysis of silicon. She concluded that, although a satisfactory general colorimetric method has been developed for most biological tissues, there are many appropriate modifications (Volk and Weintraub, 1958; Paul, 1965; Chalmers and Sinclair, 1966), one of the best being that developed by Jankowiak and LeVier (1971). Bone is a special case, and, for this tissue, emission spectrography, although it requires special techniques, is the best method available at present, particularly with small samples.

SILICON IN PLANTS

Geographic Factors

High levels of silicon are likely to be found in plants in the following regions:

- Semiarid to arid areas, especially those with alkaline soils, where little leaching has occurred, e.g., the drier areas of the great plains of North America.
- Areas of fresh volcanic ash or of highly weatherable silicate minerals such as olivine or pyroxene.
- Areas adjacent to hot springs or discharge areas of groundwater high in silicon.
- Areas containing fine-textured soils with low contents of free iron and aluminum oxides.
- Areas of poorly drained, reduced soils.

Low levels of silicon in plants are generally associated with the following:

- Areas of highly weathered soils composed mainly of oxides of iron and aluminum (oxisols). (Such soils occur principally in tropical areas on old landscapes, such as parts of India, Ceylon, Puerto Rico, and Hawaii.)
- Highly leached, coarse-textured soils.
- Soils developed from limestone of low silicon content.

Although these relationships do affect the silicon content of plants, inherent differences in the abilities of plants to exclude or to take up silicon exert a much greater influence than do the geographic factors. The silicon content of different plants grown on the same soil may vary by two orders of magnitude.

Uptake and Essentiality

Plant uptake of silicon is governed by the concentration of silicon in the soil solution and the nature of the plant. Factors that control the concentration of silicon in soil solutions were discussed previously.

Jones and Handreck (1965b) grew oats in soil and soil–oxide mixtures that had a wide range of concentrations of silicon in solution. Their data, some of which are tabulated in Table 8, showed that total silica in the plant tops was directly proportional to the concentration of monosilicic acid in the soil solution.

For rice plants, the silicon content of leaves and stems (dry wt) increased from 0.03 percent for plants grown in solutions containing no detectable silicon to 3.74 percent for plants growing in solutions containing silicon at a concentration of 47 ppm (Okuda and Takahashi, 1964). The silicon content of the plants was not directly proportional to the silicon concentration of the nutrient solution, however, as the transpiration ratio decreased from 4.5 to 3.3 at no detectable and at 47 ppm silicon, respectively. The silica content of crimson clover was approximately proportional to the concentration of silicon in solution, although the percentages of silica were very low (Handreck and Jones, 1967). Reports of increases in the silicon content of plants due to acidification of soil, addition of soluble silicates, and reduction of soil by flooding are consistent with the information on effects of these treatments on the silicon content of the soil solution (Jones and Handreck, 1967). Similarly, reports of decreases in the silicon content of plants due to liming and to additions of iron or aluminum oxides are consistent with the known effects of these treatments on the concentration of silicon in soil solutions.

Abundant data show that there are marked differences in the tendencies of various plants to accumulate silicon, although the reasons for these differences are not completely clear (Lovering, 1959; Russell, 1961; Jones and Handreck, 1967; Lewin and Reimann, 1969). Some of these data are presented in Table 9. The silicon composition of a number of plants is given in Table 10.

TABLE 8 Dissolved Silica Content of Oat Plants and Soil Solutions (as SiO_2)

Plant	SiO_2 in Solution, ppm	Total Dry Matter, g	SiO_2 in Dry Matter, %	SiO_2 in Plants, mg/plant	Water Transpired, kg/plant	SiO_2 Expected, mg/plant
A	7	7.07	0.40	28.3	3.86	27.0
B	54	6.40	2.77	177.0	3.27	176.0
C	67	6.92	3.96	274.0	3.90	261.0

SOURCE: Jones and Handreck (1965b).

TABLE 9 Removal of Silicon by Various Crops per Acre of Soil

Crop	Part of Plant	Dry Weight, lb	Total ash, lb	Si, lb	Si as Percent of Dry Weight
Wheat	Seed	1,530	30	0.3	0.02
	Straw	2,560	142	45.0	1.76
Oats	Seed	1,630	51	9.3	0.57
	Straw	2,350	140	30.6	1.30
Beans	Seed	1,610	58	0.2	0.01
	Stems and leaves	1,850	99	3.2	0.17
Meadow hay	—	2,820	203	26.6	0.94
Red clover hay	—	3,760	258	3.3	0.09
Turnips	Roots	3,130	218	1.2	0.04
	Leaves	1,530	146	2.4	0.16
Marigolds	Roots	5,910	426	4.1	0.07
	Leaves	1,650	254	4.3	0.26
Potatoes	Tubers	3,360	127	1.2	0.04

SOURCE: Russell (1961).

TABLE 10 Silicon Content of Plants

Species	Material	Si Dry Tissue, ppm	SiO_2 in Total Ash, %
Crimson clover	Aerial growth	560	1.4
Peas	Aerial growth	1,080	3.2
Alfalfa	Aerial growth	1,000– 2,400	2.5–7
Reed canary	Aerial growth	2,800–40,000	8–56
Tall fescue	Aerial growth	4,200–23,000	17–51
Coastal Bermuda	Aerial growth	3,300–32,000	10–53
Oats	Leaf	25,000	40
	Culm[a]	4,800	N.A.[b]
	Inflorescence	36,000	N.A.
Rice	Polished grain	230	9
	Bran	21,500	37
	Hulls	107,000	99
	Straw	63,500	73

[a] Blades, sheaths, and stems.
[b] Not available.
SOURCE: Van Soest (1970).

Some plants, such as the legumes, have levels close to those of animal tissues. Other plants are accumulators to a varying extent, as shown by comparison of grasses, oats, and rice. The flowering part of the plant has the highest concentration, while the endosperm is quite low in silicon. Silicon in grass varies widely, depending on the region in which the grasses are grown and on the species (Van Soest and Jones, 1968; Van Soest, 1969). Rice straw commonly contains more than 5 percent silicon, and the silicon content of *Equisetum* is similar (Okuda and Takahashi, 1964; Jones and Handreck, 1967; Lewin and Reimann, 1969).

Plants can be sorted into three classes according to their tendency to take up silicon:

• *Plants that take up silicon passively* These plants make no discrimination between H_2O and H_4SiO_4. Oats is an example of this class (Table 8). The total silica in the tops has been accounted for in terms of the silicon concentration in the soil solution and the amount of water transpired (Jones and Handreck, 1965b). The silicon concentration of the xylem sap at the base of the stem was similar to that in the soil solution (Handreck and Jones, 1968).

• *Plants that actively take up silicon (e.g., rice)* Okuda and Takahashi (1964) showed that the concentration of silicon in the xylem sap of rice plants was much greater than that in the external solution. They found that 37 h after the introduction of rice plants, the silicon concentration of a nutrient solution decreased from 47 ppm to less than 5 ppm, but the xylem sap contained 300 ppm silicon.

Okuda and Takahashi (1964) concluded that metabolic processes in the roots were closely linked to the capacity of rice to take up silicon. Tops of rice plants contained no more silicon than did tops of wheat or tomato plants, but rice roots took up silicon more than 10 times as rapidly as the tops. In wheat and tomato plants, roots contained silicon in concentrations similar to those for tops.

• *Plants that restrict the uptake of silicon* Handreck and Jones (1967) and Jones and Handreck (1969) found that the concentration of silicon in the xylem sap of whole crimson clover plants was only about 6 percent of that in the external solution. Although clover and oats had similar transpiration ratios, the silicon in clover tops was only 5–10 percent of that in oats grown in the same soils. Roots of the clover plants contained approximately eight times as much silicon as the tops. The authors concluded that crimson clover either excludes H_4SiO_4 at its external surface or binds it within the root by an unknown mechanism.

Deposition of Silicon in Plants

Silicon enters plants as H_4SiO_4 but is deposited as silica (SiO_2) (Jones and Handreck, 1967; Handreck and Jones, 1968; Lewin and Reimann, 1969). Analyses of the sap of oat plants (Jones and Handreck, 1965b) and of rice plants (Okuda and Takahashi, 1964; Jones and Handreck, 1967)

show that the concentration of H_4SiO_4 in sap can exceed that of a solution saturated with amorphous silica.

Silicon is probably deposited in plant tissue as amorphous or opaline silica with a water content commonly of 1 to 1.5 percent (Iler, 1955; Jones and Milne, 1963; Lewin and Reimann, 1969). Other reports, however, describe the occurrence of organic complexes (Engel, 1953; Heinen, 1963; Lovering and Engel, 1967). Opal is ordinarily obtained after wet ashing with acid, suggesting that it is synthesized by acid treatment (Drum, 1968). The organic forms that may occur in plants appear to be very unstable (Coombs and Volcani, 1968), particularly in the presence of traces of acid (Van Soest and Lovelace, 1969). It has been suggested that the organic bound silica is complexed by carbohydrate in plants (Van Soest and Lovelace, 1969) and in animal tissues (Johlin, 1932). Lovering and Engel (1967) suggested that it occurs in soluble complexes with polyphenols in a kind of silicone linkage. The organic complexes, if they occur in the plant, have defied isolation and characterization so far (Coombs and Volcani, 1968).

Plant silica has been shown to exist in at least two forms, based on solubility criteria (Brown, 1927; Van Soest and Lovelace, 1969). One form has a solubility of approximately 200 ppm (as silicon) in boiling water; the second has a much lower solubility—approximately 0.2–0.5 ppm. Such solutions contain organic matter, and rate-of-solution studies suggest the hydrolysis of a hydrogen-bonded silica–organic complex (M. G. Jackson, personal communication, 1973). Such association is necessary to account for the depressing effects of biogenic silica on digestibility of structural carbohydrates in straw and grasses.

Table 11 lists soluble silica for a number of species. Proportions of the two phases vary among forages; rice hulls and horsetail (*Equisetum*) show only the soluble phase, whereas all leafy graminaceous plants show two phases. The lowest proportion of hot-water-soluble silica occurs in plants from arid regions. Whether transpiration has a role in determining the amount of a given form of silica in plants remains an open question.

Detailed studies by Handreck and Jones (1968) on silica in oat plants suggested that H_4SiO_4 moved with the transpiration stream and that silica was deposited in the greatest quantities in those parts of the plant from which water was lost in the greatest quantities.

Detailed morphologic studies of plant silica show that epidermal cells are usually impregnated with silica and that it is probably deposited in intimate association with the cell wall (Yoshida *et al.*, 1962; Jones *et al.*, 1963). Jones *et al.* (1963), in their studies of oats, found that cell walls thickened with cellulose or lignin were always impregnated with silica. Yoshida *et al.* (1962) reported that silica was combined with cellulose in the epidermal cells of the leaf blade of rice.

The Role of Silicon in Plants

In plants beneficial effects attributed to silicon have been

TABLE 11 Siliceous Fractions in Forages

Material	Source	Total Silicon, % of dry matter	Potentially[a] Soluble Silicon, % of dry matter	Percent of Total Silicon, % of dry matter
Equisetum	Washington	3.9	3.9	100
Rice hulls	Texas	10.7	10.6	99
Rice straw	Arkansas	6.0	4.9	79
Buffalo grass	Kansas	3.2	2.9	88
Brome	South Dakota	2.4	2.1	87
Reed canary	South Dakota	4.1	2.9	69
Oat leaves	(hydroponic)	2.0	1.4	64
Fescue	Arizona	4.2	1.8	43
Indian rice grass	Utah	2.5	1.1	43

[a] Soluble silicon obtained by exhaustive extractions with boiling water.
SOURCE: Van Soest and Lovelace (1969).

summarized by Okuda and Takahashi (1964), Jones and Handreck (1967), and Lewin and Reimann (1969):

• Resistance to fungal disease and insect attack has been attributed to toughening of the epidermis by the deposition of SiO_2 (Sasamoto, 1958; Volk *et al.*, 1958). According to Okuda and Takahashi (1964), however, this explanation is not adequate.

• Alleviation of manganese toxicity in plants such as wheat, oats, and rice. Silicon also promotes the oxidation capacity of rice roots and thus avoids Fe^{2+} toxicity.

• Depression of the transpiration ratio.

• Promotion of the translocation of absorbed phosphorus to the panicle of rice.

• Improvement of seed retention, and, perhaps, strengthening cereal stems, thus helping to prevent lodging (bending over or breaking of the stalks).

Yields of rice are markedly higher in the presence of a generous supply of soluble silicon than in its absence, and, for degraded paddy fields, silicon-containing fertilizing materials are commonly added (Okuda and Takahashi, 1964). Some of the soils involved are highly weathered oxisols with relatively low silicon contents that have the property of binding phosphate in an unavailable form. Addition of silicates raises the pH, releases phosphate, and precipitates soluble iron, aluminum, and manganese (Russell, 1961; McKeague and Cline, 1963c; Clements, 1965; Ayres, 1966; Jones and Handreck, 1967). Silicate fertilization of sugarcane markedly increases dry-matter yield, including sugar (Clements, 1965; Alexander, 1968) and moderates toxicity of manganese at the leaf level (Clements and Awada, 1967). Application of silicate solutions to the leaves of cane increases sugar and protein content, showing that not all the effects of silicon relate to its influence on the soil (Alexander, 1968). Production of other plants that normally have a high silicon content, such as wheat, oats, and *Equisetum*, is probably improved when an adequate supply of soluble silicon is present (Okuda and Takahashi,

1964; Fox *et al.*, 1969; Lewin and Reimann, 1969). Cell division of diatoms ceases in the absence of silicon in solution (Lewin and Reimann, 1969).

Although beneficial in some plants, the question of the essentiality of silicon to plants remains unresolved (Lewin and Reimann, 1969). One viewpoint regards silicon as a passive element with respect to plant metaboism, exerting beneficial effects by modifying the soil environment (Yoshida *et al.*, 1962; Jones and Handreck, 1967). Another view is that silicon is essential to the plant—i.e., required for specific metabolic and structural functions (Alexander, 1968; Lewin and Reimann, 1969).

SILICON IN ANIMALS AND HUMANS

Plant Content of Silicon Related to Animal Feeding

Silicon has an adverse influence on the digestibility and nutritive value of fibrous plant feed and forages (Smith *et al.*, 1971a). Silicon affects digestibility by two mechanisms: first, by substituting (in volume) for nutrients because it is essentially indigestible matter; and second, by inhibiting the digestion of cellulosic carbohydrates. The mechanism of the latter effect is not understood, because the complexes between silicic acid and the carbohydrates have not been adequately characterized. Removal of the hot-water-soluble phase of silicon from plants (*in vitro*) significantly increases the fraction of digestible organic matter by increasing availability of the cellulosic carbohydrates (Van Soest and Jones, 1968; Van Soest, 1970). The precise role of the nearly insoluble phase is not known because of the difficulty in removing this fraction without disrupting the ligno–cellulose complex.

Plant silica interferes somewhat less with digestibility than does lignin. Sullivan (1959) has reported regression slopes for lignin on dry-matter digestibility for grasses in the range of 2.1–6.6 units of digestibility per unit of lignin; values for silicon range from 2 to 7.5 (Van Soest and Jones, 1968). A decline of approximately 3.4–4.2 units is observed for the chemical effects of silica on digestible dry matter and of approximately 2–3 units for the diluting effect (Van Soest and Jones, 1968). Because cumulator plants may contain as much as 5 percent silicon, the diluting effect of silicon can be of considerable importance in feed quality.

From a practical view, adverse effects of silicon in animal feeding occur only at high levels and occur in two ways: First, from feeding directly on soil minerals such as sand, clay, and vermiculite; and second, from feeding of or grazing on cumulator plants. The first situation produces adverse effects because the minerals ingested are insoluble, and thus dilute the nutritive content of the feed. These materials are not inert, however; clay and vermiculite have considerable base exchange and absorptive capacity. Denser minerals, such as quartz (in the form of sand or gravel), have been reported to influence gains in beef cattle (Van Soest, 1970); but this is probably because the passage of such materials is retarded, and the so-called gain reflects the weight of the retained quartz. The same holds for ingestion of soil by grazing animals (Van Dyne and Lofgren, 1964).

The second category, consumption of siliceous plants, presents a different set of problems. An important interaction exists between silica and lignin, as shown in Table 12, which indicates that the sum of silica and lignin shows a more consistent relation to digestibility than either

alone. The table also shows the great variation in silica content of forages from different regions. As has been previously mentioned, forage plants vary considerably among species in their ability to concentrate silica. Variation of silica content within similar forages, and even within the same species, is also very large. Climate, particularly light and temperature, is important, but the variation in soil silica provides an additional basis on which to explain geographic differences in digestibility of comparable forages (Jones and Handreck, 1967).

Balance trials in animals indicate that almost all ingested silicon is unabsorbed, passing through the digestive tract to be lost in the feces. Moreover, most of the small proportion that is absorbed is excreted in the urine. The proportion of absorbed silicon actually retained in the body is not known. The occurrence of siliceous uroliths, however, gives clear evidence that the quantity of silicon absorbed, and excreted in the urine, under conditions of high intake can be harmful. Many balance studies in herbivores have shown that more silicon is excreted than absorbed, which presumably reflects the leaching and absorption of silicon from sand and gravel lodged in pockets of the rumen or from residual plant phytoliths. In ruminants, which have multiple stomachs, the absorption of silicon occurs in the rumen (the animal's first stomach) at pH 5.5–6.5 prior to passage of ingested material through the abomasum (the fourth, true digestive stomach), which presents a much more acidic environ-

TABLE 12 Lignin and Silica Content of Forages in Relation to Digestibility

Forage	Source	SiO_2,[a] % of dry matter	Permanganate Lignin, % of dry matter	SiO_2 + Lignin, % of dry matter	Apparent percentage of Digestibility
Coastal Bermuda	Texas	2.2	4.7	6.9	56
	Arkansas	2.3	6.0	8.3	50
	South Carolina	0.7	8.5	9.2	53
	Arizona (April)	4.1	4.9	9.0	55
	Arizona (July)	6.2	4.8	11.0	47
	Louisiana	6.7	5.5	12.2	42
Fescue	Arkansas (April)	1.5	2.7	4.2	81
	Arkansas (May)	0.9	5.1	6.0	67
	Iowa (Oct)	4.0	3.5	7.5	63
	Arkansas	4.5	6.4	10.9	48
	Arizona	8.9	5.0	13.9	47
Reed canary	Pennsylvania	0.6	5.3	5.9	62
	Michigan	1.3	4.6	5.9	63
	Michigan	4.0	3.4	7.4	57
	Iowa	5.4	5.9	11.3	45
	South Dakota	8.8	4.5	13.3	50
Rice straw	Arkansas	13.1	3.1	16.2	37
Rice hulls	Texas	22.9	15.6	38.5	8
Correlation coefficients (r) with ADDM% (apparent digestibility of dry matter)		−0.877[b]	−0.743[b]	−0.916[b]	

[a] SiO_2 (silica) values are approximately twice as large as comparable Si (silicon) values.
[b] Statistically significant at the 1 percent level of probability.
SOURCE: Van Soest (1970).

ment. In many monogastric animals, ingested material encounters gastric HCl at very low pH initially—often below 1. Under these conditions, absorption of silicon from plants is greatly reduced. This behavior of plant silicon does not extend to sodium silicate or other inorganic sources of silicon.

Phytoliths as a Source of Silica

An important mechanism by which silicon enters higher animal organisms is the ingestion or inhalation of phytoliths. Phytoliths are minute bodies of amorphous silica (opal) that are formed within a great variety of plants. They have been extensively studied in grasses, where they occur in and between the epidermal cells, as well as in mineralized plant hairs, hooks, spines, and allied structures (Baker, 1959). In grasses, they range from long bamboo-like structures to small hemispheres, many of them having a minimum dimension of 10 μ or less. Twiss *et al.* (1969) give an excellent description of phytoliths in grasses and how they contribute to the content of particular soils and paleosols.

Opal phytoliths are found in the sediments of the ocean floor as well as in soils, and their mineralogical and chemical properties have been extensively studied (Jones and Beavers, 1963). Grazing animals contribute a great deal to their distribution, because, through ingestion of forage materials rich in silica, they release (by digestion) and fragment the siliceous particles, depositing them in the soil. Baker and Jones (1961) estimated that the rumen contents of a sheep fed on oaten hay contained nearly 10 g of opal phytoliths. Smaller phytoliths are readily dispersed by the wind and may be carried for great distances (Folger *et al.*, 1967).

Many of the smaller phytoliths are able to pass the intestinal epithelial barrier of animals, including man, and enter the lymphatic and blood vascular systems, thus to be distributed throughout the body. Also, many of them are inhaled and lodge permanently in the lungs or enter the pulmonary vascular systems and are carried to various internal organs and tissues. Baker and Jones (1961) observed opal phytoliths in *every* prescapular, mediastinal, mesenteric, popliteal, precrural, and bronchial lymph node taken from slaughtered sheep. "One million mineralized corpuscles were estimated to be present in one prescapular lymph node!" In a study of siliceous urinary calculi in rams, they found opal phytoliths within the calculi and concluded that these had entered the blood stream and penetrated the glomerulus (in the kidneys) to reach the urinary tract, where they acted as nuclei to initiate the precipitation of silica.

It has been suggested, particularly by Rose (1968), that phytolithicosis may be a disease entity and that the phytoliths in animal tissues may play a role in the genesis of cancer, although there are no hard data to support this view.

In addition to possibly producing pathologic effects, the presence of phytoliths throughout many tissues, especially lungs, liver, lymph nodes, and spleen, makes it virtually impossible to determine the amount of silicon in animal tissues that is actually contributing to physiologic mechanisms.

Opal phytoliths (from oats) have a hardness in the range of 4.5–6.2 on the Mohs scale, which is harder than the tooth enamel of 4-yr-old sheep (Baker *et al.*, 1959). This characteristic has caused serious problems with respect to abrasive wear of the teeth of grazing animals in the range country of the western United States, as well as in portions of Australia and South Africa. Dental erosion may be so severe that, on occasion, teeth have been capped with stainless steel to extend the useful life of selected animals. Underwood (1971, pp. 407–415) has calculated that sheep under "normal" conditions in some areas will ingest approximately 14 kg of SiO_2 during a year. This amount does not include the portion derived from the contamination of leaf surfaces by silica-containing soil—probably the major factor in tooth wear (Underwood, 1971).

Siliceous Calculi

A common problem of ruminants grazing siliceous grasses in certain arid regions is urolithiasis, i.e., urinary calculi ("stones") in the kidney, ureter, or bladder (Emerick *et al.*, 1959; Underwood, 1971). High silicon intake, coupled with a limited supply of drinking water, results in renal concentration of silicic acids to a supersaturated solution. The relation between intake of silica and its excretion in sheep under conditions of varying silicon and water intake has been studied in detail by Jones and Handreck (1965a).

Stones from animals grazing silicon cumulator plants are composed almost entirely of hydrated silica (Jones and Handreck, 1967). Urolithiasis has not been produced by the consumption of dry products very high in silicon, such as rice hulls (Whiting *et al.*, 1958), or by the feeding of soluble sodium silicate. However, supplementation of diets with tetraethyl orthosilicate ester (TES) does produce siliceous calculi in rats (Emerick *et al.*, 1959). Presumably, the organic ester and the forms present in fresh grasses are more absorbable. It seems likely that the drying of forage reduces the soluble silicon by promoting polymerization. This property of forage silicon, i.e., its ability to produce uroliths, emphasizes the peculiar nature of plant silicon as compared to inorganic sources.

Recently, Ehrhart and McCullagh (1973) have produced siliceous urinary calculi in a group of dogs fed an atherogenic diet that contained 27 percent nonnutritive bulk, of which 11 percent was $Mg_3Si_4O_{11} \cdot H_2O$ and 44 percent was $SiO_2 \cdot H_2O$.

Relatively few cases of siliceous calculi have been reported in the urinary tracts of human beings, probably because they have not been searched for carefully. Their occurrence was first described in 1960 by Herman and Goldberg (1960). As in most of the reported cases, the source of the silicon was attributed to the ingestion of magnesium trisilicate as an antacid. A recent case is described by Joekes *et al.* (1973). The original analysis did not include silicon, but showed the presence of calcium, oxalate, phosphorus, and magnesium; subsequent analysis showed the calculus to be predominantly amorphous silica.

Factors considered to be important in formation of siliceous calculi include the presence of mucoproteins (in human beings, the R-1 fraction of the nondialyzable urinary solids, uromucoid, is the important one) and acid urine, both of which are contributing factors in precipitating polymerized silicic acid (Bailey, 1970). The presence of phytoliths in the urinary tract may also be a contributing factor in urolithiasis.

Newberne and Wilson (1970) have demonstrated a significant renal lesion occurring in dogs (but not in rats) that were fed relatively large doses of sodium silicate and magnesium trisilicate over a period of 4 weeks. Major pathologic changes were focal subcapsular hemorrhages and hypertrophy of tubular epithelium, with or without degenerative changes and lowgrade inflammatory cellular reaction. They concluded that "The unusual lesions in the kidney of the dogs suggest a basic defect in the ability of this species to metabolize or excrete these compounds, . . ." reinforcing the view that comparative pathologic studies are critically important in evaluating the results of potentially toxic substances.

Essentiality and Physiologic Role in Animals

Interest in the effects of siliceous substances on animals (excluding silicosis) extends back over half a century, probably beginning with the work of Gonnermann (1918, 1919). Major interest has been focused on the deleterious aspects of silicon and its compounds, such as interference with digestibility of forage and urolithiasis. Until quite recently, however, there has been no proof that silicon plays a specific beneficial role in metabolic processes of higher animals. All animal tissues and fluids that have been adequately examined contain at least traces of silicon, but the highest levels are found in the epidermis and its appendages and in connective tissues in general. The eggs of birds (Drea, 1935; Monier-Williams, 1949; Carlisle, 1972b), milk (Kirchgessner, 1957; Carlisle, 1972b), and the fetuses of mammals (King and Belt, 1938; E. M. Carlisle, unpublished data, 1973) have small quantities. The blood of man and other mammals averages 5 ppm. In cow's milk, dietary silicon supplements have been reported (Archibald and Fenner, 1957) to have little effect on silicon concentration. However, in rat's blood, moderate increases have been obtained after feeding silicon as sodium metasilicate, and much higher levels have been reached after feeding organic silicates (E. M. Carlisle, unpublished data, 1973). The blood appears capable of maintaining a considerably higher concentration of organic than of inorganic silicate. The silicon content of parenchymal tissues such as liver, heart and muscle, for example, range from 2 to 10 ppm. The consistently low concentration of silica in most organs does not appear to

vary appreciably during life except in the lungs, which ordinarily accumulate large amounts of silicon from long-continued inhalation of finely particulate silica.

Adequate silicon intakes for growth in chicks (2–3 weeks) and rats (3–7 weeks) are 2 mg (silicon) and 4 mg (silicon), respectively, administered as sodium silicate. From recent *in vivo* findings, Carlisle (1973, 1974, 1976) concluded that silicon is associated with mucopolysaccharide synthesis in the formation of cartilage matrix and connective tissue and showed that the site of action of silicon is in the mucopolysaccharide–protein complexes of the ground substance. In higher animals, the mucopolysaccharides, hyaluronic acid, chondroitin sulfates, and keratin sulfate are found to be linked covalently to proteins as components of the extracellular, amorphous ground substance that surrounds the collagen and elastic fibers and the cells of connective tissues.

Silicon appears to be covalently bound to the polysaccharide matrix in these mucopolysaccharides, most likely in ester linkage: C-O-Si (Schwarz, 1973; Carlisle, 1974). Additional evidence for the possibility of a polysaccharide–silicon–ester linkage existing in biologic material may be found in the literature (Carlisle, 1974).

These findings are important in terms of biological structure and function, and human health and disease, because the mucopolysaccharides are involved in induction of calcification in general (bone formation in particular); in the maintenance of fibrous, elastic, osseous, and cartilaginous tissues; and in those restorative processes that require the production of new collagen or bone, e.g., repair of wounds and fractures. Moreover, conditions such as atherosclerosis, osteoarthritis, and the overall processes of aging are associated with significant changes in the mucopolysaccharides. Quite a different role of mucopolysaccharides is their contribution to the control of metabolites, ions, and water, since mucopolysaccharides are intimately involved with bound water and its constituents. The full extent to which silicon affects these many processes through its influence on mucopolysaccharides remains to be determined.

In vitro studies beginning in 1964, using electron microprobe analyses (Carlisle, 1969, 1970a), have shown the unique localization of silicon in active calcification sites in young bone. Furthermore, in the earliest stages of calcification in these sites, when the calcium content of osteoid tissue is very low, a direct relation exists between silicon and calcium. Subsequent *in vivo* experiments with weanling rats (Carlisle, 1970b) have also shown a relation between silicon and calcium in bone formation and have demonstrated that dietary silicon increases the rate of mineralization; this effect was particularly apparent under conditions of low calcium intake. A somewhat similar mechanism of action (in bone formation) has been demonstrated for vitamin D (Muller *et al.*, 1966).

Skeletal changes have also been observed in silicon-deficient weanling rats (Schwarz and Milne, 1972). Compared with controls, the skulls were shorter and the bone structure surrounding the eye appeared distorted. Pigmentation of the incisors was also affected, suggesting a disturbance of enamel development. Significant improvement in pigmentation could be produced by dietary supplements of silicon (59 mg/100 g of diet), as well as by administering tin, vanadium, or fluorine at physiologic levels.

The importance of silicon for skeletal development has also been demonstrated in the chick (Carlisle, 1972a, 1972b). On a silicon-deficient diet, the leg bones of the chicks were shorter, smaller in circumference, and had thinner cortices than those of the controls. The metatarsals were more flexible, and the femur and tibia fractured more easily. Skull development was also affected, the cranial bones being flatter than normal, giving the skull an appearance much like that of a serpent's head. Other tissue changes were observed, particularly in the comb and in the skin.

Skeletal and other abnormalities involving mucopolysaccharide synthesis in the formation of the cartilage matrix and connective tissue are also found to be associated with silicon deficiency in the chick (Carlisle, 1973, 1974, 1976). The tibial–metatarsal and tibial–femoral joints were smaller in the silicon-deficient chicks. The ends of the bones had less articular cartilage and were less well formed. Analyses of the bones showed no difference in dry weight, amount of organic matter, or the absolute amount or percentage of ash between the groups. However the bones of the deficient group contained 35 percent less water. This, in addition to the presence of less articular cartilage, is strongly suggestive of mucopolysaccharide involvement. Analysis revealed a significantly decreased total amount and proportion of mucopolysaccharides in the cartilage of the silicon-deficient group of chicks. This relation, established between silicon and mucopolysaccharide synthesis in cartilage formation, was confirmed in another type of connective tissue, the cock's comb. These findings indicate that silicon plays an essential role in the formation of cartilage and connective tissue.

Carlisle (1974) has established an inverse relationship between silicon content and aging of certain tissues, notably skin, aorta, and thymus; see also Brown (1927), MacCardle *et al.* (1943), and Loeper *et al.* (1966). This has been found to be true of several species (rat, rabbit, pig, and chicken). In contrast, other tissues such as heart, kidney, muscle, and tendon showed no significant changes in silicon content with aging. In an earlier study, Leslie *et al.* (1962) reported that a 60 percent decrease of silicon content occurs in rat skin between 5 weeks and 30 months of age. No significant change in silicon was observed in muscle and tendon.

With respect to aging it is probably quite important that the level of silicon in the arterial wall was found to decrease with the development of atherosclerosis (Loeper *et al.*, 1966). It is also of interest that organosilicon compounds have been considered as prophylactic and therapeutic agents for a variety of human diseases (Garson and Kirchner, 1971).

Food Sources of Silicon for Humans

Rich natural sources of silicon for human beings include whole-grain cereals, particularly unpolished rice and oats, legumes in general, sorghum, and beer (husks of hops have a high silicon content). Animal skin and connective tissues are also good sources. Much more work needs to be done to determine the bioavailability of the silicon contained in the various foods because there is a wide variation in the amount absorbed.

An important medicinal source of silicon is the group of antacids that contain magnesium trisilicate. There are many preparations containing approximately 0.5 mg of magnesium trisilicate per tablet or a 4-ml unit of dosage. Excessive intake of magnesium trisilicate has led to formation of siliceous calculi (see the previous section on this topic).

Optimal intake figures are not available for human beings. Almost certainly, however, young, rapidly growing individuals, and those involved in stress situations (such as repair of wounds or fractures), have much higher requirements than the normal adult. It is estimated that man assimilates 20 to 30 mg of SiO_2 daily. This figure correlates well with the report of Goldwater (1936) that man excretes 20 mg of SiO_2 in the urine daily. Obviously, much more work needs to be done in this area.

EFFECTS ON HEALTH

Silicon deficiency has not been recognized in human beings. If it occurs, it will probably be recognized first in conjunction with particular situations of stress, as was the case with zinc deficiency or in rapidly growing (i.e., young) individuals, under conditions of low silicon intake. Except for siliceous calculi, no pathologic effects from excessive dietary intake of silicon in humans have been reported.

A number of factors could contribute to inadequate or excessive silicon intake. As certain regions are significantly lower or higher in available silicon than others, geographical factors should be considered. In addition, there is the possibility that otherwise available silicon may be bound by certain substances or antagonized by other elements in the diet. Obviously, genetic faults and/or acquired metabolic defects could interfere with silicon utilization, although no such entities have yet been recognized.

Effects from high natural dietary intake of silicon in grazing animals have already been described. Aside from its influence on digestion and absorption of food, the principal untoward effect is the occurrence of siliceous calculi, although excessive wear of teeth can also be a problem. Other pathologic aspects are to be considered, however, although they are less well understood. Varying effects on the growth of sheep have been reported as a result of the administration of high levels of sodium silicate (400 ppm, silicon) in drinking water. Sex-linked differences in effects on growth have been noted in both lambs and rats; males show a positive growth response, females a slight growth depression (Smith *et al.*, 1971b; Smith *et al.*, 1972b). Diminished reproductive performance of female rats has also been reported (Smith *et al.*, 1972a). Some of these effects may be related to the sodium component of the sodium silicate administered.

Pneumoconioses

Most published reports on the pathologic effects of silicon and its inorganic compounds in human beings concern the pneumoconioses, a group of diseases characterized by chronic fibrous reaction in the lungs resulting from long, heavy occupational exposure to a variety of dusts. Inhalation of finely particulate silica (especially in the range of $0.3-3\ \mu$) and the group of fibrous silicates collectively termed asbestos causes the most marked reactions. Asbestos comprises fibrous forms of five amphiboles—amosite, $(Mg,Fe)_7Si_8O_{22}(OH)_2$; crocidolite, $Na_2Fe_3Fe_2Si_8O_{22}(OH)_2$; tremolite, $Ca_2Mg_5Si_8O_{22}(OH)_2$; anthophyllite, $(Mg,Fe)_7Si_8O_{22}(OH,F)_2$; and actinolite, $Ca_2(Mg,Fe)_5(Si_8O_{22})(OH,F)_2$—and one serpentine—chrysotile, $Mg_3(Si_2O_5)(OH)_4$ (National Institute of Occupational Safety and Health, 1972; Goodwin, 1974). Over the years, such reactions result in development and progressive expansion of fibrous pulmonary nodules and linear deposits to the point that many of them coalesce; in the course of this process, numerous lymph and blood vessels, as well as small air passages, become obstructed. Ultimately, much of the lung may be converted to dense fibrous tissue, associated with emphysema, congestion, and pulmonary hypertension, which, in turn, cause profound systemic effects. Silicosis, in particular, increases susceptibility to various respiratory infections, notably tuberculosis, and it is often difficult to separate the effects of silica from those of associated infectious organisms.

Slight to moderate pulmonary fibrosis results from inhalation of most silicates, such as kaolin, fire clay, feldspar, olivine, sericite, and mica (Bryson and Bischoff, 1967). This group represents the so-called benign pneumoconioses, because functional effects are usually minimal. Processed diatomaceous earth and bentonite (a quasigeneric term applied to a group of clays) are also said to be capable of producing silicosis (Bryson and Bischoff, 1967; Phibbs *et al.*, 1971).

Coal miners' pneumoconiosis, which differs greatly depending on where it occurs, dramatically illustrates that the geochemical environment may be the primary factor in determining the character of a given disease. As Harington (1972) remarked in speaking of reactions to coal dust, "Geology is the basis of the real differences that are known to exist. . . ."

Although silicosis is closely tied to occupation, it is difficult to assess the degree of risk of silicosis in particular occupations unless one knows the natural products involved, the details of the processing, and the individual's specific work. In a brickyard, for example, the most common clay mineral used in making refractory bricks is kaolinite ($H_4Al_2Si_2O_9$), which contains varying quantities of fine-grained quartz, along with minute quantities of a number of other materials. When this material is fired, mullite ($Al_6Si_2O_{13}$) is produced. At high temperatures, the SiO_2 reacts with various cations, e.g., sodium, calcium, potassium, or iron, to form a siliceous glass. Thus, depending on the natural mineral present, more or less fine-grained quartz may be present; moreover, depending on the firing temperature and, to some extent, the kind and availability of cations present, silica is released in varying proportions that react to form a siliceous glass. In some operations, a special-purpose silica brick is made, which provides a much more hazardous environment. With this basic source of contamination, various workers are exposed to varying degrees of contamination, influenced a great deal by the protective devices that are in operation. The individuals at greatest risk are those who work in the grinding shed where bricks are ground to meet very close tolerances. Lesser danger exists for those who work in the holding bins, and there is practically no danger to those who work in the firing kiln or with the wet clay. It is, therefore, inappropriate to make a general statement concerning the risk of pneumoconiosis faced by brickworkers (Walter D. Keller, personal communication, 1975).

Environmental Factors in Respiratory Disease, edited by Douglas H. K. Lee (1972), gives an excellent up-to-date review of the pneumoconioses, as does *Occupational Lung Disorders*, especially chapters 7 and 9, by W. R. Parkes (1974).

Silicosis

Many hypotheses have been advanced to explain the fibrogenic properties of silica (see historical review by Zaidi, 1969), among which the solubility theory was most widely accepted until the past several years. According to this theory, silicic acid exerted its effects *directly* and depended on both the concentration and form of silicic acid. Although there is no direct correlation between solubility and fibrogenic effect, this concept was supported by observations that, in general, more highly soluble silica compounds have a predominantly necrotizing effect, whereas less soluble silica (quartz) is primarily fibrogenic. The lack of direct correlation was explained (in part) by Engelbrecht and Burger (1961), who found that ". . . monosilicic acid is more toxic than either colloidal silicic acid or partially polymerized silicic acid when allowed to polymerize in contact with tissue fractions." More recently, Allison *et al.* (1966) have proposed that once silica of proper particle size becomes arrested in the lungs and "available" for phagocytosis, it is phagocytosed just as any other electronegative particle. The particles do not lie free in the cytoplasm of the phagocyte, however. They are encapsulated within the cell in a phagosome that soon becomes converted into a lysosome (digestion vacuole) through acquisition of a variety of potent enzymes. (A detailed discussion appears in *Lysosomes*, edited by J. T. Dingle, 1972.) Protective substances adsorbed onto the silica particles (e.g., plasma proteins) are stripped off by these enzymes, exposing "naked" polymeric silicic acid. This material, now in a highly reactive state, acts as a hydrogen donor to form hydrogen-bonded complexes with active groups of the lipid membrane, such as quaternary and phosphate ester groups, and with secondary amide (peptide) groups of proteins. This reaction causes the lysosomal membrane to become permeable, allowing its enzymes to leak into the cytoplasm and destroy the cell (Nash *et al.*, 1966). With cellular dissolution, the cell contents, including active lysosomal enzymes, along with the ingested silica, are released into the tissue interstices (Figure 7). Thus it is hypothesized that intracellular reaction to silicic acid is the first of a two-stage process in which the major fibrogenic stimulus comes from action of cellular enzymes rather than directly from silicic acid per se. Moreover, the freed particles of silica are again phagocytosed by other macrophages, and the chain of events starts all over again.

Asbestosis

Not nearly so much information is available on asbestosis as on silicosis, principally because asbestosis has been recognized relatively recently. The first detailed description of asbestosis was published by Cooke in 1927, at which time the term *asbestosis* was introduced. Moreover, asbestosis affects many fewer persons than does silicosis, and the causative agent is much more complex than silica. Although there are many points of similarity between the pneumoconiosis caused by silica and that caused by asbestos fibers, there are also important differences: (a) much larger particles are involved in asbestosis, (b) the fibrosis develops more rapidly (with heavy exposure) and produces greater pulmonary disability, (c) there is not the marked predisposition to tuberculosis that occurs with silicosis, and (d) there is a clearly discernible causal relationship between asbestosis and cancer. Excellent references on the pathology of silicosis and asbestosis include *Medicine in the Mining Industries* (Rogan, 1972) and *The Biological Effects of Asbestos* (Bogovski *et al.*, 1973).

Within the past few years, great concern has developed over nonoccupational exposure to minute submicroscopic airborne or waterborne fibers of asbestos that may be inhaled or ingested. This concern has arisen because of the evidence that asbestos is carcinogenic and the fear that exposure to amounts far less than those required to produce asbestosis may significantly increase the risk of developing cancer. Asbestos is, indeed, commonly contained in the lungs of urban dwellers having no known occupational exposure to asbestos. Selikoff *et al.* (1972) found asbestos bodies present in lungs in 48.3 percent of 3,000 consecutive autopsied individuals in New York. Electron microscopic studies of a sample group of these positive cases showed the presence of chrysotile fibers or fibrils in every instance. Pooley *et al.* (1970) found that the lungs of almost 80 percent of the individuals they studied in London contained chrysotile asbestos.

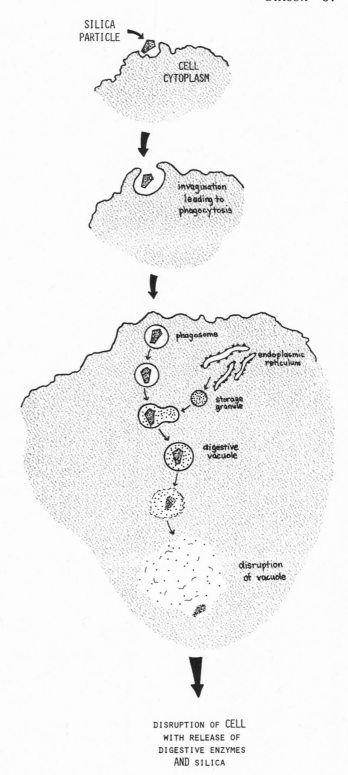

FIGURE 7 Suggested mechanism of silicosis.

Recently, asbestos has become generally recognized as a common and widespread contaminant of food, beverages, drinking water (Cunningham and Pontefract, 1971) and medicinals—even parenteral solutions (Nicholson *et al.*, 1972). This realization, coupled with the relatively high incidence of gastrointestinal cancers among asbestos workers, in addition to lung cancer and mesothelioma, has heightened concern that asbestos is an important carcinogen that affects a very large segment of the population. This general subject has been extensively reviewed (Committee on Biologic Effects of Atmospheric Pollutants, 1971; Cralley, 1972; National Institute for Occupational Safety and Health, 1972).

In an extensive five-part article that appeared in *The New Yorker*, Brodeur (1973) gives a dramatic account of some of the politicoeconomic factors involved in exposure of human beings to asbestos. This important aspect is also discussed by Wagner (1974).

Asbestos fibers are common in lymph nodes draining the intestinal tract, and there is little question but that ingested asbestos can penetrate the intestinal wall and gain access to the circulatory system. Recently, Cunningham and Pontefract (1973) have shown that transintestinal passage in rats can involve fibers as large as $23 \times 1/2 \, \mu$ and that ingested asbestos reaches a variety of internal organs and tissues in addition to lymph nodes.

Asbestos bodies are readily seen in tissue sections as refractile bodies, ordinarily $10–25 \, \mu$ in length, their walls containing much iron. Suzuki and Churg (1969) have demonstrated that these structures represent a cellular (macrophagic) reaction in which the phagocytosed asbestos fiber becomes primarily coated with hemosiderin, and that, in addition to these asbestos bodies, great numbers of sub-light-microscopic fibers (very thin and less than $1 \, \mu$ in length) are invariably present. Suzuki and Churg support the view that it is the submicroscopic fibers that are responsible for the major serious effects of asbestos. One could speculate that the fibers of the asbestos bodies have been effectively encapsulated as a defensive measure and, for all practical purposes, are no longer in contact with body tissues or fluids.

Studies to determine precisely the behavior of asbestos in terms of carcinogenicity are complicated because asbestos comprises a group of chemical compounds and because the fibers contain varying amounts of iron, nickel, chromium, cobalt, and scandium (Holmes *et al.*, 1971). Recently, P. Gross (personal communication, 1975) established that most of these metals come from milling balls and are not inherent in the asbestos fibers. Carcinogenicity of asbestos containing these metals and asbestos free of these metals is approximately equal. Further confusion arises because asbestos fibers may contain a variety of polycyclic hydrocarbons (Hilborn *et al.*, 1974). It appears that carcinogenicity is associated with a fibrous shape. Fibrous asbestos that is definitely carcinogenic can be rendered quite innocuous by grinding it into roughly spherical particles. Mesothelioma cannot be produced by injecting fibrils less than $1/2 \, \mu$ long into the pleural space, whereas longer fibers will produce

mesothelioma in a high percentage of animals (J. Churg, personal communication, 1975). The physical state of asbestos seems to be important in carcinogenesis, although it is not known why and how.

Many of the cancers associated with asbestos in human beings have been related to chrysotile, but this may simply reflect a greater exposure because approximately 95 percent of the asbestos in general use in the United States is chrysotile. Recently it has been shown that amosite also is carcinogenic (Selikoff *et al.*, 1972).

The role of asbestos in causing cancer is very complex. In some instances, at least, it appears to act as a cocarcinogen. Selikoff *et al.* (1968), in a study of 370 asbestos insulation workers, found no special predisposition to carcinoma of the lung among the 87 members of the group who were not cigarette smokers, but the other 283 had ". . . about 92 times the risk of dying of bronchogenic carcinoma as men who neither work with asbestos nor smoke cigarettes" (see also the report by Berry *et al.*, 1972).

Much additional work is required to clarify the nature of the carcinogenic actions of asbestos, as well as to evaluate the extent of risk with respect to the amount and kind of exposure. Some of the problems involved in such studies are considered in a recent paper by Enterline *et al.* (1972). An excellent, though somewhat out-of-date, consideration of silicate-induced neoplasms is that of Bryson and Bischoff (1967).

RECOMMENDATIONS FOR RESEARCH

To develop an adequate understanding of the importance as well as the dangers of silicon compounds, studies are needed to accomplish the following:

1. Provide further information on the silicon content of rocks, waters, soils, and air, with special concern for its availability to plants and animals, particularly those used as food.

2. Clarify our understanding of the physiologic mechanisms of uptake and transport of silicon in plants and animals, including man.

3. Determine the precise chemical structure of silicon compounds and/or complexes in plants, the ways in which these interrelate with other ions, and the effects that the silicon compounds and complexes have on the digestibility of forage plants.

4. Establish whether silicon is essential for higher plants and, if so, for which species.

5. Provide further information on silicon content, including its form and availability, in foods consumed by animals, including humans.

6. Provide more information on the physiologic mechanisms of transport of silicon in animals, its chemical forms in various tissues, and its precise location in the tissues.

7. Determine the precise chemical structure of the silicon compounds or complexes found in animals and the

ways in which these interrelate with other ions (e.g., fluorine, calcium, phosphorus, manganese, and magnesium) affecting their absorption and utilization.

8. Establish silicon requirements for humans at different ages under both normal conditions and states of stress.

9. Develop better analytical methods for silicon, particularly as it occurs in low concentrations in complex biological materials such as bone.

10. Determine the influence of geographic factors on silicon uptake and absorption in man and other animals.

11. Clarify the role of (inorganic) silicates as carcinogens and further elaborate the pathogenesis of silicosis and asbestosis.

12. Characterize the nature of opaline phytoliths distribution within various tissues of animals and their extent, with special concern for their possible carcinogenicity.

13. Determine the incidence of siliceous urolithiasis in various animals, including man, and its etiology and pathogenesis.

14. Determine what diseases or adverse conditions (e.g., the Balkan nephropathy, atherosclerosis, dental caries, etc.) may be caused by excesses, deficiencies, or imbalances of silicon.

REFERENCES

Acquaye, D. K., and J. Tinsley. 1964. Soluble silica in soils. *In* Experimental pedology, E. G. Hallsworth and D. V. Crawford [eds.]. Butterworths, London. pp. 126–148.

Alexander, A. G. 1968. *In vitro* effects of silicon on the action patterns of sugarcane acid invertase. J. Agric. Univ. P.R. 52:311–322.

Allison, A. C., J. S. Harington, and M. Birbeck. 1966. An examination of the cytotoxic effects of silica on macrophages. J. Exp. Med. 124:141–154.

Archibald, J. G., and H. Fenner. 1957. Silicon in cow's milk. J. Dairy Sci. 40:703–706.

Ayres, A. S. 1966. Calcium silicate slag as a growth stimulant for sugarcane on low-silicon soils. Soil Sci. 101:216–224.

Bailey, C. B. 1970. The precipitation of polymerized silicic acid by urine protein: A possible mechanism in the etiology of silica urolithiasis. Can. J. Biochem. 50:305–311.

Baker, G. 1959. Fossil opal–phytoliths and phytolith nomenclature. Aust. J. Sci. 21(9):305–306.

Baker, G., and L. H. P. Jones. 1961. Opal in the animal body. Nature 189(4765):682–683.

Baker, G., L. H. P. Jones, and I. D. Wardrofs. 1959. Cause of wear in sheep's teeth. Nature 184:1583–1584.

Beckwith, R. S., and R. Reeve. 1963. Studies on soluble silica in soils. 1. The sorption of silicic acid by soils and minerals. Aust. J. Soil Res. 1:157–168.

Berry, G., M. L. Newhouse, and M. Turok. 1972. Combined effect of asbestos exposure and smoking on mortality from lung cancer in factory workers. Lancet 2(7775):476–479.

Bogovski, P., J. C. Gilson, V. Timbrell, and J. C. Wagner [eds.]. 1973. Biological effects of asbestos. Proceedings of a Working Conference held at the International Agency for Research on Cancer, Lyon, France, October 2–6, 1972. IARC Scientific Publ. No. 8. World Health Organization, Lyon, 346 pp.

Brodeur, P. 1973. Annals of industry casualties of the workplace. Parts I–V. The New Yorker, October 29:44–106; November 5:92–142; November 12:131–177; November 19:87–149; November 26:126–179.

Brown, H. 1927. The mineral content of human skin. J. Biol. Chem. 75:789–794.

Bryson, G., and F. Bischoff. 1967. Silicate-induced neoplasms. Prog. Exp. Tumor Res. 9:77–164.

Carlisle, E. M. 1969. Silicon localization and calcification in developing bone. Fed. Proc. 28:374. (Abstract)

Carlisle, E. M. 1970a. Silicon a possible factor in bone calcification. Science 167:279–280.

Carlisle, E. M. 1970b. A relationship between silicon and calcium in bone formation. Fed. Proc. 29:565. (Abstract)

Carlisle, E. M. 1972a. Silicon an essential element for the chick. Fed. Proc. 31:700. (Abstract)

Carlisle, E. M. 1972b. Silicon an essential element for the chick. Science 178:619–621.

Carlisle, E. M. 1973. A skeletal alteration associated with silicon deficiency. Fed. Proc. 32:930. (Abstract)

Carlisle, E. M. 1974. Silicon as an essential element. Fed. Proc. 33:1758–1766.

Carlisle, E. M. 1976. *In vivo* requirements for silicon in articular cartilage and connective tissue formation in the chick. J. Nutr. 106:478–484.

Chalmers, R. A., and A. G. Sinclair. 1966. Analytical applications of β-heteropolyacids. II. The influence of complexing agents on selective formation. Anal. Chim. Acta 34:412–418.

Clements, H. F. 1965. The roles of calcium silicate slags in sugar cane growth. Report of 24th Annual Meeting, Hawaii Sugar Cane Technology. pp. 103–126.

Clements, H. F., and M. Awada. 1967. Experiments on the artificial induction of flowering in sugar cane. Proceedings of the International Society of Sugar Cane Technologies, 12th Congress, San Juan, Puerto Rico, March 28 to April 10, 1965, Jaime Bague [ed.]. Elsevier, Amsterdam, pp. 795–812.

Committee on Biologic Effects of Atmospheric Pollutants. 1971. Asbestos: The need for and feasibility of air pollution controls. National Academy of Sciences, Washington, D.C. 40 pp.

Cooke, W. E. 1927. Pulmonary asbestosis. Brit. Med. J. 2:1024–1025.

Coombs, J., and B. E. Volcani. 1968. Studies on the biochemistry and fine structure of silica-shell formation in diatoms. Planta 82:280–292.

Cralley, L. V. [ed.] 1972. Industrial environmental health: The worker and the community. Academic Press, New York. 544 pp.

Cunningham, H. M., and R. Pontefract. 1971. Asbestos fibres in beverages and drinking water. Nature 232:332–333.

Cunnginham, H. M., and R. Pontefract. 1973. Asbestos fibers in beverages, drinking water, and tissues. Their passage through the intestinal wall and movement through the body. J. Assoc. Off. Anal. Chem. 56:976–986.

Davis, S. N. 1964. Silica in streams and ground water. Am. J. Sci. 262:870–891.

Davis, S. N. 1969. Silica in streams and ground water of Hawaii. Tech. Rep. No. 29. Water Resources Research Center, University of Hawaii, Honolulu. 31 pp.

Dingle, J. T. [ed.] 1972. Lysosomes: A laboratory handbook. American Elsevier, New York. 247 pp.

Drea, W. F. 1935. Spectrum analysis of hen eggs and chick tissues. J. Nutr. 10:351–355.

Drum, R. W. 1968. Silification of *Betula* woody tissue *in vitro*. Science 161:175–176.

Ehrhart, L. A., and K. G. McCullagh. 1973. Silica urolithiasis in dogs fed an atherogenic diet. Proc. Soc. Exp. Biol. Med. 143:131–132.

Elgawhary, S. M., and W. L. Lindsay. 1972. Solubility of silica in soils. Soil Sci. Soc. Am. Proc. 36:439–442.

Emerick, R. J., L. B. Embry, and O. E. Olsen. 1959. Effect of sodium silicate on the development of urinary calculi and the excretion of various urinary constituents in sheep. J. Anim. Sci. 18:1025.

Engel, W. 1953. Untersuchungen über die Kieselsäureverbindungen im Roggenhalm. Planta 41:358–390.

Engelbrecht, F. M., and F. J. Burger. 1961. The toxicity of silicic acid. S. Afr. J. Lab. Clin. Med. 7(1):16–21.

Enterline, P., P. DeCoufle, and V. Henderson. 1972. Mortality in relation to occupational exposure in the asbestos industry. J. Occup. Med. 14(12):897–903.

Fanning, K. A., and M. E. Q. Pilson. 1971. Interstitial silica and pH in marine sediments: Some effects of sampling procedures. Science 173:1228–1231.

Folger, D. W., L. H. Bunkle, and B. C. Heezen. 1967. Opal phytoliths in a North Atlantic dust fall. Science 155(3767):1243–1244.

Fox, R. L., J. A. Silva, D. L. Plucknett, and D. Y. Teranishi. 1969. Soluble and total silicon in sugar cane. Plant Soil 30(1):81–92.

Garrels, R. M., and C. L. Christ. 1965. Solutions, minerals, and equilibria. Harper & Row, New York. 450 pp.

Garson, L., and L. Kirchner. 1971. Organosilicon entities as prophylactic and therapeutic agents. J. Pharmacol. Sci. 60:1113–1127.

Goldwater, L. J. 1936. The urinary excretion of silica in non-silicotic humans. J. Ind. Hyg. 18:163–166.

Gonnermann, M. 1918. Beiträge zur Kenntnis der Biochemie der Kieselsäure and Tonerde. Biochem. Z. 88:401–415.

Gonnermann, M. 1919. The quantitative excretion of silicic acid in human urine. Biochem. Z. 94:163–173.

Goodwin, A. [comp.]. 1974. Proceedings of the Symposium on Talc, Washington, D.C., May 8, 1973. U.S. Bureau of Mines Information Circular 8639. U.S. Department of the Interior, Washington, D.C. 102 pp.

Handreck, K. A., and L. H. P. Jones. 1967. Uptake of silica by Trifolium incarnatum. Aust. J. Biol. Sci. 20:483–485.

Handreck, K. A., and L. H. P. Jones. 1968. Studies of silica in the oat plant. 4. Silica content of plant parts in relation to stage of growth, supply of silica and transpiration. Plant Soil 29:449–459.

Harington, J. S. 1972. Investigative techniques in the laboratory study of coal workers' pneumoconiosis: Recent advances at the cellular level. Ann. N.Y. Acad. Sci. 200:816–834.

Heinen, W. 1963. Siliciumverbindungen. In Moderne Methode der Pflanzenalalyse, M. V. Tracey and H. F. Linskens [eds.], Vol. 6. pp. 4–20.

Herman, J. R., and A. S. Goldberg. 1960. New type of urinary calculus caused by antacid therapy. J. Am. Med. Assoc. 174:128–129.

Hilborn, J., K. S. Thomas, and R. C. Lao. 1974. The organic content of international reference samples of asbestos. Sci. Total Environ. 3:129–140.

Hingston, F. J., R. J. Atkinson, A. M. Posner, and J. P. Quirk. 1967. Specific adsorption of anions. Nature 215:1459–1461.

Hingston, F. J., R. J. Atkinson, A. M. Posner, and J. P. Quirk. 1968. Specific adsorption of anions on goethite. Trans. 9th. Int. Congr. Soil Sci. (Adelaide, Australia) 1:669–678.

Holmes, A., A. Morgan, and F. J. Sandalls. 1971. Determination of iron, chromium, cobalt, nickel, and scandium in asbestos by neutron activation analysis. Am. Ind. Hyg. Assoc. J. 32:281–286.

Iler, R. K. 1955. The colloid chemistry of silica and silicates. Cornell Univ. Press, Ithaca, New York. 324 pp.

Jankowiak, M. E., and R. R. LeVier. 1971. Elimination of phosphorus interference in the colorimetric determination of silicon in biological material. Anal. Biochem. 44:462–472.

Joekes, A. M., G. A. Rose, and J. Sutor. 1973. Multiple renal silica calculi. Brit. Med. J. 1:146–147.

Johlin, J. M. 1932. Occurrence of a silico-carbohydrate derivative in animal tissue. Proc. Soc. Exp. Biol. Med. 29:760–761.

Jones, B. F., S. L. Rettig, and H. P. Eugster. 1967. Silica in alkaline brines. Science 158:1310–1313.

Jones, L. H. P., and K. A. Handreck. 1965a. The relation between the silica content of the diet and the excretion of silica by sheep. J. Agric. Sci. 65:129–134.

Jones, L. H. P., and K. A. Handreck. 1965b. Studies of silica in the oat plant. III. Uptake of silica from soils by the plant. Plant Soil 23:79–96.

Jones, L. H. P., and K. A. Handreck. 1967. Silica in soils, plants, and animals. Adv. Agron. 19:107–149.

Jones, L. H. P., and K. A. Handreck. 1969. Uptake of silica by Trifolium incarnatum in relation to the concentration in the external solution and to transpiration. Plant Soil 30:71–80.

Jones, L. H. P., and A. A. Milne. 1963. Silica in the oat plant. I. Chemical and physical properties of SiO_2. Plant Soil 18:207–220.

Jones, L. H. P., A. A. Milne, and S. M. Wadham. 1963. Studies of silica in the oat plant. II. Distribution of silica in the plant. Plant Soil 18:358–370.

Jones, R. L., and A. H. Beavers. 1963. Some mineralogical and chemical properties of plant opal. Soil Sci. 95(6):375–379.

Kelley, W. P., and S. M. Brown. 1939. An unusual alkali soil. J. Am. Soc. Agron. 31:41–43.

King. E. J. 1939. The biochemistry of silicic acid. VIII. The determination of silica. Biochem. J. 33:944–954.

King, E. J., and T. H. Belt. 1938. The physiological and pathological aspects of silica. Physiol. Rev. 18:329–365.

King, E. J., B. D. Stacy, P. F. Holt, D. M. Yates, and D. Pickles. 1955. The colorimetric determination of silicon in the microanalysis of biological material and mineral dusts. Analyst 80:441–452.

Kirchgessner, M. 1957. Der mengen-und spurenelementgehalt von Rinderblut. Z. Tierphysiol. Tierernachr. Futtermittelk. 12:156–169.

Kobayashi, J. 1960. A chemical study of the average quality and characteristics of river waters of Japan. Ber. Ohara Inst. Landwirtsch. Biol. Okayama Univ. 9(3):313–358.

Lee, Douglas H. K. [ed.] 1972. Environmental factors in respiratory disease. Academic Press, New York. 256 pp.

Leslie, J. G., T. K. Kung-Ying, and T. H. McGavack. 1962. Silicon in biological material. II. Variations in silicon contents in tissues of rats at different ages. Proc. Soc. Exp. Biol. Med. 110:218–220.

Lewin, J., and B. E. F. Reimann. 1969. Silicon and plant growth. Annu. Rev. Plant Physiol. 20:289–304.

Loeper, J., J. Loeper, and A. Lemaire. 1966. Étude du silicium en biologie animale et au cours de l'atherome. Presse Med. 74:865–868.

Lovering, T. S. 1959. The significance of accumulator plants in rock weathering. Geol. Soc. Am. Bull. 70:781–800.

Lovering, T. S., and C. Engel. 1967. Translocation of silica and other elements from rock into Equisetum and three grasses. Geol. Surv. Prof. Pap. 594-B. U.S. Government Printing Office, Washington, D.C. 16 pp.

MacCardle, R. C., M. F. Engman, Jr., and M. F. Engman, Sr.

1943. XCIV. Mineral changes in neurodermatitis revealed by microincineration. Arch. Dermatol. Syphilol. 47:335–372.

McKeague, J. A., and M. G. Cline. 1963a. Silica in soil solutions. I. The form and concentration of dissolved silica in aqueous extracts of some soils. Can. J. Soil Sci. 43:70–82.

McKeague, J. A., and M. G. Cline. 1963b. Silica in soil solutions. II. The adsorption of monosilicic acid by soil and by other substances. Can. J. Soil Sci. 43:83–96.

McKeague, J. A., and M. G. Cline. 1963c. Silica in soils. Adv. Agron. 15:339–396.

Monier-Williams, G. W. 1949. Trace elements in foods. Chapman-Hall, Ltd., London. 511 pp.

Morey, G. W., R. O. Fournier, and J. J. Rowe. 1962. The solubility of quartz in water in the temperature interval from 25° to 300° C. Geochim. Cosmochim. Acta 26:1029–1043.

Muller, S. A., A. S. Posner, and H. E. Firschein. 1966. Effect of vitamin D deficiency on the crystal chemistry of bone mineral. Proc. Soc. Exp. Biol. Med. 121:844–846.

Nash, T., A. C. Allison, and J. S. Harington. 1966. Physico-chemical properties of silica in relation to its toxicity. Nature 210(5033):259–261.

National Institute for Occupational Safety and Health. 1972. Criteria for a recommended standard-occupational exposure to asbestos. Public Health Service Publ. HSM 72-10267. Health Services and Mental Health Administration, U.S. Department of Health, Education, and Welfare, Washington, D.C. pp. III 1–24.

Newberne, P. M., and R. B. Wilson. 1970. Renal damage associated with silicon compounds in dogs. Proc. Natl. Acad. Sci. 65(4):872–875.

Nicholson, W. J., C. J. Maggiore, and I. Selikoff. 1972. Asbestos contamination of parenteral drugs. Science 177:171–173.

Obihara, C. H., and E. W. Russell, 1972. Specific adsorption of silicate and phosphate by soils. J. Soil Sci. 23:105–117.

Okuda, A., and E. Takahashi. 1964. The role of silicon. *In* The mineral nutrition of the rice plant. The Johns Hopkins Press, Baltimore. pp. 123–146.

Parkes, W. R. 1974. Occupational lung disorders. Butterworths, London. 528 pp.

Paul, J. 1965. Simultaneous determination of arsenic, phosphorus and silicon. Mikrochim. Acta (Wien):836–841.

Phibbs, B. P., R. E. Sundin, and R. S. Mitchell. 1971. Silicosis in Wyoming bentonite workers. Am. Rev. Resp. Dis. 103(1):1–17.

Pooley, F. D., P. D. Oldham, C. H. Um, and J. C. Wagner. 1970. The detection of asbestos in tissues. *In* Pneumoconiosis. Proceedings of an International Conference, 1969. H. A. Shapiro [ed.]. pp. 108–116.

Ponnamperuma, F. N. 1964. Dynamic aspects of flooded soils and the nutrition of the rice plant. *In* The mineral nutrition of the rice plant. Johns Hopkins Press, Baltimore. pp. 295–328.

Raupach, M. 1957. Investigations into the nature of soil pH. Soil Publ. No. 9. Commonwealth Scientific and Industrial Research Organization, Melbourne. 35 pp.

Rochow, E. 1951. An introduction to the chemistry of the silicones, 2d ed. Wiley, New York. 213 pp.

Rogan, J. M. [ed.] 1972. Medicine in the mining industries. F. A. Davis Company, Philadelphia. 397 pp.

Rose, E. F. 1968. Phytolithicosis—A disease entity? Trans. N.Y. Acad. Sci. 30:1196–1200.

Russell, E. W. 1961. Soil conditions and plant growth, 9th ed. Longmans, London. 688 pp.

Sasamoto, K. 1958. Studies on the relation between the silica content in the rice plant and the insect pests. VI. On the injury of a silicated rice plant caused by the rice stem borer and its feeding behavior. Jpn. J. Appl. Entomol. Zool. 2:88–92.

Schwarz, K. 1973. A bound form of silicon in glycosaminoglycans and polyuronides. Proc. Natl. Acad. Sci. 70(5):1608–1612.

Schwarz, K., and D. B. Milne. 1972. Growth-promoting effects of silicon in rats. Nature 239:333–334.

Selikoff, I. J., E. C. Hammond, and J. Churg. 1968. Asbestos exposure, smoking, and neoplasia. J. Am. Med. Assoc. 204(2):104–112.

Selikoff, I. J., E. C. Hammond, and J. Churg. 1972. Carcinogenicity of amosite asbestos. Arch. Environ. Health 25:183–186.

Siever, R. 1957. The silica budget in the sedimentary cycle. Am. Mineral. 42:821–841.

Siever, R. 1972. Geochemistry of silicon: Low temperature. *In* Handbook of geochemistry, K. H. Wedepohl [ed.], Vol. II-1, Sec. 14, Silicon. Springer-Verlag, New York–Berlin–Heidelberg.

Siever, R., and R. A. Scott. 1963. Organic geochemistry of silica. *In* Organic geochemistry, I. A. Breger [ed.]. Pergamon Press, Oxford. pp. 579–595.

Siever, R., and N. Woodford. 1973. Sorption of silica by clay minerals. Geochim. Cosmochim. Acta 37:1851–1880.

Slavnyy, Y. A., and E. S. Vorob'yeva. 1962. Precipitation of silica on freezing of soil solutions. Sov. Soil Sci. 9:958–961.

Smith, G. S., A. B. Nelson, and Eloy J. A. Boggino. 1971a. Digestibility of forages *in vitro* as affected by content of "silica." J. Anim. Sci. 33:466–471.

Smith, G. S., A. B. Nelson, E. C. Smith, and A. L. Neumann. 1971b. Effects of "soluble silica" on growth of lambs. J. Anim. Sci. 32:246. (Abstract)

Smith, G. S., A. L. Neumann, V. H. Gledhill, and C. A. Arzola. 1972a. Effects of "soluble silica" on growth, nutrient balance and reproductive performance of albino rats. J. Anim. Sci. 36:271–278.

Smith, G. S., A. L. Neumann, A. B. Nelson, and E. E. Ray. 1972b. Effects of "soluble silica" upon growth of lambs. J. Anim. Sci. 34:839–845.

Sosman, R. B. 1965. The phases of silica. Rutgers University Press, New Brunswick, New Jersey. 388 pp.

Sowden, F. J. 1972. Effect of silicon on automated methods for the determination of phosphate in water. Can. J. Soil Sci. 52:237–243.

Sullivan, J. T. 1959. A rapid method for the determination of acid-insoluble lignin in forages and its relation to digestibility. J. Anim. Sci. 18:1292–1298.

Suzuki, Y., and J. Churg. 1969. The structure and development of the asbestos body. Am. J. Pathol. 55:79–107.

Tiller, K. G. 1967. Silicic acid and the reaction of zinc with clays. Nature 214:852. (Abstract)

Tiller, K. G. 1968. The interaction of some heavy metal cations and silicic acid at low concentrations in the presence of clays. Trans. 9th Int. Congr. Soil Sci. (Adelaide, Australia) 2:567–575.

Twiss, P. C., E. Suess, and R. M. Smith. 1969. Morphological classification of grass phytoliths. Soil Sci. Soc. Am. Proc. 33(1):125–128.

Underwood, E. J. 1971. Trace elements in human and animal nutrition, 3d ed. Academic Press, New York. pp. 407–415.

Van Dyne, G. M., and G. P. Lofgren. 1964. Comparative digestion of dry annual range forage by cattle and sheep. J. Anim. Sci. 23:823–832.

Van Soest, P. J. 1969. The chemical basis for the nutritive evaluation of forages. Proc. Natl. Conf. Forage Qual. Eval. Util. Nebraska Center for Continuing Education, Lincoln. U-1-19.

Van Soest, P. J. 1970. The role of silicon in the nutrition of plants and animals. Proc. 1970 Cornell Nutr. Conf. Cornell University, Ithaca, New York, pp. 103–109.

Van Soest, P. J., and L. H. P. Jones. 1968. Effect of silica upon digestibility. J. Dairy Sci. 51:1644–1648.

Van Soest, P. J., and F. E. Lovelace. 1969. Solubility of silica in forages. J. Anim. Sci. 29:182. (Abstract)

Volk, R. J., and R. L. Weintraub. 1958. Microdetermination of silicon in plants. Anal. Chem. 30:1011–1014.

Volk, R. J., R. P. Kahn, and R. L. Weintraub. 1958. Silicon content of the rice plant as a factor influencing its resistance to infection by the blast fungus, *Piricularia oryzae*. Phytopathology 48:179–184.

Wagner, C. 1974. Disputes on the safety of asbestos. New Scientist 61 (888):606–609.

Weaver, R. M., M. L. Jackson, and J. K. Syers. 1971. Magnesium and silicon activities in matrix solutions of montmorillonite-containing soils in relation to clay mineral stability. Soil Sci. Soc. Am. Proc. 35:823–830.

Weitz, E., H. Franck, and M. Schuchard. 1950. Silicic acid and silicates. Chem. Ztg. 74:256–257.

Whiting, F., R. Connell, and S. A. Forman. 1958. Silica urolithiasis in beef cattle. Can. J. Comp. Med. 22:332–337.

Yoshida, S., Y. Ohnishi, and K. Kitagishi. 1962. Histochemistry of silicon in rice plant. II. Localization of silicon within rice tissues. Soil Sci. Plant Nutr. (Tokyo) 8:36–41.

Zaidi, S. H. 1969. Experimental pneumoconiosis. The Johns Hopkins Press, Baltimore. pp. 94–120.

VII

Strontium

ROBERT H. WASSERMAN, *Chairman*

Evan M. Romney, Marvin W. Skougstad, Raymond Siever

Strontium is not an essential element for growth and development of either plants or animals, including man, even though it is an almost universal component of soil, water, and plant life.

The great similarity in the chemical properties of calcium and strontium made the separation and identification of the small amounts of strontium usually present in soil, water, and biological materials a difficult task for early investigators (Robinson *et al.*, 1917; Noll, 1931). This analytical problem was solved by the development of sensitive and specific methods of emission spectroscopy (Mitchell, 1948), neutron activation (Bowen and Dymond, 1955), and X-ray fluorescence (Vose and Koontz, 1959). Applications of these methods have increased our understanding of the sources and amounts of stable strontium in soil, water, and biological materials. Intensive work on radioactive strontium during the past two decades has increased understanding of the behavior and movement of strontium in the biosphere.

Although this chapter is concerned primarily with stable strontium, we cannot minimize the environmental impact of radioactive strontium. There is no doubt about the potential health hazards that could result from indiscriminate dissemination of the long-lived strontium-90 fission product into the natural environment. No such threat is apparent from stable strontium because of its low toxicity to plant and animal life.

Much of the work pertaining to the strontium-90 fallout problem has been documented in reviews, symposia proceedings, and government hearings, of which the following are examples: Joint Committee on Atomic Energy Hearings (1957, 1959a, 1959b), Caldecott and Snyder (1960), U.N. Scientific Committee on Effects of Atomic Radiation (1962, 1964), Food and Agriculture Organization (1960, 1964), Frere *et al.* (1963), Schultz and Klement (1963), Aleksakhin (1963), Hungate (1965), Fowler (1965), Russell (1966), Åberg and Hungate (1967), Benson and Sparrow (1971).

GEOCHEMISTRY AND OCCURRENCE

Geochemical Cycle

Figure 8 shows the general outline of the routes traveled by strontium as it is weathered from rock terrains and dissolved in natural waters, ultimately to be transported to the oceans, from which it is precipitated in sediments. Strontium is strongly coupled to calcium in primary source rocks and is most abundant in calcium-rich rocks like basalt and limestones. Strontium occurs primarily in plagioclase feldspars in igneous and metamorphic rocks, and in calcite or dolomite in limestones.

Weathering of limestone supplies the bulk of the strontium that appears in both groundwaters and surface waters and thus the major part of the flux to the oceans. The content of strontium in groundwaters is highly variable because of its dependence on the type of rock through which the infiltrating water passes. Thus, in limestone terrains the strontium content in groundwater may climb

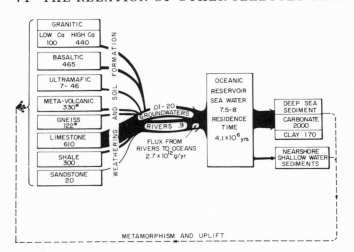

FIGURE 8 The geochemical cycle of strontium in rocks and waters at the surface of the earth. All figures are averages or composites in parts per million, with the exception of the asterisked values, which are specific samples. These data were taken from various sources including Turekian (1964), Dasch (1969), Livingstone (1963), White *et al.* (1963), and Turekian and Schutz (1965).

to 20 ppm or more, whereas in calcium-poor rocks the groundwater strontium levels may be in the parts-per-billion range. The average strontium content for near-surface groundwaters low in dissolved solids is probably of the same order of magnitude as the average for major North American rivers—0.9 ppm (Livingstone, 1963).

The strontium dissolved from continental terrains enters the oceans, where it participates in the general steady-state (or homeostatic) reservoir. The entering flux from all rivers is estimated at 2.7×10^{12} g/yr (Turekian and Schutz, 1965), which is assumed to be equal to the rate at which strontium is deposited by sedimentation at the bottom of the oceans. The residence time in the ocean is about 4.1×10^6 yrs, which is one of the shorter residence times and somewhat comparable to that of calcium, with which it coprecipitates as carbonate. Most strontium is precipitated as biogenic carbonate, largely in the form of the shells of pelagic foraminifera.

The strontium content of invertebrate shell material is a function of mineralogy, temperature, and species (Lowenstam, 1964). Strontium is accommodated in solid solution in the aragonite (orthorhombic) crystal structure in much greater abundance than in the calcite (rhombohedral) structure; thus, aragonite shells will have a higher strontium content than those primarily formed of calcite. One species, *Mytilus edulis* (the common mussel, a bivalve mollusk or pelecypod), secretes aragonite containing moderate amounts of strontium along the inner part of its shell and calcite containing very low amounts of strontium on the outer shell surface. The amount of strontium incorporated into either aragonite or calcite is also a function of temperature, both in pure laboratory systems and apparently in some animal and plant species, though recent work has established independence of tempera-

ture in corals (Thompson and Livingston, 1970). Sr:Ca ratios of organisms in waters of constant temperature at Palau tend to be close to those of seawater (about 0.018 on a ppm basis, or 8.5×10^{-3} on an atomic basis) except for the higher mollusca, which show lower Sr:Ca ratios. In comparison to the relatively abundant information on composition of shell materials, little is known about strontium in the soft tissue of marine plants or animals.

Because most of the carbonate sediment in the ocean is deposited in the deep sea as the shells of pelagic foraminifera, coccoliths, and other organisms, and because strontium is so strongly localized in carbonate, most of the strontium dropped out of the ocean by sedimentation is in deep-sea sediments, relatively isolated from rapid mixing with surface environments. This is the sink for strontium in the surface geochemical cycle. Ultimately, the strontium in sediments is reintroduced at the surface of the continents by the geological cycle of burial and mountain building and uplift, normally accompanied by some degree of metamorphism.

Soils

Figure 9 shows the nature of the cyclical system of ion and water transfer in the complex soil system. The primary agent of chemical attack on soil is rainwater infiltrating into the top of the soil and carrying with it dissolved carbon dioxide (CO_2) and other atmospheric gases, after depositing a thin layer of aerosol and particulate material at the surface. The mineral decomposition in a stratified soil is made up of three indistinctly separated processes. At the bottom of the C-horizon, fresh rock is dissolved and strontium may be leached from its primary host rock, ultimately leaving behind an altered product, generally

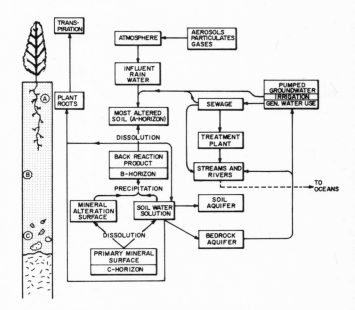

FIGURE 9 Cyclical movement of strontium solutions through soil–plant–water systems.

clay. The soil solution at this point is enriched in strontium. In the overlying B-horizon, soil waters may participate in a back reaction with the altered minerals produced in the C-horizon, thus modifying the soil solution. The most altered or leached soil is in the A-horizon. The soil profile advances downward with time as solutions continue to move downward. Depending on the parent rock materials, the rate of water flow through the soil, and the type of vegetation, the amounts of strontium available to plant roots may vary widely. A precise knowledge of these amounts in natural soil systems is needed to understand more completely the availability of strontium to plants. Future sampling and analysis should concentrate on the amounts of strontium, on the Sr:Ca ratios in the various solids and solutions in the soil profile in relation to their position, and on the penetration of the roots of various types of plants.

Total Strontium

Although early workers encountered difficulties in chemically determining the small amounts of strontium associated with the relatively large amounts of calcium in soils, some of the data reported fall in line with results from recently developed analytical methods. For example, Robinson (1914) found from 80 to 900 ppm strontium in surface and subsurface samples of 13 soils. Thomas (1923) found the strontium content of surface and subsurface soils to be 110 ppm and 260 ppm, respectively.

Using the sensitive and specific methods of emission spectroscopy, Mitchell (1944, 1948) found from 50 to 5,000 ppm of total strontium in the soils of Scotland. Later work on Scottish soils reported by Swaine (1955) showed total strontium contents ranging from 50 to 1,000 ppm in 43 surface soils and 118 subsoils; the average total strontium content was 400 ppm. Results reported on Russian soils by Vinogradov (1945) indicated total strontium contents within the same range as Scottish soils. Samples taken along the fortieth meridian contained from 130 to 2,600 ppm total strontium. Later, Vinogradov (1954) reported additional results of 10 to 2,000 ppm in 14 surface soils and from 30 to 2,800 ppm in 28 subsoils.

Total stable strontium contents of soils in Florida have been reported to range from less than 10 to more than 1,000 ppm. This range represents work reported by Rogers *et al.* (1939), Allison and Gaddum (1940), and Carrigan and Rogers (1940) for 88 surface soils and 71 subsoils. For 10 New Jersey surface soils, Prince (1957) reported the strontium content to vary from 36 to 142 ppm. Vanselow (1966) found strontium contents ranging from 100 to 1,000 ppm in 12 surface soils and subsoils from California. Preliminary results from a geochemical survey of Missouri indicate a range of strontium contents from 25 to 360 ppm in 1,140 samples of agricultural soils, with a mean concentration of 110 ppm (U.S. Geological Survey, 1972).

Shacklette *et al.* (1971) reported data on the elemental composition of surficial materials from 862 sites in the conterminous United States. Results showed that the surficial materials of the western half of the United States generally contain more strontium than do those of the eastern half (Figure 10).

Available Strontium

Plant uptake of strontium is determined primarily by the amount of available soil strontium, but chemical extraction methods to estimate availability are not completely satisfactory for all conditions encountered. An extracting solution of neutral 1 N ammonium acetate appears to have the widest application. Using this procedure, Mitchell (1937) reported exchangeable strontium levels of 3.5–17.5 ppm in seven surface soils from Scotland and 0.2–19.7 ppm at two depths in their subsoils. In four other soils of Scotland, Glentworth (1944) reported 4.4–8.8 ppm in surface samples and 3.7–17.5 ppm in subsoil samples. Viro (1951) reported 0.1–7.7 ppm exchangeable strontium in the surface of 47 Finnish soils. Bowen and Dymond (1955) reported levels of exchangeable strontium in seven soils of England ranging from 0.5 to 10 ppm; two strontium-rich soils near celestite mines have exchangeable levels as high as 1,500 ppm. The vegetation growing on these two soils contained exceptionally large quantities of strontium, but neither the plants nor the animals feeding upon them appeared to be affected. For 93 soils from 11 states in the United States, Menzel and Heald (1959) found from 0.5 to 32 ppm of exchangeable strontium by the ammonium acetate extraction method.

Water

Alexander *et al.* (1954) determined by emission spectrography the concentrations of stable strontium in waters provided for human consumption in 50 United States cities. The range of dissolved strontium concentrations in raw waters was from 5.8 to 1,900 ppb, with a mean of 280 ppb. Tap-water concentrations ranged from 9.4 to 680 ppb, with a mean of 160 ppb. In several cities, as much as 75 percent of the strontium was removed by routine water-softening procedures. The strontium concentrations of raw surface waters were related to the predominant soil types in their drainage basins. The lowest values observed were for waters from noncalcareous soils; the highest values were associated with waters from regions in which the soils contain a large proportion of easily leached, soft limestone. Nichols and McNall (1957) found one groundwater source in eastern Wisconsin with an unusually high content of 39,000 ppb, reflecting the high mineral content of the aquifer.

Skougstad and Horr (1963) analyzed samples from 75 major rivers of the conterminous United States and found concentrations of strontium that ranged from 0.007 to 13.7 ppm; average concentrations (in ppm) are shown in Figure 11. The proportion of strontium in the dissolved solids carried by these rivers ranged from a few hundred to 3,700 ppm. The greatest strontium concentration, both in the water and in the dissolved material, was found to occur in the saline streams of the Southwest. In this area,

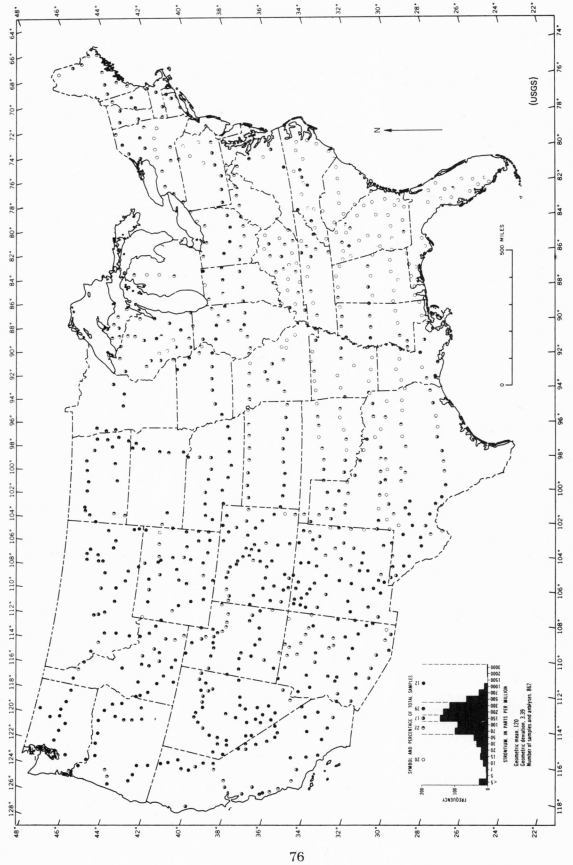

FIGURE 10 Strontium content of surficial materials in the conterminous United States. Values are in parts per million (Shacklette *et al.*, 1971).

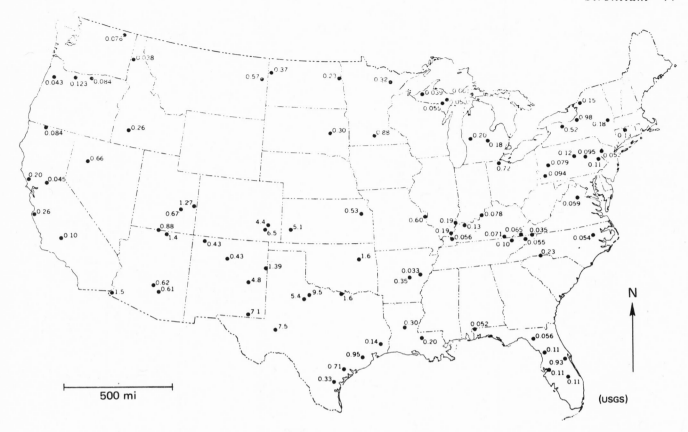

FIGURE 11 Average strontium concentration in surface waters. Values are in parts per million (Skougstad and Horr, 1963).

characterized by relatively low annual rainfall, high evaporation rate, and low physical relief, the concentration of strontium is generally two to three times as great as in most other streams in the nation. The report of this study also includes analyses of more than 175 groundwater samples, 60 percent of which contained less than 200 ppb of strontium, although 50,000 ppb was found in one sample of potable groundwater.

Vanselow (1966) found strontium in water ranging in concentration from less than 100 ppb to more than 1,000 ppb. More recently, Skougstad (personal communication, 1973) compiled data obtained from the analysis of more than 600 samples of municipal water supplies, including both treated water collected at the treatment plant and water from the distribution system. Fifty-eight percent of the samples contained between 46 and 220 ppb of strontium. Only 1 percent of the samples contained more than 2,200 ppb and only 1 percent less than 10 ppb strontium.

ANALYTICAL METHODS

Some earlier data on the concentrations of strontium in natural materials may not be completely reliable because of the difficulties inherent in the separation of traces of strontium from calcium, which is much more abundant

and chemically similar. Present instrumental methods not only provide satisfactory data but also are sufficiently sensitive to detect and measure strontium at the concentration levels at which the element commonly occurs in most natural substances. Flame photometry, atomic-absorption spectrophotometry, emission spectrometry, and X-ray-fluorescence spectrometry have been used with considerable success in a wide variety of materials. These techniques seem adequate for present needs.

PLANTS

The higher plants are unable to discriminate strontium (nonessential) from calcium (essential) because of the great similarity in the chemical properties of the two elements. In fact, the ratio of strontium to calcium in plants is essentially equal to the ratio of their concentrations in the culture media or to the ratio of their available amounts in the soil (Collander, 1941; Schroeder, 1941; Rediske and Selders, 1953; Knauss and Porter, 1954; Menzel, 1954; Menzel and Heald, 1955; Bowen and Dymond, 1955). Strontium can replace calcium, but not completely, in higher plants (Hasselhoff, 1893, 1898; Molisch, 1896; Loew, 1903, 1911). Plants vary in their tolerance for strontium. Scharrer and Schropp (1937) reported that cereals are more tolerant of high concentrations of strontium than legumes. Cotton is less tolerant of strontium than is Swiss chard. Hurd-Karrer (1937, 1939) demonstrated a low toxicity of strontium to higher plants when the calcium concentration is sufficiently high (Sr:Ca ratio < 1) to suppress the toxic action of strontium salts.

The low toxicity of strontium to higher plants and its partial replacement of calcium when the plants are grown on soil was demonstrated by Baird and Mehlich (1950). Swiss chard grown on soil containing 90 percent exchangeable strontium and 10 percent exchangeable calcium gave almost as high a yield as when the soil contained 80 percent exchangeable calcium and 20 percent exchangeable strontium. The strontium content of the crop grown on high-strontium soil was more than twice the calcium content, whereas the sum of the calcium and strontium contents was practically independent of the cations. In this instance, strontium appeared, at least in part, to replace calcium.

The replacement of calcium by strontium seems to be more apparent in aquatic plants for which calcium is a micronutrient. Walker (1953) reported that the calcium requirement of *Chlorella pyrenoidosa* can be satisfied on

essentially an equimolar basis by strontium. In a subsequent report, Walker (1956) indicated that the green algae constitute three groups: One group cannot tolerate excess strontium, another has a low sensitivity to strontium but cannot accept it as a replacement for calcium, and the third can accept strontium as a replacement for calcium.

Table 13 summarizes some representative tissue-analysis values reported in the literature indicating the status of strontium in vegetation of concern to grazing animals and man. Because strontium is not an essential nutrient, and no natural occurrence of strontium toxicity has been reported, one would not expect to find strontium indicator plants. A few species, such as *Arabis stricta* and *Carex humilis*, apparently can accumulate strontium in greater amounts in relation to calcium uptake (Bowen and Dymond, 1955).

Robinson *et al.* (1917) reported strontium contents ranging from 8 to 210 ppm for 38 plants, with a mean of 70 ppm. Mitchell (1948) reported strontium levels in pasture plants ranging from 11 to 137 ppm. Bowen and Dymond (1955) reported that the strontium content of plants grown on normal soils varied between 1 and 169 ppm, with a mean value of 36 ppm for some 40 different species of

TABLE 13   Strontium Concentrations in Vegetation of Concern to Grazing Animals and Man

Plants	Strontium, ppm (dry wt)	Source
Legumes and grasses		
Alfalfa	50–1,500	Vanselow (1966)
Clover	180–850	Vanselow (1966)
Meadow fescue	25–40	Vanselow (1966)
Orchard grass	20–25	Vanselow (1966)
Rye grass	18–23	Vanselow (1966)
Grain plants		
Barley grain	3[a]	Bowen and Dymond (1955)
Corn leaf	9–26	Prince (1957)
Corn leaf	15	Shimp *et al.* (1957)
Wheat straw	36.2	Champion *et al.* (1966)
Oat straw	13.1	Champion *et al.* (1966)
Oat grain	3.2	Champion *et al.* (1966)
Rye grass	13.3	Champion *et al.* (1966)
Vegetables, edible parts		
Beets	107	Bowen and Dymond (1955)
Broad beans	116	Bowen and Dymond (1955)
Cabbage	41	G. V. Alexander, personal communication (1959)
Carrots	34–131	G. V. Alexander, personal communication (1959)
Green beans	142	Bowen and Dymond (1955)
Lettuce	48	G. V. Alexander, personal communication (1959)
Mustard greens	138	G. V. Alexander, personal communication (1959)
Onions	209[a]	Bowen and Dymond (1955)
Peas	7	G. V. Alexander, personal communication (1959)
Parsnips	23	Bowen and Dymond (1955)
Snap beans	6–67	G. V. Alexander, personal communication (1959)
Spinach	45–70	Albrecht and Schroeder (1942)
Tomatoes	7–91	G. V. Alexander, personal communication (1959)

[a]Bowen and Dymond (1955) show these values in their Table, even though their text gives a range of 1–169 ppm for plants grown on normal soils.

native and pasture plants. On strontium-rich soils near celestite mines, the natural vegetation contained strontium in amounts up to at least 26,000 ppm dry weight. Even at these high concentrations, there appeared to be no adverse effect on the plants or on the animals feeding on them. Gerloff *et al.* (1964) reported the results of an extensive survey of the mineral content of native plants of Wisconsin in which the strontium concentrations were found to range from 1 to 115 ppm. In pasture plants of California, Vanselow (1966) found strontium levels ranging from 18 to 850 ppm, with occasionally higher values of up to 1,500 ppm in alfalfa. Vose and Koontz (1959) found strontium levels in grasses ranging from 75 to 160 ppm. In an extensive survey of perennial vegetation of the northern Mojave Desert, in southern Nevada, Wallace and Romney (1971, 1972) found strontium levels ranging from 10 to 800 ppm, with most species normally containing from 50 to 200 ppm.

Some information on the strontium content of vegetables and grains common to the human diet has been reported (Table 13). Numerous reports in the literature (Albrecht and Schroeder, 1942; Prince, 1957; Shimp *et al.*, 1957; G. V. Alexander, personal communication, 1959; Champion *et al.*, 1966) indicate that strontium does not concentrate in the edible grain and tubers of food crop plants. Champion *et al.* (1966) reported strontium concentrations of 9.3 ppm in milk powder.

ANIMALS AND HUMANS

In general, the metabolism and distribution of strontium mimics that of calcium. The major site of retention of both elements is the skeleton and teeth, in which more than 95 percent of body calcium and strontium is found (with the next greatest concentration occurring in the aorta, and lesser amounts in other soft tissues and body fluids). As with calcium, ingested strontium is absorbed in moderate amounts (5–25 percent in various species) and enters the exchangeable pool, from which it is incorporated into the mineral phase of bones and teeth (Vaughan, 1970), is transferred into intracellular compartments, traverses the placenta and mammary gland, and is handled like calcium by the kidney (Comar and Wasserman, 1964). Some strontium is excreted in sweat.

Strontium–Calcium Discrimination at the Physiological Level

During the transfer of strontium and calcium from one compartment to another, a preferential movement of calcium produces a decrease in the ratio of strontium to calcium. Prominent sites of strontium–calcium discrimination are the intestine, kidney, placenta, and mammary gland. If the strontium to calcium ratio in the diet is normalized to unity, the typical strontium:calcium ratio of the absorbed minerals is, in the adult human, 0.4. The ratio in bone is about 0.25; in milk, 0.1; in the fetus, 0.16;

and in urine, 0.87 (Figure 12; Comar and Wasserman, 1964).

In the kidney, discrimination occurs during the reabsorption step, leading to an increase in the Sr:Ca ratio of urine as compared to the ratio in the glomerular filtrate. Walser and Robinson (1963) indicated that the following exponential relationship existed between strontium and calcium in the filtrate and the urine:

$$\left(\frac{Sr_u}{Sr_f}\right) = \left(\frac{Ca_u}{Ca_f}\right)^{0.7}$$

where:

Sr_u = strontium excreted in urine
Ca_u = calcium excreted in urine
Sr_f = strontium filtered
Ca_f = calcium filtered

This equation assumes that reabsorption follows first-order kinetics and indicates that the strontium rate constant is 0.7 that for calcium (cf. Walser and Robinson, 1963, for theoretical development).

When the gastrointestinal absorptive process was considered in the same way, the relation was found to be:

$$\left(\frac{Sr_t}{Sr_0}\right) = \left(\frac{Ca_t}{Ca_0}\right)^{0.72}$$

where:

Sr_t and Ca_t = strontium and calcium in the lumen at time t
Sr_0 and Ca_0 = strontium and calcium in the lumen at time zero (i.e., before absorption proceeded)

The rate constant for strontium absorption is, therefore,

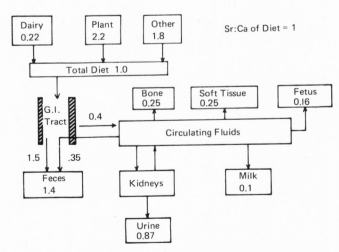

FIGURE 12 Typical strontium:calcium ratios in diet, tissues, and excretions of adult humans. Values are normalized to a strontium:calcium ratio of 1 in the total diet (Comar and Wasserman, 1964).

0.72 that for calcium absorption (Marcus and Wasserman, 1965). In addition, the rate constant of entry of strontium into mitochondria was found to be 0.63 that for calcium (Mraz, 1962; and Marcus and Wasserman, 1965).

Strontium–Calcium Discrimination at the Molecular Level

Diamond and Wright (1969) reviewed and extended the theory of ion selectivity based on the relative energy (enthalpy) of hydration of various cations as compared to the energy of interaction between these various cations and the binding site. This theory, largely after Eisenmann (1961), has a high degree of predictability with regard to discerning the expected selectivity sequence of ions in a given series. Some of the relative interactions or effects of the alkaline-earth series on various biological systems are given below; calcium is taken as unity (Diamond and Wright, 1969):

- Blocking negative (anionic) charges on the gall bladder: Ca(1.00)   Sr(0.61)
- Ability to restore intestinal permeability after EDTA treatment: Ca(1.00)   Sr(0.50)
- Effect on rise of spike of barnacle muscle: Ca > Sr
- Peak transient membrane currents in the alga, *Chara australis*: Ca(1.00)   Sr(0.8)
- Ability to reactivate Taka-amylase-A: Sr(1.04)   Ca(1.00)
- Apparent binding constant to G-actin: Ca(1.00)   Sr(0.03)
- Amount of cation bound to reconstituted collagen: Ca(1.00)   Sr(0.79).

The relative binding affinity of calcium and strontium to the vitamin-D-dependent calcium-binding protein is about 10:1 (calcium to strontium) (Wasserman and Taylor, 1972). These findings illustrate that, in most systems, calcium is bound preferentially over strontium. In contrast, however, the binding affinity of strontium to alginates exceeds that of calcium by factors of 1.5 to 4.3, depending on the guluronic acid (the uronic acid derived from L-gulose) content of the alginate (Triffitt, 1968).

Metabolism

Physiological and nutritional variables that affect strontium metabolism are similar to those that affect calcium metabolism and usually operate in the same direction (Comar and Wasserman, 1964; Wiseman, 1964). Younger people absorb both calcium and strontium to a greater extent than older people, and the stresses of pregnancy and lactation are associated with a greater efficiency of absorption of both elements. The intestinal site of greatest absorption of both calcium and strontium is the duodenum, whereas the site of most *effective* (efficiency × residence time) absorption is the ileum. Factors tending to enhance calcium and strontium absorption are vitamin D, lactose, and specific amino acids, such as lysine and arginine.

Parathyroid hormone accelerates the resorption of bone strontium, as it does bone calcium (Catsch, 1967; Vanderborght *et al.*, 1972). The inclusion of alginates in the diet has a greater depressing effect on strontium absorption than on calcium absorption (Waldron-Edward *et al.*, 1964).

Levels in the Human Diet

Table 14 lists values for strontium and calcium content of various food constituents of the diet in the United Kingdom given by Bryant *et al.* (1958). Values for institutional

TABLE 14   Average Daily Intake by Adults of Strontium and Calcium in Various Foods in the United Kingdom

	Daily Intake		
	Sr, mg/day	Ca, g/day	Sr:Ca, mg/g
Milk and milk products	0.193	0.667	0.290
Flour, bread, etc.	0.714[a]	0.332	2.150
Potatoes	0.062	0.019	3.240
Carrots	0.026	0.006	4.310
Cabbage (and greens)	0.031	0.014	2.190
Peas (and beans)	0.015	0.005	3.050
Lettuce	0.005	0.002	2.720
Other vegetables and fruit	0.174	0.058	3.000
Other cereals	0.033	0.011	3.000
Meat	0.016	0.027	0.600
Eggs	0.024	0.024	1.000
Fish	0.095	0.019	5.000
Other foods	0.045	0.015	3.000
TOTAL	1.433	1.199	Ratio of Totals=1.195

[a] In England, bread is fortified with calcium (and therefore also with strontium).
SOURCE: Bryant *et al.* (1958).

diets for eight different sections of the United States are presented in Table 15 (from Strong *et al.*, 1972). The range of values for the eight locations, in terms of strontium intake per day, is 1.03–2.67 mg/day, and the range of ratios of Sr:Ca (mg/g) is 0.89–2.41. It is apparent, then, that the variation is not great from region to region in the sampled areas and does not exceed a factor of 3. This uniformity of strontium concentrations in diet might reflect the interchange of dietary components from region to region. Strong *et al.* (1972) show a trend in dietary strontium:calcium ratios in the east–west profile across the United States (Figure 13). The eastern region (Atlantic Ocean to the Ohio and Mississippi rivers) was noted as having a Sr:Ca ratio of less than 1 mg/g; the middle region (including the states bordering the Gulf of Mexico), a ratio of 1–2 mg/g; the southern far-western area, values of more than 2 mg/g. This general trend of strontium with respect to calcium in different diets reflects the concentration of strontium in surface material as given in Figure 10.

Content of Human and Animal Skeletons

The strontium:calcium ratio in the skeleton of humans of various ages, as derived by Bryant *et al.* (1958) for the United Kingdom, is given in Table 16. Because the calcium content of the skeleton, on a dry fat-free basis, is about 26 percent, the strontium values were calculated in parts per million, and these were also included in Table 16. Values for six regions of the United States given in the report of Strong *et al.* (1972) are summarized in Table 17.

Again, the uniformity of the values, like those for the diet, is striking, and the range, in terms of strontium:calcium ratio or strontium in parts per million, does not exceed a factor of 2. Strong *et al.* (1972) also suggested that the strontium concentration of the human skeleton seemed to vary little with age, but tended toward a higher concentration in older people. The strontium:calcium ratio also did not vary greatly with the type of bone sampled.

Martin (1969) recorded a mean strontium:calcium ratio

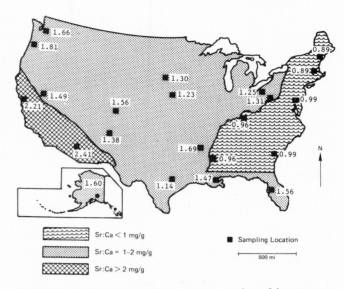

FIGURE 13 Strontium:calcium ratios at selected locations in the Institutional Total Diet Sampling Network (Strong *et al.*, 1972).

of 0.51 for 75 human skeletons in Wisconsin, with a range of 0.29–2.20. The strontium content had a mean value of 136 and a range of 77–587 ppm (dry fat-free basis). These values are apparently higher than those for most regions in the United States and might be attributed to the relatively high strontium:calcium ratios in Wisconsin drinking water (Strong *et al.*, 1972).

Thurber *et al.* (1958) have provided strontium data on a worldwide basis, although the number of samples in any given region was small (Table 18). The data again indicate the relative constancy of the strontium:calcium ratio and of the stable strontium content of human bone from region to region.

Additional data on the strontium and calcium content of human bone tissue is available from the report of Alexander and Nusbaum (1959).

The strontium content of portions of the skeletons of two diverse animal species, mule deer and freshwater fish, are presented in Table 19 (Farris *et al.*, 1967) and Table 20 (Ophel and Judd, 1967), respectively. The stron-

TABLE 15 Daily Intake of Strontium and Calcium in Eight Localities in the United States, and the Mean and Range of 22 Localities

	Sr, mg/day	Ca, g/day	Sr:Ca, mg/g
Boston, Massachusetts	1.18 ± 0.08	1.33 ± 0.06	0.89 ± 0.05
Pittsburgh, Pennsylvania	1.54 ± 0.12	1.18 ± 0.12	1.31 ± 0.07
Charleston, South Carolina	1.03 ± 0.12	1.07 ± 0.22	0.99 ± 0.14
Omaha, Nebraska	2.44 ± 0.26	1.99 ± 0.15	1.23 ± 0.10
Austin, Texas	1.83 ± 0.45	1.64 ± 0.44	1.14 ± 0.09
Phoenix, Arizona	2.63 ± 0.50	1.11 ± 0.28	2.41 ± 0.33
San Francisco, California	2.67 ± 0.28	1.25 ± 0.14	2.21 ± 0.41
Seattle, Washington	1.84 ± 0.14	1.14 ± 0.22	1.66 ± 0.16
Mean of 22 localities	1.74 ± 0.26	1.30 ± 0.18	1.40 ± 0.21
Range of 22 localities	1.03 – 2.77	0.67 – 2.03	0.89 – 2.41

SOURCE: Strong *et al.* (1972).

TABLE 16 Strontium:Calcium Ratio and Strontium Content in Human Skeleton in the United Kingdom

Age	Sr:Ca, mg/g	Sr,[a] ppm, dry wt, fat-free basis
<1 month	0.190 ± 0.024	50.7
1 month–4 yr	0.205 ± 0.029	54.7
5–18 yr	0.276 ± 0.021	73.7
> 18 yr	0.271 ± 0.014	72.4

[a] Calculated by the present authors, assuming skeleton contains 26.7 percent calcium on a dry weight, fat-free basis.
SOURCE: Bryant *et al.* (1958).

TABLE 17 Strontium and Calcium Content in Human Bone from Six Regions of United States

Region	No. of Samples	Average Age	Sr, mg/g ash	Ca, g/g ash	Sr:Ca, mg/g	Sr, ppm dry, fat-free basis
New England	23	13	0.07 ± 0.02	0.35 ± 0.02	0.02 ± 0.07	53.4
Mid-Atlantic	50	15	0.08 ± 0.03	0.35 ± 0.03	0.22 ± 0.08	58.7
South	28	17	0.09 ± 0.04	0.37 ± 0.01	0.25 ± 0.11	66.8
Great Lakes	8	18	0.08 ± 0.03	0.36 ± 0.02	0.23 ± 0.07	61.4
Plains	9	18	0.09 ± 0.02	0.36 ± 0.01	0.27 ± 0.06	72.1
West	14	18	0.08 ± 0.02	0.37 ± 0.02	0.21 ± 0.06	56.1

SOURCE: Strong *et al.* (1972).

tium content and the strontium:calcium ratio of bone from the deer and the fish were quite similar. As a reference, the strontium:calcium ratios in the lake waters are also given in Table 20.

Content of Human Soft Tissues

Tipton *et al.* (1965) published an extensive report on the concentrations of various mineral elements in a number of human soft tissues. Table 21 summarizes the values for strontium and calcium in aorta, heart, and kidney. The aorta contains considerably more strontium than the other soft tissues and has a commensurately high relative calcium content. The strontium:calcium ratios cited in Table 21, ranging from 0.57–1.30, are generally higher than those of the human skeleton or bone as given in Tables 16–18, a relationship that was not expected. The overall discrimination between strontium and calcium at the blood–bone interface is usually considered to be close to unity (Comar and Wasserman, 1964).

Compartmental Model

Dolphin and Eve (1963) summarized a large body of human adult data in diagrammatic form (Figure 14). For this model, it is assumed that 1.5 mg of strontium is ingested per day, and, of this, 0.3 mg/day is absorbed and 0.04 mg/day is endogenously secreted into the digestive tract and excreted with the unabsorbed portion (1.5–0.30

= 1.2) in the feces. Other routes of excretion are the kidney (0.24 mg/day in urine) and skin (0.02 mg/day in sweat). The net retention of strontium is therefore zero, indicating that the model individual is in the steady state or in balance. The various compartments contributing to the exchangeable strontium pool are also given in Figure 14, as is the rate of interchange of skeletal strontium with exchangeable strontium (0.12 mg/day).

Essentiality for Animals

An unconfirmed report by Rygh (1949) indicated that rats and guinea pigs fed a diet relatively free of strontium had a high incidence of dental caries and poor growth, conditions corrected by the addition of strontium to the diet. Shorr and Carter (1950) suggested that strontium, as an adjunct to calcium therapy, might be beneficial in the remineralization of bone in persons with osteoporosis; this approach has not been widely used or accepted.

Toxicity

There is no evidence that stable strontium at levels encountered in the biosphere exerts any deleterious effects on animals, including humans. However, strontium at quite high levels in the diet was shown several years ago to produce the syndrome known as "strontium rickets" in experimental animals (Shipley *et al.*, 1922). The skeleton failed to mineralize, even in the presence of adequate vitamin D. The early explanation for the toxicity was that strontium formed an insoluble phosphate complex, thereby producing phosphate-deficiency rickets (Jones,

TABLE 18 Strontium Content and Strontium:Calcium Ratio of Human Bone from Various Worldwide Regions

Region	No. of Localities	Sr, ppm, dry, fat-free basis Mean	Range	Sr:Ca, mg/g
North America	9	109	73–141	0.41
South America	5	144	111–238	0.54
Europe	5	125	95–175	0.47
Asia	3	136	130–143	0.51
Africa	2	181	135–225	0.68

SOURCE: Thurber *et al.* (1958).

TABLE 19 Strontium in Deer Metacarpal Bone from Colorado

Age, months	Sr, mg/g ash[a]	Ca, g/g ash	Sr:Ca, mg/g
4–5	0.244 ± 0.014	0.370	0.66
28–29	0.276 ± 0.019	0.370	0.75
64–77	0.283 ± 0.017	0.370	0.76
88–113	0.300 ± 0.015	0.370	0.81

[a] Value is mean ±SEM (standard error of the mean).
SOURCE: Farris *et al.* (1967).

TABLE 20 Strontium and Calcium in Lake Waters and Fish Bones

Lake Waters	Sr, μg/ml	Ca, mg/ml	Sr:Ca, mg/g
Lake Huron	0.11	0.0266	4.14
Perch Lake (Chalk River)	0.032	0.00625	5.12
Fish Bones	Sr, mg/g ash[a]	Ca, g/g ash[a]	Sr:Ca, mg/g
Ribs			
Lake Huron perch (10)[b]	0.294 ± 0.0494	0.368 ± 0.0226	0.80
Lake huron redhorse sucker (5)	0.360 ± 0.0197	0.352 ± 0.0083	1.02
Perch Lake perch (10)	0.399 ± 0.0388	0.370 ± 0.0161	1.07
Anterior vertebrae			
Lake Huron perch (9)	0.277 ± 0.0385	0.370 ± 0.0207	0.75
Lake Huron redhorse sucker (5)	0.321 ± 0.0287	0.358 ± 0.0108	0.90
Perch Lake perch (10)	0.373 ± 0.0247	0.364 ± 0.0163	1.02

[a] Values of the mean ±SD (standard deviation).
[b] Number of samples.
SOURCE: Ophel and Judd (1967).

1938). Another proposal was that strontium directly inhibited the calcification mechanism (Sobel, 1954; Weber *et al.*, 1963), and still another was that the high levels of dietary strontium in some fashion depressed the intestinal calcium absorptive mechanism. The latter idea was suggested by the studies of Bartley and Reber (1961) in which it was shown that dietary strontium depressed the retention of a subsequent oral dose of radiostrontium.

The problem of strontium toxicity was examined in more detail using chicks (Corradino and Wasserman, 1970; Corradino *et al.*, 1971; Corradino, 1972). Replacing most of the calcium in the diet with stable strontium was found to depress radiocalcium and radiostrontium absorption within a few days, as measured by the ligated duodenal loop technique *in situ*. This effect is illustrated in Table 22, in which the duodenal absorption of [85]Sr by

TABLE 21 Strontium and Calcium in Ash of Human Soft Tissues from Four Different Geographic Regions

Tissue	Median Values for Males Aged 20–59			
	Sr, ppm in ash	Sr, % in ash (× 10³)	Ca, % in ash	Sr:Ca, mg/g
Aorta				
United States	34	3.4	5.0	0.68
Africa	32	3.2	3.3	0.97
Near East	35	3.5	2.9	1.21
Far East	22	2.2	2.4	0.92
Heart				
United States	2.5	0.25	0.38	0.66
Africa	3.6	0.36	0.40	0.90
Near East	3.1	0.31	0.45	0.69
Far East	3.0	0.30	0.43	0.70
Kidney				
United States	4.9	0.49	0.82	0.60
Africa	6.6	0.66	0.78	0.85
Near East	7.8	0.78	0.60	1.30
Far East	6.8	0.68	1.2	0.57

SOURCE: Tipton *et al.* (1965).

rachitic chicks is compared to that of chicks on a normal diet (1.2 percent calcium), those on a low-calcium diet (0.1 percent calcium), and those on a low-calcium diet supplemented with stable strontium (2.62 percent strontium, the molar equivalent of 1.2 percent calcium). From Table 22, we see that the presence of strontium in the diet (IV) depressed [85]Sr absorption to a level comparable to vitamin-D-deficiency rickets (I). Those chicks on the low-calcium diet (III) underwent an adaptation that enhanced the efficiency of [85]Sr absorption as compared to the control group (II). It should be recognized that the technique used for determining [85]Sr absorption measures alterations in the process of absorption per se rather than direct dietary influences.

Later investigations revealed that the inhibitory effect of stable dietary strontium (2.62 percent) on intestinal activity in chicks appeared at less than 2 days and reached a maximum at about 4 days. High dietary strontium also inhibited calcification; a significant reduction in percentage of ash in tibia (dry, fat-free basis) was evident at 3–4 days of feeding. Plasma calcium levels were also reduced.

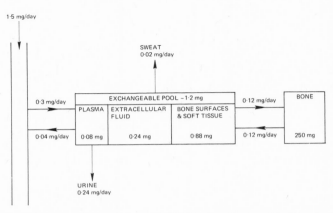

FIGURE 14 Compartment model for strontium metabolism in human adults (Dolphin and Eve, 1963).

TABLE 22   Effect of Dietary Calcium Level, Vitamin D_3, and Strontium on ^{85}Sr-Labeled, Calcium Absorption by Ligated Duodenal Segment *in Situ*.

Group	Previous Diet	Experimental Diet	% Dose ^{85}Sr Absorbed[a]
I	Rachitogenic	Normal Ca, $- D_3$	6.02 ± 1.08[b]
II	Normal	Normal Ca, $+ D_3$	23.12 ± 2.35
III	Normal	Low Ca, $+ D_3$	70.50 ± 3.22
IV	Normal	Low Ca, $+ Sr, + D_3$	7.88 ± 0.88[b]

[a] Each value represents the mean $\pm SE$ (standard error) of five chicks per group.
[b] These values were the only pair not significantly different at 1 percent level by Student's t-test.
SOURCE: Corradino and Wasserman (1970).

When chicks, previously fed the high-strontium diet (2.62 percent) for 7 days, were given a normal diet, rapid recovery of all measured parameters quickly ensued. ^{85}Sr absorption began to increase within 1–2 days, as did the intestinal level of calcium-binding protein. Bone ash and plasma calcium also began to return to normal but were still less than the control group after 7 days.

The mechanism by which strontium exerts its effect is still not absolutely clear, although there is evidence for interference with vitamin D metabolism. According to the reviews of Wasserman and Corradino (1971) and Wasserman and Taylor (1972), it has been demonstrated that cholecalciferol (D_3) is hydroxylated in the liver to form 25-hydroxycholecalciferol [25-$(OH)D_3$], and is subsequently further hydroxylated in the kidney to $1,25$-dihydroxycholecalciferol [$1,25$-$(OH)_2D_3$]. The $1,25$-$(OH)_2D_3$ is the most biologically active form of D_3 and is considered to be the form that affects calcium absorption through an intermediate protein synthetic event at the intestinal level. Omdahl and DeLuca (1971) observed in a similar experimental procedure to that given above that the defect in calcium absorption could be rectified by the administration of $1,25$-$(OH)_2D_3$, but not by 25-$(OH)D_3$ or vitamin D_3 as such. This suggested that the ingestion of a diet with a high strontium content reduces or inhibits the ability of kidney enzymes to convert 25-$(OH)D_3$ to $1,25$-$(OH)_2D_3$. Studies with radioactively labeled vitamin D_3 or 25-$(OH)D_3$ tended to confirm this hypothesis (Omdahl and DeLuca, 1972).

In a series of studies using different strontium:calcium dietary ratios (on a molar basis), it was observed that at 3 days a ratio of 0.2 yielded a significant reduction in the synthesis of the vitamin-D-dependent calcium-binding protein, CaBP (Corradino, 1972). At 14 days, there was a "spontaneous" recovery in this group in which intestinal CaBP levels became equal to those in the control group. No spontaneous recovery occurred if the dietary Sr:Ca ratio was 2 or greater.

Mraz *et al.* (1967) fed laying hens diets containing varying levels of supplemental $SrCO_3$ and a calcium level of 2.8 percent. At a strontium level of 0.5 percent (Sr:Ca molar ratio 0.08), no deleterious effect was noted. At the 1 percent strontium level (Sr:Ca molar ratio 0.16), the hens lost more weight than the controls, but egg production and egg-shell strength were not adversely affected, and there was only a slight increase in late embryonic mortality.

Because strontium:calcium molar ratio encountered by humans is usually of the order of 6×10^{-4}, it is very unlikely that usual levels of dietary strontium would be sufficiently high to elicit any of the responses noted above.

RECOMMENDATIONS FOR RESEARCH

1. More extensive information on stable strontium concentrations in plant materials, animal and human diets, and animal and human tissues is required.

2. More information on the state of strontium in soils, its availability for plants, and the strontium–calcium selectivity in the soils-to-plant transfer step is required.

3. Information about the movement of stable strontium through the biosphere, particularly for comparison with ^{90}Sr movement, is required.

4. The long-term toxic effects of moderate levels of stable strontium should be studied.

5. The interrelation of strontium to various elements (calcium and fluorine, for example) in water, soil, and biological tissue and to various disease states should be studied. Regions of low or high strontium should be selected for the study of animal, human, and plant health.

REFERENCES

Åberg, B., and F. P. Hungate [eds.] 1967. Radioecological concentration processes. Pergamon Press, Oxford. 1040 pp.

Albrecht, W. A., and R. A. Schroeder. 1942. Plant nutrition and the hydrogen ion. Soil Sci. 53:313–327.

Aleksakhin, R. M. 1963. Radioactive contamination of soils and plants. Academy of Sciences of the USSR, Moscow. [Translation, AEC-tr-6631, U.S. Department of Commerce, Springfield, Virginia. 108 pp.]

Alexander, G. V., and R. E. Nusbaum. 1959. The relative retention of strontium and calcium in human bone tissue. J. Biol. Chem. 234:418–421.

Alexander, G. V., R. E. Nusbaum, and N. S. MacDonald. 1954. Strontium and calcium in municipal water supplies. J. Am. Water Works Assoc. 46:643–654.

Allison, R. V., and L. W. Gaddum. 1940. Trace elements of some important soils. Soil Sci. Soc. Am. Proc. (Fla.) 2:68–91.

Baird, G. B., and A. Mehlich. 1950. The effect of soil-exchangeable cations on Swiss chard and cotton. Soil Sci. Soc. Am. Proc. 15:201–205.

Bartley, J. C., and E. F. Reber. 1961. Metabolism of radiostrontium in young pigs and in lactating rats fed stable strontium. J. Dairy Sci. 44:1754–1762.

Benson, D. W., and A. H. Sparrow [eds.] 1971. Survival of food crops and livestock in the event of nuclear war. U.S. AEC Rep. CONF-700909. Technical Information Center, Oak Ridge, Tennessee. 745 pp.

Bowen, H. J. M., and J. A. Dymond. 1955. Strontium and barium in plants and soils. Proc. R. Soc. London Ser. B 144:355–368.

Bryant, F. J., A. C. Chamberlain, G. S. Spicer, and M. S. W. Webb. 1958. Strontium in the diet. Brit. Med. J. 1:1371–1375.

Caldecott, R. S., and L. A. Snyder [eds.] 1960. Radioisotopes in the biosphere. University of Minnesota, Center for Continuation Study, Minneapolis. 597 pp.

Carrigan, R. A., and L. H. Rogers. 1940. Trace-element content of certain Florida soils and related plants. Soil Sci. Soc. Am. Proc. (Fla.) 2:92–103.

Catsch, A. 1967. Removal of radiostrontium from the mammalian body. *In* Strontium metabolism, J. M. A. Lenihan, J. F. Lontit and J. H. Martin [eds.]. Academic Press, New York. pp. 265–281.

Champion, K. P., J. C. Taylor, and R. N. Whittem. 1966. Rapid X-ray fluorescence determination of traces of strontium in samples of biological and geological origin. Anal. Chem. 38(1):109–112.

Collander, R. 1941. Selective absorption of cations by higher plants. Plant Physiol. 16:691–720.

Comar, C. L., and R. H. Wasserman. 1964. Strontium. *In* Mineral metabolism: An advanced treatise, C. L. Comar and F. Bronner [eds.]. Academic Press, New York. pp. 523–572.

Corradino, R. A. 1972. Strontium inhibition of the vitamin D-induced calcium-binding protein and the intestinal calcium-absorptive mechanism. Proceedings of the 2d International Conference on Strontium Metabolism, Glasgow and Strontian, August 16–19, 1972. pp. 277–287.

Corradino, R. A., and R. H. Wasserman. 1970. Strontium inhibition of vitamin D3-induced calcium-binding protein (CaBP) and calcium absorption in chick intestine. Proc. Soc. Exp. Biol. Med. 133:960–963.

Corradino, R. A., J. G. Ebel, P. H. Craig, A. N. Taylor, and R. H. Wasserman. 1971. Calcium absorption and the vitamin D3-dependent calcium-binding protein. I. Inhibition by dietary strontium. II. Recovery from dietary strontium inhibition. Calcium Tissue Res. 7:81–102.

Dasch, E. J. 1969. Strontium isotopes in weathering profiles, deep-sea sediments, and sedimentary rocks. Geochim. Cosmochim. Acta 33:1521–1552.

Diamond, J. M., and E. M. Wright. 1969. Biological membranes: The physical basis of ion and nonelectrolyte selectivity. Annu. Rev. Physiol. 31:581–646.

Dolphin, G. W., and I. S. Eve. 1963. The metabolism of strontium in adult humans. Phys. Med. Biol. 8:193–203.

Eisenman, G. 1961. On the elementary atomic origin of equilibrium ionic specificity. *In* Symposium on membrane transport and metabolism, A. Kleinzeller and A. Kotyk [eds.]. Academic Press, New York. 608 pp.

Farris, G. C., F. W. Whicker, and A. H. Dahl. 1967. Effect of age on radioactive and stable strontium accumulation in mule deer bone. *In* Strontium metabolism, J. M. A. Lenihan, J. F. Loutit and J. H. Martin [eds.]. Academic Press, New York. 93 pp.

Food and Agriculture Organization. 1960. Radioactive materials in food and agriculture. United Nations, Food and Agriculture Organization, Rome, Italy. 123 pp.

Food and Agriculture Organization. 1964. Agricultural and public health aspects of radioactive contamination in normal and emergency situations. United Nations, Food and Agriculture Organization, Rome, Italy. 421 pp.

Fowler, E. B. [ed.]. 1965. Radioactive fallout, soils, plants, foods, man. Elsevier, New York. 317 pp.

Frere, M. H., R. G. Menzel, K. H. Larson, R. Overstreet, and R. F. Reitemeier. 1963. The behavior of radioactive fallout in soils and plants. NAS–NRC Publ. 1092. National Academy of Sciences, Washington, D.C. 32 pp.

Gerloff, G. C., D. D. Moore, and J. T. Curtis. 1964. Mineral content of native plants of Wisconsin. Univ. Wis. Res. Rep. 14. University of Wisconsin, Madison. 27 pp.

Glentworth, R. 1944. Studies on the soils developed on basic igneous rocks in central Aberdeenshire. Trans. R. Soc. Edinburgh 61:149–170.

Haselhoff, E. 1893. Versuche über den Ersatz des Kalkes durch Strontium bei der Pflanzenernahrung. Landwirtsch. Jahrb. 22:851–867.

Haselhoff, E. 1898. Versuche über den Ersatz des Kalkes durch Strontium und den Kalibedarf einiger Planzen. Arb. Dtsch. Landwirtsh. Ges. (Berlin):34.

Hungate, F. P. [ed.] 1965. Hanford Symposium on Radiation and Terrestrial Ecosystems. Health Phys. 11:1255–1675.

Hurd-Karrer, A. M. 1937. Rubidium and strontium toxicity in plants inhibited by potassium and calcium, respectively. J. Wash. Acad. Sci. 27:351–353.

Hurd-Karrer, A. M. 1939. Antagonism of certain elements essential to plants toward chemically related toxic substances. Plant Physiol. 14:9–29.

Joint Committee on Atomic Energy Hearings. 1957. The nature of radioactive fallout and its effects on man. May 27–29 and June 3, 1957. U.S. Government Printing Office, Washington, D.C.

Joint Committee on Atomic Energy Hearings. 1959a. Fallout from nuclear weapons tests. May 5–8, 1959. U.S. Government Printing Office, Washington, D.C.

Joint Committee on Atomic Energy Hearings. 1959b. Biological and environmental effects of nuclear war. June 22–26, 1959. U.S. Government Printing Office, Washington, D.C.

Jones, J. H. 1938. The metabolism of calcium and phosphorus as influenced by the addition to the diet of salts of metals which form insoluble phosphates. Am. J. Physiol. 124:230–237.

Knauss, H. J., and J. W. Porter. 1954. Absorption of inorganic ions by *Chlorella pyrenoidosa*. Plant Physiol. 29:229–234.

Livingstone, D. A. 1963. Data of geochemistry, 6th ed. Chapter G: Chemical composition of rivers and lakes. U.S. Geol. Surv. Prof. Pap. 440-G. U.S. Government Printing Office, Washington, D.C. 64 pp.

Loew, O. 1903. The toxic action of salts of magnesium, strontium and barium on plants. Landwirtsch. Jahrb. 32:509–515.

Loew, O. 1911. Concerning the effect of strontium salts on algae. Flora (Jana) 2:96.

Lowenstam, H. A. 1964. Sr/Ca ratio of skeletal aragonites from the recent marine biota at Palau and from fossil gastropods. *In* Isotopic and cosmic chemistry, H. Craig, S. L. Miller and G. J. Wasserburg [eds.]. North Holland, Amsterdam. pp. 114–132.

Marcus, C. S., and R. H. Wasserman. 1965. Comparison of intestinal discrimination between calcium 47, strontium 85 and barium 133. Am. J. Physiol. 209:973–977.

Martin, A. 1969. Content and distribution of stable strontium and 226Ra in human skeletons from Wisconsin decedents 1957–1961. Brit. J. Radiol. 42:295–298.

Menzel, R. G. 1954. Competitive uptake by plants of potassium, rubidium, cesium, and calcium, strontium, barium from soils. Soil Sci. 77:419–425.

Menzel, R. G., and W. R. Heald. 1955. Distribution of potassium, rubidium, cesium, calcium and strontium within plants given in nutrient solutions. Soil Sci. 80:287–293.

Menzel, R. G., and W. R. Heald. 1959. Strontium and calcium contents of crop plants in relation to exchangeable strontium and calcium in the soil. Soil Sci. Soc. Am. Proc. 23:110–112.

Mitchell, R. L. 1937. Base-exchange equilibria in soil profiles. J. Agric. Sci. 27:557–568.

Mitchell, R. L. 1944. The distribution of trace elements in soils and grasses. Proc. Nutr. Soc. 1:183–189.

Mitchell, R. L. 1948. The spectrographic analysis of soils, plants and related materials. Commonw. Bur. Soil Sci. Tech. Commun. No. 44. Herald Printing Works, York, England. 183 pp.

Molisch, H. 1896. The nutrition of the algae. Sitzungsber. Akad. Wiss. Wien. 105 pp.

Mraz, F. R. 1962. Calcium and strontium uptake by rat liver and kidney mitochondria. Proc. Soc. Exp. Biol. Med. 111:429–431.

Mraz, F. R., P. L. Wright, and T. M. Ferguson. 1967. Effect of dietary strontium on reproductive performance of the laying hen. *In* Strontium metabolism, J. M. A. Lenihan, J. F. Loutit, and J. H. Martin [eds.]. Academic Press, New York.

Nichols, M. S., and D. R. McNall. 1957. Strontium content of Wisconsin municipal waters. J. Am. Water Works Assoc. 49:1493–1498.

Noll, W. 1931. On the determination of strontium in mineral and rock analysis. Z. Anorg. Allg. Chem. 199:193–208.

Omdahl, J. L., and H. F. DeLuca. 1971. Strontium-induced rickets: Metabolic basis. Science 174:949–950.

Omdahl, J. L., and H. F. DeLuca. 1972. Rachitogenic activity of dietary strontium. I. Inhibition of intestinal calcium absorption and 1,25-dihydroxycholecalciferol synthesis. J. Biol. Chem. 247:5520–5526.

Ophel, I. L., and J. M. Judd. 1967. Skeletal distribution of strontium and calcium and strontium/calcium ratios in several species of fish. *In* Strontium metabolism, J. M. A. Lenihan, J. F. Loutit, and J. H. Martin [eds.]. Academic Press, New York.

Prince, A. L. 1957. Trace element delivering capacity of ten New Jersey soil types as measured by spectrographic analysis of soils and mature corn leaves. Soil Sci. 84:413–418.

Rediske, J. H., and A. A. Selders. 1953. Absorption and translocation of strontium by plants. Plant Physiol. 28:594–605.

Robinson, W. O. 1914. Inorganic composition of some important American soils. U.S. Dep. Agric. Bull. 122. U.S. Department of Agriculture, Washington, D.C.

Robinson, W. O., L. A. Steinkoenig, and C. F. Miller. 1917. The relation of some of the rarer elements in soils and plants. U.S. Dep. Agric. Bull. 600. U.S. Department of Agriculture, Washington, D.C.

Rogers, L. H., O. E. Gall, L. W. Gaddum, and R. M. Burnette. 1939. Distribution of the macro- and microelements in some soils. Univ. Fla. Agric. Exp. Stn. Bull. 341. University of Florida, Gainesville. 31 pp.

Russell, R. S. [ed.] 1966. Radioactivity and human diet. Pergamon Press, Oxford. 552 pp.

Rygh, O. 1949. Recherches sur les oligo-elements. I. De l'importance du strontium du baryum et du zinc. Bull. Soc. Chim. Biol. 31:1052.

Scharrer, K., and W. Schropp. 1937. The effect of barium and strontium ions upon the growth of some plants. Bodenk. Pflanzenernaehr. 3:369–385.

Schroeder, R. A. 1941. Some effects of calcium and pH upon spinach. Proc. Am. Soc. Hortic. Sci. 38:482–486.

Schultz, V., and A. W. Klement, Jr. [eds.] 1963. Radioecology. AIBS and Reinhold, Washington, D.C. 746 pp.

Shacklette, H. T., J. G. Hamilton, J. G. Boerngen, and J. M. Bowles. 1971. Elemental composition of surficial materials in the conterminous United States. U.S. Geol. Surv. Prof. Pap. 574-D. U.S. Government Printing Office, Washington, D.C. 71 pp.

Shimp, N. F., Jane Conner, A. L. Prince, and F. E. Bear. 1957. Spectrochemical analysis of soils and biological materials. Soil Sci. 83:51–64.

Shipley, P. G., E. A. Park, E. V. McCollum, N. Simmonds, and E. M. Kinsey. 1922. Studies in experimental rickets. XX. Bull. Johns Hopkins Hosp. 33:216–220.

Shorr, E., and A. C. Carter. 1950. The value of strontium as adjuvant to calcium in the remineralization of the skeleton in osteoporoses in man. *In* Metabolic interrelations, Trans. 2d Conf. Josiah Macy, Jr., Found., New York. pp. 144–154.

Skougstad, M. W., and C. A. Horr. 1963. Chemistry of strontium in natural water. U.S. Geol. Surv. Water Supply Pap. 1496. U.S. Geological Survey, Washington, D.C. 97 pp.

Sobel, A. E. 1954. Local factors in the mechanism of calcification. Ann. N.Y. Acad. Sci. 60:713–732.

Strong, A. B., C. R. Porter, and B. Kahn. 1972. Stable strontium:calcium ratios in U.S. bone and total diet samples. Second International Conference on Strontium Metabolism, Glasgow and Strontian, August 16–19, 1972. pp. 513–520.

Swaine, D. J. 1955. The trace-element content of soils. Commonw. Bur. Soil Sci. Tech. Commun. No. 48. Herald Printing Works, York, England. 167 pp.

Thomas, W. 1923. Ultimate analysis of the mineral constituents of a Hagerstown silt clay loam, and occurrence in plants of some of the elements found. Soil Sci. 15:1–18.

Thompson, G., and H. D. Livingston. 1970. Strontium and uranium concentrations in aragonite precipitated by some modern corals. Earth Planet. Sci. Lett. 8:439–442.

Thurber, D. L., J. L. Kulp, E. Hodges, P. W. Gast, and J. M. Wampler. 1958. Common strontium content of the human skeleton. Science 128:256–257.

Tipton, I. H., H. A. Schroeder, H. M. Perry, Jr., and M. J. Cook. 1965. Trace elements in human tissue. Health Phys. 11:403–451.

Triffitt, J. T. 1968. Binding of calcium and strontium by alginates. Nature 217:457–458.

Turekian, K. K. 1964. The marine geochemistry of strontium. Geochim. Cosmochim. Acta 28:1479–1496.

Turekian, K. K., and D. F. Schutz. 1965. Trace element economy in the oceans. *In* Symposium on marine geochemistry. Occasional Publication No. 3. University of Rhode Island, Narragansett Marine Laboratory, Kingston. pp. 41–89.

U.N. Scientific Committee on Effects of Atomic Radiation. 1962. Supplement No. 16 A/5216. United Nations, New York. 442 pp.

U.N. Scientific Committee on Effects of Atomic Radiation. 1964. Supplement No. 14 A/5810. United Nations, New York. 120 pp.

U.S. Geological Survey. 1972. Environmental geochemistry: Geochemical survey of Missouri. U.S. Geol. Surv. Open-File Rep. Nos. 1658–1661, 1706, 1800, and 1982. U.S. Geological Survey, Denver and Reston.

Vanderborght, O., D. Keslev, van Puymbroeck, and J. Colard. 1972. Combined influence of diet, alginate, parathyroid hormone and vitamin D on ^{85}Sr and ^{47}Ca mobilization from bone. Second International Conference on Strontium Metabolism, Glasgow and Strontian, August 16–19, 1972. pp. 397–401.

Vanselow, A. P. 1966. Strontium. *In* Diagnostic criteria for plants and soils, H. D. Chapman [ed.]. Division of Agricultural Sciences, University of California, Riverside. 793 pp.

Vaughan, J. M. 1970. The physiology of bone. Clarendon Press, Oxford, England. 325 pp.

Vinogradov, A. P. 1945. The chemical study of the biosphere. Pochvovedenie (Leningrad) 8:348–353.

Vinogradov, A. P. 1954. Geochemie seltener und nur in Spuren vorhandener chemischer Elemente in Boden. Academic-Verlag, Berlin. 249 pp.

Viro, P. J. 1951. Nutrient status and fertility of forest soil. I. Pine stands. Commun. Inst. For. (Finland) 39(4):1–54.

Vose, P. B., and H. N. Koontz. 1959. Uptake of strontium by plants and its possible significance in relation to the fallout of radiostrontium. Nature 183:1447–1448.

Waldron-Edward, D., T. M. Paul, and S. C. Skoryna. 1964. Studies on the inhibition of intestinal absorption of radioactive strontium. Can. J. Med. Assoc. 91:1006–1010.

Walker, J. B. 1953. Inorganic micronutrient requirements of *Chlorella*. I. Requirements of calcium, strontium, copper and molybdenum. Arch. Biochem. Biophys. 46:1–11.

Walker, J. B. 1956. Strontium inhibition of calcium utilization by green algae. Arch. Biochem. Biophys. 60:264–265.

Wallace, A., and E. M. Romney. 1971. Some interactions of Ca, Sr, and Ba in plants. Agron. J. 63:245–248.

Wallace, A., and E. M. Romney. 1972. Radioecology and ecophysiology of desert plants at the Nevada Test Site. USAEC Rep. TID-25954. U.S. Department of Commerce, Springfield, Virginia. 439 pp.

Walser, M., and B. H. B. Robinson. 1963. Renal excretion and tubular reabsorption of calcium and strontium. *In* The transfer of calcium and strontium across biological membranes: Proc. Conf. Ithaca, NY, May 13–16 1962, sponsored by U.S. AEC and others. R. H. Wasserman [ed.]. Academic Press, New York. pp. 305–326.

Wasserman, R. H., and R. A. Corradino. 1971. Metabolic role of vitamins A and D. Annu. Rev. Biochem. 40:501–532.

Wasserman, R. H., and A. N. Taylor. 1972. Metabolic roles of fat-soluble vitamins D, E, and K. Annu. Rev. Biochem. 41:179–202.

Weber, C. W., Z. R. Doberenz, R. W. G. Wyckoff, and B. L. Reid. 1963. Strontium metabolism in chicks. Poult. Sci. 47:1218–1323.

White, D. E., J. D. Hem, and G. A. Waring. 1963. Data of geochemistry, 6th ed. Chapter F: Chemical composition of subsurface waters. U.S. Geol. Surv. Prof. Pap. 440-F. U.S. Government Printing Office, Washington, D.C. 67 pp.

Wiseman, G. 1964. Absorption from the intestine. Academic Press, New York. 564 pp.

VIII

Tin

KENNETH C. BEESON, *Chairman*

Wallace R. Griffitts, David B. Milne

Tin occurs widely in nature in such disparate modes as insoluble oxides (such as cassiterite, SnO_2) and as organic complexes in peats and coals. Tin and its alloys have been economically important for man since the beginning of the bronze age. Although tin, in minor quantities, appears to be a common constituent of plants, it has no known function in them. In 1970, however, Schwarz *et al.* (1970) found that various tin compounds have a significant effect on the growth of rats if trace element contamination from the environment is rigidly excluded.

GEOCHEMISTRY

Rocks

Tin is dispersed in very small and generally uniform amounts in most silicate rocks. Among igneous rocks, it is found in concentrations of 0.3–0.5 ppm in ultramafic rocks; 1–1.2 ppm in basalts, gabbros, and other mafic rocks; and about 3.5 ppm in granites (Wedepohl, 1969). Some granites that are associated with tin deposits may contain locally much more tin, 120 ppm having been found in a sample of one granite on the Seward Peninsula, Alaska (Sainsbury *et al.*, 1968).

Granitic micas generally have two to four times the concentrations of tin found in the associated feldspars, biotite, and other dark minerals. Sphene contains far more tin than other minerals. In basic rocks, tin is found in plagioclase, and in smaller amounts in olivine, pyroxenes, and hornblende. Perhaps most important from a geochemical standpoint is the occurrence of tin in minerals that are moderately susceptible to weathering—the feldspars, biotite and other dark minerals (Wedepohl, 1969). Thus tin may be mobilized as rocks disintegrate.

Sedimentary rocks seem more varied. Shales average about 5 ppm of tin, but few data are available for limestones and sandstones because they generally contain tin in concentrations below the limits of detection of most analytical methods.

Soils

A review by Hamaguchi and Kuroda (1969) of early work on the distribution of tin concludes that variations in the concentration of tin in soils are largely related to the bedrock from which the soils are developed. Gordon (1953) reported 30–300 ppm of tin in the ash of 50 peats in Finland. Pinta and Oliat (1961) rarely found more than traces of tin in the tropical soils of Dahomey.

Soils underlain by crystalline rocks or shales may be expected to contain a few parts per million of tin. This generally low level of tin is substantiated by the analyses of nearly 900 soil samples taken throughout the United States. Only 1 percent of the samples contained more than 10 ppm of tin, and the maximum found was 20 ppm. A similar number of soil samples collected across the state of Missouri also yielded about 1 percent with 10 ppm or more of tin (personal communication, W. R. Griffitts, 1974). Tin districts may exhibit exceptional tin levels in soil. For example, soils in the Lost River tin district of

Alaska contain as much as 1,500 ppm tin (Sainsbury *et al.,* 1968).

Water

Public water supplies in 42 cities in the United States were found to contain 1.1 to 2.2 ppb of tin; water from 175 sources in west-central Arizona contained 0.8 to 30 ppb. Seawater contains 0.2 to 0.3 ppb of tin (Wedepohl, 1969).

ANALYTICAL METHODS

Most workers with tin in the biological field appear to have assumed that their ashing and solution methods have been satisfactory. However, a recent report by Hutner (1972) notes that wet-ashing methods are inadequate for tin; Schroeder *et al.* (1964) state that drying tissues at 110° C and subsequent ashing at 450° C may incur losses, especially from the organotin compounds.

Jeltes (1969) analyzed air samples for tin by atomic absorption spectroscopy. The air was filtered through glass-fiber filters, which were then extracted with methylisobutyl ketone. He reports a detection limit of 0.1 mg of tin per cubic meter of air. Maienthal and Taylor (1968) achieved a sensitivity of 50 ppm tin in water using polarography. Kahn (1968) used an atomic absorption spectrograph with an air–hydrogen flame for the determination of tin in water and reached a detection limit of 0.02 ppm.

Atomic absorption has a sensitivity of about 1 ppm for 1 percent absorption and a detection limit of 0.1 ppm in a water solution using an air–hydrogen flame, according to Capacho-Delgado and Manning (1966). The use of an argon–hydrogen or air–hydrogen flame increases the sensitivity of this method for tin by about two and a half times over that achieved in an acetylene–air flame. However, the atomic absorption method, particularly with the air–hydrogen flame, is subject to a large number of interferences (Harrison and Juliano, 1969; Juliano and Harrison, 1970). Interferences that depress absorption are: sulfuric acid (H_2SO_4); phosphoric acid (H_3PO_4) if above a 5 percent concentration; lithium; sodium; and aluminum. Most organic solvents almost obliterate tin absorption, even when present at levels of less than 10 percent of the sample by volume (Harrison and Juliano, 1969). Absorption is enhanced by the presence of cesium, rubidium, potassium, strontium, magnesium, calcium, cobalt, copper, and titanium (Juliano and Harrison, 1970).

D. B. Milne and his associates (personal communication, 1974) state that dry ashing is inadequate for the determination of tin, particularly in the case of most organotin compounds. Thus, tin chloride ($SnCl_4$)—volatile at 114° C—may be formed in the ashing process, and SnO_2, which is relatively insoluble and is nonreactive toward color reagents, may also be a product. In experiments with dry ashing, only 20 percent of the $SnCl_4$ and none of the organic tin as tryphenyltin were recovered. Using a wet-ashing technique with a nitric–sulfuric acid digestion (Thompson and McClellan, 1962), nearly 100 percent of the inorganic tin and about 80 percent of the organotin compounds were recovered.

Perchloric acid in the digestion mixture could form $SnCl_4$ (Gorsuch, 1970). Probably any chloride in the sample in significant amounts could lead to some losses of the tin as $SnCl_4$. Sandell (1959) describes a procedure in which the tin is converted to the chloride and distilled, but this is time consuming and requires the use of specialized glassware and equipment. The colorimetric method described by Thompson and McClellan (1962) seems to be the most satisfactory for tin in biological samples, and, in one experiment, as little as 0.2 μg of tin in a 3-g sample (or about 0.067 ppm) was detected. This procedure, however, has numerous sources for error, including a pH-sensitive color development and losses during the extraction process. Obviously, a critical aspect of health research in tin is development of suitable analytical methods.

PLANTS

Content and Accumulation

Tin is not known to be essential for plants. Peterson (1971) reported that generally concentrations of tin were higher in lichens than in mosses. Lounamaa (1956) also noted the higher accumulation of tin in lichens. Peterson (1971) reported 20 ppm of tin in the ash of one sample of an herb, *Silene cucubates* (a *Caryophyllaceae*). Sarosiek and Klys (1962) reported the average tin concentration of a number of native plants in the Sudeten mountains of Czechoslovakia on peat soils to be 46 ppm in the ash and roughly 5–10 ppm in the moisture-free tissue. Bardyuk and Ivashov (1969) reported a threefold to tenfold higher concentration of tin in plants growing over tin-ore deposits than those growing in soils outside the deposits. Sedge and mosses were found to be the best accumulators. Bowen (1966) has reviewed much of the early work on tin in higher plants, marine organisms, and animal tissue.

Dobrovol'skii (1963) states that the ratio of the concentration of tin in the plant to that in the soil (a coefficient of biological accumulation) is greater than 1 and that tin is severalfold more concentrated in plant ash than in the topmost layers of the soil. Glazovskaya (1964), however, believes that the coefficient is close to 1.

Sainsbury *et al.* (1968) found that the concentration ratio based on ash of bulk tundra vegetation is above 1 where the tin content of the soil is below about 100 ppm, but the ratio is below 1 where the tin content of the soil is above about 150 ppm. Porutskii *et al.* (1962) report more tin in the plastids than in the sap of apple and pear trees and potato vines. Paribok and Kuznetsova (1963) report a higher accumulation of tin in the roots of young bean, barley, and tomato plants than in the aboveground plants.

Curtin *et al.* (1974) found 23–80 ppm tin in the ashed residue of vapor transpired from several coniferous trees.

The twig ash was also high, 6–40 ppm. Very little tin was found in the needles or the A- and B-horizons of the underlying soil.

Foods

There appears to be little recent work on tin in unprocessed foods. Zook *et al.* (1970) determined the tin contents of wheats and wheat products. Although there were differences in concentration, they seemed largely unrelated to the geographical origin of the samples. Monier-Williams (1949) found that the tin concentration in canned foods and drinks is usually less than 100 ppm, although higher levels may be present in certain products after prolonged storage in closed, nonlacquered cans, or after some days of storage in open cans.

Schroeder and co-workers (1964) analyzed a number of natural foods, many of them from a garden soil reported to have an appreciable concentration of tin. In vegetables, they found tin contents that ranged from below the detectable limit to 8.5 ppm on a fresh-weight basis, or, roughly, up to 40 ppm in the dry material. In general, grains contained less tin than vegetables. Higher levels were found in foods preserved in tin cans, but there was no direct comparison with fresh foods from the same sources.

Aerial Contamination

There may be a danger of contamination from the use of the organotin fungicides such as triphenyltinacetate, although Klimmer (1968) reported that experiments with radioactive materials show no systemic effect in plants sprayed with triphenyltinacetate, and Brueggmann *et al.* (1964) had earlier reported no damage to animals from eating beet leaves also sprayed with this material. Dust sediments from industrial regions may contain from 10 to 10,000 ppm of tin, according to Morik and Morlin (1959). The possibility of obtaining high toxic levels of tin from leafy vegetables, but not necessarily from roots, seems likely although it has received little attention.

ANIMALS AND MAN

Essentiality and Availability

Trace amounts of tin are widely distributed in biological tissues and nutrients, but the element has generally been considered an "environmental contaminant" instead of an essential nutrient (Underwood, 1962). A recent review (Schroeder *et al.*, 1964) on tin in man and foods, for instance, treated tin as an abnormal trace metal and concluded that "measurable tin is not necessary for life or health." This conclusion was based mainly on the fact that, with the prevailing inadequate methods of analysis, "zero" levels of tin were found in the newborn and in organs of natives of some foreign countries. However, Schwarz *et al.* (1970) found that various tin compounds exert a significant effect on the growth of rats if trace

element contamination from the environment is rigidly excluded.

In rats maintained on purified amino acid diets in trace element controlled isolators, trimethyltin hydroxide, dibutyltin maleate, stannic sulfate, and potassium stannate were found to enhance growth at dose levels supplying 1 ppm of tin to the diet. When supplied as stannic sulfate, 0.5, 1, and 2 ppm of tin in the diet increased growth in rats by 24, 53, and 59 percent, respectively (Schwarz *et al.*, 1970). Tin as stannic sulfate or in the form of many trialkyltin derivatives of the type R_3SnX (R denotes any organic side chain in which tin is covalently linked to a carbon atom, and X is an anionic group such as Cl^-, OH^-, and acetate) also improved growth rates and significantly increased pigmentation of rat incisors (Milne *et al.*, 1972). From the above data, the requirement for, or optimum concentration of, tin in the rat diet would be between 1 and 2 ppm. If the requirement for a 100-g rat is approximately 10 μg/day, one could reasonably assume that a 75-kg man would require approximately 7.5 mg/day. Schroeder *et al.* (1964) found the average composition of an institutional diet to contain 1.41 ppm tin on a wet basis. They calculated that a normal human intake of tin from food might range from 1 to about 40 mg/day.

Values relating to balance studies in man vary greatly. Kehoe *et al.* (1940) found an average of 18 ppm tin in 30 urine samples from the United States and an average of 9 ppm in 30 urine samples from Mexico. They measured the content of tin in the food, fluids, and feces of a normal male from the United States for 28 days, with the following results (mean/day ± SEM [Standard Error of the Mean]):

Intake, mg	Output, mg	
Food and Beverages 17.14 ± 1.9	Feces	22.88 ± 1.9
	Urine	0.14 ± 0.001

Kent and McCance (1941), on the other hand, found that with a dietary intake of tin of 14.4 mg/day for 7 days, their subject excreted 7.2 mg/day in the urine and 6.6 mg/day in the feces. Almost all of an injected dose of tin was excreted via the kidney. Twenty-three years later, Schroeder *et al.* (1964) estimated that the average daily human intake of tin was slightly over 4 mg/day, with the bulk being excreted in the feces. The discrepancies in these results are illustrative of the inconsistencies found in the tin literature.

Tipton *et al.* (1966) reported that, in a 30-day study of food intake, about 1.5–2.5 mg of tin was ingested per day, and most of this was excreted. A later work (Tipton *et al.*, 1969) reported a daily intake in the United States ranging from 0.10 to 100 mg of tin with an average of 5.8 mg per day. The wide variation of the daily intake of tin probably arises because most of this element enters the diet from tin-coated cans and utensils.

Organotin compounds, particularly of the type R_3SnX, have been used widely over the past 20 yr as fungicides, bactericides, and insecticides (Barnes and Stoner, 1959; Poller, 1970). At tin levels of 1–2 ppm, many of these compounds showed an effect on growth in rats (Milne and Schwarz, unpublished observations). Trimethyl-, ethyl-, propyl-, butyl-, phenyl-, and, benzyltin chloride were quantitatively compared to stannic sulfate as a standard. All compounds of this type elicited positive growth responses in the isolated rats, but the ethyl and propyl derivatives were distinctly less effective than the others. An inverse relation appears to exist between the growth effect of tin compounds and their antimicrobial potential, indicating that the growth effect in rats is dissimilar to effects on microorganisms. Also, inorganic tin salts that have no antimicrobial effect were most effective in supplying the tin requirement of the rat.

The chemical nature of tin found in nutrients and tissues is unknown. However, much of the tin may be present in the form of low-molecular-weight compounds because it can be stored in fats (Schroeder *et al.*, 1964).

TOXICITY

Most of the information on the biological action of tin compounds has been from the toxicological point of view, since many of these compounds (particularly of the type R_3SnX) are useful as fungicides, bactericides, insecticides, and for their antihelminthic (antiintestinal worm) activity. The toxicities of organotin compounds have been reviewed thoroughly by Barnes and Stoner (1959), and more recently by Poller (1970). A striking feature is that, unlike the toxicity of lead, mercury, or arsenic, tin toxicity is manifest only when the tin is part of certain organic (organotin) compounds, such as the trialkyltin salts (Barnes and Stoner, 1959). Inorganic tin is relatively nontoxic. De Groot *et al.* (1973) noted that stannous chloride, orthophosphate, sulfate, oxalate, or tartrate had no toxic effect on rats fed these compounds over a 13-week period at levels of 450–650 ppm tin in the diet. Stannous or stannic oxides, and stannous sulfide and oleate, had no effect on rats at three times that level. There is little or no evidence that tin contributed by canned foods has any significant biological effect.

CONCLUSIONS AND RECOMMENDATIONS

1. Most of the available data on the distribution of tin in soils, plants, and animal tissues are scanty or questionable because of analytical procedures that are inadequate, both from the point of view of recoveries and of the sensitivity of the method. Before any conclusive study on the distribution of tin in biological systems or in the geochemical environment can be made, a more sensitive, reliable, rapid, and standardized method of tin analysis needs to be developed.

2. At present we have no evidence that tin is either beneficial or detrimental to plants. Investigations of these matters do not appear to merit a high priority at this time.

3. Before tin can be conclusively considered as an essential trace element, effects should be shown in other species besides the rat. Careful experiments involving several generations of animals may be needed.

4. Tin deficiencies or tin toxicities of natural origin are unknown in man and animals (except the rat), but these possibilities should not be ignored.

REFERENCES

Bardyuk, V. V., and P. V. Ivashov. 1969. The accumulation of trace elements in plants on a tin ore deposit in the southern part of the Soviet Far East. Tr. Buryat. Inst. Estestv. Nauk Buryat. Fil. Sib. Otd. Akad. Nauk SSSR 2:83–93; Chem. Abstr. 75:150847 (1971).

Barnes, J. M., and H. B. Stoner. 1959. The toxicology of tin compounds. Pharmacol. Rev. 11:211.

Bowen, H. J. M. 1966. Trace elements in biochemistry. Academic Press, New York. 241 pp.

Brueggmann, J., O. R. Klimmer, and K. H. Niesar. 1964. Residues of triphenyl acetate of tin in plants and animals and their importance in hygiene and toxicology. Zentralbl. Veterinaermed. Reihe A 11(1):40–48.

Capacho-Delgado, L., and D. C. Manning. 1966. Determination of tin by atomic absorption spectroscopy. Spectrochim. Acta 22:1505.

Curtin, G. C., H. D. King, and E. L. Mosier. 1974. Movement of elements into the atmosphere from coniferous trees in subalpine forests of Colorado and Idaho. J. Geochem. Explor. 3:245–263.

De Groot, A. P., V. J. Feron, and H. P. Til. 1973. Short-term toxicity studies on some salts and oxides of tin in rats. Food Cosmet. Toxicol. 11:19–30.

Dobrovol'skii, V. V. 1963. Distribution of trace elements between the soil forming ground layer, soil, and vegetation under conditions of the Moscow region. Nauchn. Dokl. Vyssh. Shk. Biol. Nauki (3): 193–198.

Glazovskaya, M. A. 1964. Biological cycle of elements in various landscape zones of the Urals. Fiz. Khim. Biol. Mineralog. Pochv. SSSR: 148–157 (in Russian); Chem. Abstr. 62:4562 (1965).

Gordon, M. 1953. Trace elements in peats. Torfnachrichten 3:12. Chem. Abstr. 47:4533 (1953).

Gorsuch, T. T. 1970. The destruction of organic matter. Pergamon Press, Oxford. pp. 92–93.

Hamaguchi, H., and R. Kuroda. 1969. Biogeochemistry. *In* Handbook of geochemistry. K. H. Wedepohl [ed.]. Vol. II:1–5. Springer–Verlag, New York–Berlin–Heidelberg.

Harrison, W. W., and P.O. Juliano. 1969. Effects of organic solvents on tin absorbance in an air–hydrogen flame. Anal. Chem. 41:1016.

Hutner, S. H. 1972. Inorganic nutrition. Annu. Rev. Microbiol. 26:313–346.

Jeltes, R. 1969. Determination of bistributyltin oxide in air by atomic absorption spectroscopy or pyrolysis gas chromatography. Ann. Occup. Hyg. 12(4):203–207; Chem. Abstr. 72: 35456 (1970).

Juliano, P. O., and W. W. Harrison. 1970. Atomic absorption interferences of tin. Anal. Chem. 42:84.

Kahn, Herbert L. 1968. Principles and practices of atomic absorption. *In* Trace inorganics in water. Adv. Chem. Ser. 73:183–229. American Chemical Society, Washington, D.C.

Kehoe, R. A., J. Cholak, and R. V. Story. 1940. A spectrochemical study of the normal ranges of concentration of certain trace metals in biological materials. J. Nutr. 19:579.

Kent, N. L., and R. A. McCance. 1941. The absorption and excretion of "minor" elements by man. 2. Cobalt, nickel, tin, and manganese. Biochem. J. 35:877.

Klimmer, O. R. 1968. Toxicological viewpoint on the application of organotin fungicides in agriculture. Pflanzenschutzberichte 37(4/5/6):57–66.

Lounamaa, J. 1956. Trace elements in plants growing wild on different rocks in Finland. A semiquantitative spectrographic survey. Ann. Bot. Soc. Zool. Bot. Fennicae Vanamo 29:4; Chem. Abstr. 53:5418.

Maienthal, E. J., and J. K. Taylor. 1968. Polarographic methods in determination of trace inorganics in water. *In* Trace inorganics in water. Adv. Chem. Ser. 73:172–182. American Chemical Society, Washington, D.C.

Milne, D. B., K. Schwarz, and R. F. Sognnaes. 1972. Effect of newer essential trace elements on rat incisors pigmentation. Fed. Proc. 31:700.

Monier-Williams, G. W. 1949. Trace elements in foods. Chapman-Hall, Ltd., London. 511 pp.

Morik, J., and Z. Morlin. 1959. Pollution of the air of industrial regions by metals. Nepegerzsegugy 40:288–293; Chem. Abstr. 57:8840 (1962).

Paribok, T. A., and G. N. Kuznetsova. 1963. Effect of soil temperature on the absorption and distribution of trace elements in plants. Tr. Bot. Inst. Akad. Nauk SSSR, Ser. 4, Eksperim. Botan. 16:27–48.

Peterson, P. J. 1971. Unusual accumulations of elements by plants and animals. Sci. Prog. (Oxford) 59:505–526.

Pinta, M., and C. Oliat. 1961. Physicochemical research on trace elements in tropical soils. 1. Some soils in Dahomey. Geochim. Cosmochim. Acta 25:14; Chem. Abstr. 55:23896b.

Poller, R. C. 1970. The chemistry of organotin compounds. Academic Press, New York. pp. 271.

Porutskii, G. E., V. P. Golovchenko, and S. V. Cherednichenko.

1962. Content of trace elements in various plant organs. Dokl. Akad. Nauk. SSSR 146:1223–1226.

Sainsbury, C. L., J. C. Hamilton, and C. Huffman, Jr. 1968. Geochemical cycle of selected trace elements in the tin–tungsten–beryllium district, western Seward Peninsula, Alaska—A reconnaissance study. U.S. Geol. Surv. Bull. 1242-F., U.S. Government Printing Office, Washington, D.C. 42 pp.

Sandell, E. B. 1959. Colorimetric determination of traces of metals, Vol. III, 3d ed. Wiley-Interscience, New York. pp. 852–867.

Sarosiek, J., and B. Klys. 1962. Observations on the tin content of the plants and soils of the Sudetes. Acta Soc. Bot. Pol. 31(4):737–752.

Schroeder, H. A., J. J. Balassa, and I. H. Tipton. 1964. Abnormal trace elements in man: Tin. J. Chron. Dis. 17:483–502.

Schwarz, K., D. B. Milne, and E. Vinyard. 1970. Growth effect of tin compounds in rats maintained in a trace element controlled environment. Biochem. Biophys. Res. Commun. 40:22.

Thompson, M. H., and G. McClellan. 1962. The determination of microgram quantities of tin in foods. J. Assoc. Off. Anal. Chem. 45:979.

Tipton, I. H., P. L. Stewart, and J. Dickson. 1969. Patterns of elemental excretion in long-term balance studies. Health Phys. 16:455-462.

Tipton, I. H., P. L. Stewart, and P. V. Martin. 1966. Trace elements in diets and excreta. Health Phys. 12:1683.

Underwood, E. J. 1962. Trace elements in human and animal nutrition, 2d ed. Academic Press, New York. 429 pp.

Wedepohl, K. H. [ed.] 1969. Handbook of geochemistry. Springer–Verlag, New York–Berlin–Heidelburg.

Zook, E. G., F. E. Greene, and E. R. Morris. 1970. Nutrient composition of selected wheats and wheat products. VI. Distribution of manganese, copper, nickel, zinc, magnesium, lead, tin, cadmium, chromium, and selenium, as determined by atomic absorption spectroscopy and colorimetry. Cereal Chem. 47(6):720–731.

IX

Vanadium

LEON L. HOPKINS, JR., *Chairman*

Helen L. Cannon, Alfred T. Miesch, Ross M. Welch, Forrest H. Nielsen

Vanadium could possibly affect human health in two ways. First, because vanadium has been shown to be essential for laboratory animals, it is presumably essential for man, and thus inadequate dietary intake may affect human health. Second, and more unlikely, toxic effects from the intake of relatively high levels of vanadium from the diet and air may occur.

Weathering of rocks and localized pollution provide a continuous source of vanadium to the soil, but for the most part, soil vanadium is bound by organic and clay particles. Except for a few vanadium accumulator plants, the small amount of vanadium taken up by plants from the soil is largely retained in the roots and only smaller amounts reach the aerial parts. Except for a few root crops, feed and foods come from the aerial portions of plants. Overall, man ingests relatively small amounts of vanadium from food, low-to-moderate amounts from drinking water, and, except from localized industrial exposure and the burning of fossil fuel (oil and coal), little that is airborne.

Vanadium is not highly toxic when taken orally by mammals, because it is poorly absorbed, rapidly excreted, and not excessively accumulated. Vanadium toxicity in humans has been reported only rarely, and generally as a result of industrial exposure to high concentrations of airborne vanadium. Vanadium appears to be neither a mutagen nor a carcinogen. Indeed, vanadium transport through the food chain appears to benefit rather than harm man because of the essentiality of vanadium in nutrition.

Of greater concern than toxicity is the possible effect of borderline vanadium deficiency in man. It has been estimated (but not shown) that the dietary vanadium requirement is around 100 ppb for laboratory animals consuming a purified diet and probably higher when they are consuming a natural diet. Limited data indicate that most food levels appear to be below this level, although water may contribute minor amounts of highly available vanadium. Food processing and refinement further reduce the amounts of available vanadium. If one can extrapolate animal data to man, the observation of elevated blood–lipid levels in vanadium-deficient animals makes one wonder if marginal vanadium deficiency occurs in man and if it is in part responsible for the increased serum–lipid concentrations that occur in some people. For a more extensive review, see Chapter 7 in *Vanadium*, Report of the Panel on Vanadium (Committee on Biologic Effects of Atmospheric Pollutants, 1974).

GEOCHEMISTRY, DISTRIBUTION, AND AVAILABILITY

Rocks

To obtain some estimate of the variation present in the major rock types of the near-surface environment, the computer-based Rock Analysis Storage System (RASS) file of the U.S. Geological Survey was searched and the retrieved data were passed through several programs to provide some general descriptors of the frequency distributions (Table 23). The RASS file contains modern analytical data (mostly 1968 or later) on a large number of rock, soil, and plant samples collected from all parts of the United States by numerous Survey investigators in a wide range of research programs, and analyzed in the Survey laboratories. No samples containing a substantial amount of vanadium were included.

Vanadium in igneous rocks is found largely in titaniferous magnetites (titanium-bearing Fe_3O_4) and ilmenite ($FeTiO_3$). The tendency for vanadium to accumulate in these minerals and in ferromagnesian minerals accounts for its abundance in the basic, as compared to the silicic, igneous rocks.

Vanadium is present in igneous rocks largely as V^{3+} and V^{4+}. It is oxidized by weathering to V^{5+}, which forms a great number of soluble complexes. It is largely in this form that vanadium is dispersed into the surface environment. The dominant factor controlling the sedimentary deposition of vanadium appears to be the presence of organic matter, although it can also be precipitated in aluminum and iron hydroxides. The importance of organic matter in the reduction and precipitation of vanadium is emphasized by an observation that all known sandstone-type vanadium ore deposits are in rocks of late Paleozoic age or younger and that this coincides with the evolutionary development of land plants (Fischer, 1973). The association of vanadium with organic matter accounts for its relatively high concentrations in black shales of late Cretaceous age, in phosphate deposits, and in many of the sandstone-type uranium–vanadium ores of western Colorado and southeastern Utah.

Tourtelot *et al.* (1960) have shown that vanadium tends to be associated with organic carbon in shales and claystones of the Pierre Shale of late Cretaceous age in the northern Great Plains. Shales and claystones with more than 1 percent organic matter average 400 ppm vanadium; whereas, those with less carbon than this, average less than 200 ppm vanadium.

Vanadium also occurs in relatively high concentrations in fine-grained sedimentary rocks that are older than late Paleozoic and poor in organic matter. This type of occurrence probably results from its accumulation in the clays that form by weathering of other minerals without being converted to a soluble form. Carbonates (limestones and dolomites) and coarse-grained sedimentary rocks (sandstones and conglomerates) usually contain about 20–35 ppm vanadium, respectively; the finer-grained sedimentary rocks (shales and clays) usually contain somewhat in excess of 100 ppm (Table 23).

Vine and Tourtelot (1969) summarized a great deal of data on vanadium and other elements in black shales and reported a median vanadium content of 1,000 ppm.

Ores

Although 75 percent of the world production of vanadium is from titaniferous magnetite deposits, about 90 percent of the U.S. production is from deposits formed at relatively low temperatures in sandstones (Fischer, 1973). The principal titaniferous magnetite deposits are in Australia, Canada, South Africa, and the USSR. The sandstone-type deposits in the United States are chiefly in western Colorado and southeastern Utah, where the ore generally occurs within several hundred feet of the surface and averages about 7,000 ppm vanadium in the more vanadiferous deposits. These deposits are in sandstones of Jurassic age in the eastern part of the uranium–vanadium mining area identified on Figure 15; the less vanadiferous deposits are in sandstones of Triassic age, averaging about 600 ppm vanadium. Vanadium accumulation in the sandstone-type deposits is commonly believed to have been by reduction and precipitation from groundwater solutions through the action of organic debris (leaf imprints and carbonized fossil logs and leaves are abundant in these deposits). The geologic source of the vanadium is unknown.

Vanadium is also present in relatively high concentrations in certain carbonaceous beds of the Phosphoria Formation in Idaho, Wyoming, Utah, Montana, and Nevada; it is recovered as a by-product of phosphate

TABLE 23 Median and Percentile Concentration of Vanadium (in ppm) in Some Broad Categories of Rock Type

Percentile	Basic and Ultrabasic (Mafic) Igneous Rocks	Intermediate and Silicic (Felsic) Igneous Rocks	Coarse-Grained Sedimentary Rocks	Fine-Grained Sedimentary Rocks	Limestones and Dolomites
10th	27	—	7	47	—
20th	54	—	13	70	5
50th (median)	150	24	35	130	21
80th	300	82	78	260	61
90th	380	140	140	580	120
SAMPLES	2,498	3,219	1,785	2,166	1,177

SOURCES: The estimates in this table are based on retrievals of data from the computer-based Rock Analysis Storage System (RASS) file of the U.S. Geological Survey; the samples represented are from widely scattered localities throughout the United States. The medians given here agreed closely with average values published by Turekian and Wedepohl (1961), Rankama and Sahama (1950), and Vinogradov (1959).

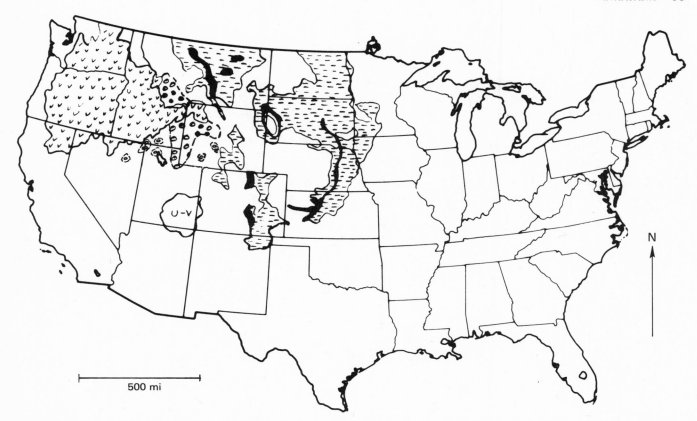

FIGURE 15 Broad regions of the conterminous United States that are underlain by some rock units that generally contain more than 100 ppm vanadium.

 Volcanic rocks of the Columbia Plateau–Snake River Plain region.

Black shales of late Cretaceous age. Dark-colored areas represent parts that are more vanadium-rich, as identified by H. A. Tourtelot (personal communication, 1973).

Region of phosphate deposits of the Phosphoria, Park City, and Embar formations (after V. E. McKelvey, 1949; H. L. Cannon, personal communication, 1976).

Principal region of sandstone-type uranium–vanadium deposits of the Colorado Plateau. Deposits in the eastern part of this region are distinctly more vanadiferous than those in the western part.

mining. Jaffé (1961) reported that the phosphate rock contains 70–4,000 ppm vanadium and cited vanadium pentoxide reserve estimates of 550,000 tons. Seventeen analyses of phosphate rock from the western conterminous United States and Alaska, stored in the RASS system, show 20–700 ppm vanadium, with a median of 150 ppm.

Studies of all the major coal provinces of the United States (Zubovic *et al.*, 1961, 1964, 1967; Zubovic, 1966) showed that the raw coals average less than 50 ppm vanadium and seldom exceed 100 ppm. Of 197 coal samples represented in the RASS system, 193 show less than 1,000 ppm vanadium in the coal ash and, presumably, less than about one-tenth of this amount in the whole coal.

Other ores in the United States in which vanadium is present in relatively high concentrations are restricted in geographic occurrence. These ores include the sedimen

tary iron ores of the Lake Superior region, which contain 100–1,000 ppm vanadium (Fischer, 1959); some copper, lead, and zinc deposits of the Southwest; a gold–quartz vein in Colorado averaging about 10,000 ppm vanadium (Lovering and Goddard, 1950); and a deposit in altered argillic rock in Arkansas that averages about 5,000 ppm vanadium according to Fischer (1973). There has been some commercial production of vanadium from each of these sources.

An extensive search for a number of metals, including vanadium, in other types of ore deposits and smelter products was made by the U.S. Geological Survey in the 1940's at the request of the War Production Board. The analytical results, given by Kaiser *et al.* (1954), have been studied by Fischer (1959). They show that vanadium concentrations in most ores, mill heads, tailings, ore con centrates, flue dusts, and slags are less than 50 ppm. Of 775 samples of these materials that were analyzed, 36 contained vanadium in the range of 500–3,000 ppm; most of these higher concentrations were found in association with manganese, chromium, gold, and copper, but the general indication of the study was that vanadium does not tend to be a prominent constituent of most types of ore or smelter products.

Geographic Distribution

Some general observations on what might be called the first-order characteristics of the distribution of vanadium could be useful in broad-scale epidemiology. Most of the central interior of the United States (the Midwest) and the central Great Plains, as well as the intermountain portions of the West, are underlain by sedimentary rocks or unconsolidated sedimentary deposits. Except where shales are abundant, the vanadium in these deposits ranges, for the most part, from a few to 200–300 ppm. The concentrations in the shales range up to about 600 ppm. Values this high may be common in certain shales of late Cretaceous age in the northern Great Plains region (Figure 15), with even higher vanadium content in black shales as noted by Vine and Tourtelot (1969).

Another broad region of the United States underlain by strata that are relatively rich in vanadium includes most of Oregon, Washington, Idaho, and nearby areas. Strata here consist of volcanic rocks, mostly basalts. Judging from the concentrations of vanadium found in mafic igneous rocks, the vanadium concentrations in the basalts may range above 400 ppm. The relatively high concentrations of vanadium in the soils of this part of the country (Shacklette et al., 1971) probably reflect this occurrence. Another area underlain by volcanic rocks rich in vanadium is the entire State of Hawaii.

The largest natural accumulations of vanadium in the United States are concentrated in ore bodies that occur in sandstones within several hundred feet of the surface in western Colorado and southeastern Utah (Figure 15). Vanadium concentrations in these rocks range up to several percent.

The vanadiferous areas of the conterminous United States shown on Figure 15 include only those of broad geographic extent: some smaller areas surrounding known or unknown vanadium-bearing mineral deposits are higher in vanadium content.

Soils

Content The vanadium content of a soil is dependent on the rocks and minerals from which the soil parent materials were derived and the action of soil-forming factors (physical, chemical, and biological) to which the parent materials were subjected during soil formation and development (Mitchell, 1964, 1971). The more mature and fully developed a soil is, the less direct is the effect parent rock has on its vanadium content.

Swaine (1955) lists the normal range of total soil vana-

dium concentrations as 20–500 ppm. Mitchell (1964) reported a range of 20–1,000 ppm vanadium in various soils of Scotland. The average vanadium content of soils worldwide has been reported to be approximately 100 ppm (Swaine, 1955; Cannon, 1963; Mitchell, 1964; Bowen, 1966). Shacklette et al. (1971) report the total range of vanadium concentrations in 862 samples from throughout the conterminous United States to be from less than 7 to about 500 ppm. The approximate percentile concentrations of vanadium in soil (in parts per million), interpolated from their histogram, are tenth percentile, 20; twentieth percentile, 30; fiftieth percentile (median), 60; eightieth percentile, 100; and ninetieth percentile, 130. The median concentration of vanadium in the surficial materials (soils and regoliths) of the United States, therefore, is found to be intermediate between mafic (150) and felsic (24) igneous rocks from which the vanadium ultimately was derived and is closer to the median for the felsic rocks, which are the more abundant of the two broad types.

Figure 16 is a map of soil vanadium in the conterminous United States (from Shacklette et al., 1971). It shows that soils and other surficial materials in Florida, which were derived largely from underlying carbonate rocks, are vanadium-poor.

Mitchell (1964, 1971) has stated that vanadium is usually highest in soils developed on olivine gabbro and has listed the total vanadium contents in 10 Scottish soils developed on various parent materials derived from different rock types as follows: serpentine, 100 ppm vanadium; olivine gabbro, 200 ppm; andesite, 100 ppm; trachyte, 60 ppm; granite, 20 ppm; granite gneiss, 250 ppm; quartz mica schist, 200 ppm; shale, 200 ppm; sandstone, 60 ppm; and quartzite, 250 ppm. Cannon (1963) has reported vanadium concentrations in soils overlying uranium–vanadium deposits to be as high as 1,500 ppm.

Exceptionally high concentrations of vanadium (430 ppm) were also reported for the anomalous selenium-rich alkaline soils from certain regions in Kansas (Goldschmidt, 1954). Vanadium does not necessarily occur in selenium soils. Uranium, vanadium, and selenium occur together in uranium ores, but high-selenium ash falls embedded in shales are not always high in uranium or vanadium.

Availability to Plants

Vanadium availability to plants may be best measured by its extractable (soluble plus exchangeable) content in soils. However, the content of extractable vanadium in different soil types depends on a number of complex factors (Mitchell, 1964, 1971). These include soil parent, soil physical factors (e.g., structure, texture, aeration), soil chemical factors (e.g., oxidation–reduction potential, pH, organic matter content, amount of hydrous manganese and iron oxides), and soil biological factors (e.g., microbial activity and plant root exudates). Mitchell (1964, 1971) has reported ranges in Scottish soils of from 0.1 to 1.0

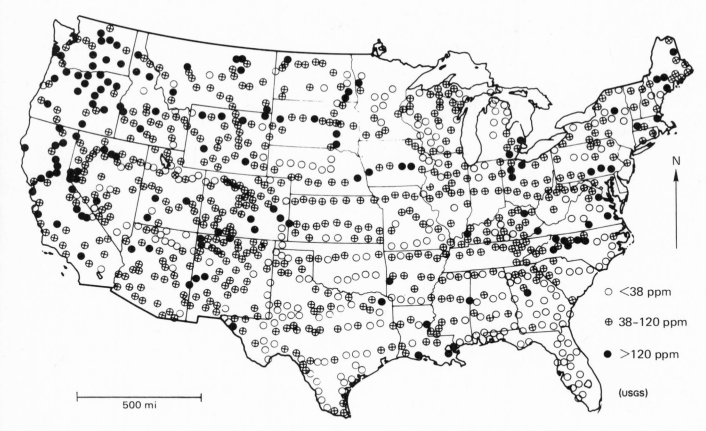

FIGURE 16   Vanadium in soils and other surficial materials of the conterminous United States (modified from Shacklette *et al.*, 1971).

ppm vanadium extractable with 2.5 percent acetic acid solutions (pH 2.5) and from 0.2 to 5 ppm vanadium extractable with EDTA (ethylenediaminetetraacetic acid). Swaine (1955) reported soil levels of extractable vanadium up to 0.5 ppm with 0.5 N acetic acid and less than 0.2 ppm using neutral 1 N ammonium acetate. Acetic acid-extractable vanadium levels in most soils are less than 0.1 ppm. Impeded soil drainage and accompanying anaerobic conditions can result in large increases in extractable vanadium from soils (from 0.08 to 4.3 ppm vanadium after flooding) (Mitchell *et al.*, 1957; Mitchell, 1964; Ng and Bloomfield, 1962). Apparently, increasing the organic-matter content of waterlogged soils also increased the extractable (either water or acetic acid extraction) level of vanadium in those soils. Mitchell (1964) concluded that vanadium may be associated with hydrous iron oxides in soils and thus can be mobilized on anaerobic incubation with organic matter. Reoxidation of hydrous oxide coatings on soil particles of those soils may immobilize some of the vanadium released during anaerobic incubation (Ng and Bloomfield, 1962). There appears to be a fairly good correlation between total and extractable vanadium in soils that are freely drained, but not in those that are poorly drained (Mitchell, 1964).

Not only do the total and extractable vanadium levels

vary greatly among the different soil types, but large differences in total and extractable vanadium concentrations are found in various soil fractions (Swaine and Mitchell, 1960; Mitchell, 1964, 1971). In general, sand particles contain most of the vanadium (usually greater than 60 percent) in the soil, except in poorly drained soils. Clays from poorly drained soils may have up to five times the total vanadium content of clays from freely drained soils (Mitchell, 1964).

During the course of soil development, vanadium may be released from rocks and minerals as a complex anion (Goldschmidt, 1954) and either precipitated with calcium (or other polyvalent cations) or lost in drainage waters. The vanadium anion may also be bound to organic complexes, adsorbed on anion-exchange materials in soil (Mitchell, 1964), or coprecipitated with or adsorbed by sesquioxide surface coatings on soil particles [chiefly hydrous, amorphous ferric oxide (Fe_2O_3) or goethite]. In precipitation tests, vanadium was almost completely coprecipitated with Fe_2O_3. Extraction of a soil with ammonium oxalate solutions readily releases trace elements bound to amorphous Fe_2O_3 and organic matter, but not from Al_2O_3. Ammonium oxalate extractions of a soil released 40 percent of its total vanadium content; over 600 times as much vanadium was extracted from a soil with ammonium oxalate as was extracted with acetic acid. The occurrence of vanadium in sesquioxides has been reviewed by Mitchell (1964).

Under reducing soil conditions, trivalent or tetravalent vanadium may be chelated to organic matter complexes; the trivalent form may also be precipitated as sulfides (Mitchell, 1964). Depletion of vanadium from surface horizons has been found for various soil types throughout the world (Mitchell, 1964). Table 24 shows the total and acetic acid (2.5 percent) extractable vanadium content at different depths of a peaty podzol with a thin iron pan. Apparently, vanadium was mobilized in surface horizons, transported, and then accumulated in the A_1A_2 horizon.

Mitchell (1964) has summarized the range of vanadium concentrations in commonly used fertilizers. Several contained less than 1 ppm, but rock phosphate, superphosphate, and basic slag contained 10–1,000 ppm, 50–2,000 ppm, and 1,000–5,000 ppm, respectively. Thus, these fertilizers may be a problem if the vanadium in them is available to plants because of the fairly high phytotoxicity of vanadium (Pratt, 1966). The forms in which soil vanadium is available to plants are not known. Mitchell (1964) has suggested that vanadium chelates may be the form most readily taken up by plants. Welch (1973) has studied the uptake of vanadium by roots from radioactively labeled (^{48}V) solutions. The accumulation of vanadium by excised barley roots was shown to be a passive process requiring no metabolic energy; a 0.2 μM (micromolar) concentration of Ca^{2+} ions was found to be required for maximum rates of uptake. The rate of uptake was also dependent on the form of vanadium in solution that, in turn, was controlled by pH level and the oxidation–reduction potential of the solution. Within the normal pH range found in most freely drained soils (pH's from 5 to 8), the metavanadate ion (VO_3^-) was the predominant ionic species taken up. At high pH levels (where the orthovanadate ions are formed), rates of absorption dropped rapidly, with virtually no vanadium taken up between pH 9 and 10. At pH levels below 4, the rate of vanadium absorption increased rapidly, corresponding to the rise in solution concentrations of the dioxovanadium ion (VO_2^+). These results suggest that the VO_2^+ species may be taken up more rapidly than the VO_3^- ionic species. It appears that under normal conditions, the form of vanadium taken up by roots from soil solutions is either the VO_3^- or VO_2^+ ionic species. Under reducing conditions and acid pH's, the tetravalent oxovanadium species (VO^{2+}) may also be taken up. Both VO_2^+ and VO^{2+} ions are capable of being chelated and could contribute substantially to vanadium uptake by plants. Vanadium uptake by roots from solution was not appreciably affected by the addition of various other cations and anions in the culture (i.e., Na^+, NH_4^+, HPO_4^{2-}, $HAsO_4^{2-}$, MoO_4^{2-}, SeO_4^{2-}, CrO_4^{2-}, Cl^-, NO_3^-, BO_3^{3-}, and SeO_3^{2-}) at 10 times the concentration of the vanadium (i.e., 5 μM vanadium compared to 50 μM concentrations of the other elements studied).

Except in accumulator-plant species (see Cannon, 1963; Pratt, 1966), vanadium is not readily translocated to plant tops and even less so to seeds (Pratt, 1966). In this context, Mitchell et al. (1957) have shown that extractable soil vanadium is poorly correlated to vanadium concentrations in plant tops.

Water

Vanadium occurs in water in several chemical forms, of which the dioxovanadium cation (VO_2^+) and the vanadate anion (VO_4^-) are readily soluble. The mobility of these compounds in water, however, is probably low because of easy adsorption on clay and precipitation with organic matter. Relatively high concentrations of vanadium have been reported (Kuroda, 1939) in nine Japanese hot springs, five of which contained from 22–79 ppb and three contained 208, 220, and 247 ppb vanadium. Seawater contains about 0.3 ppb vanadium (Mason, 1958).

Twenty-one samples of river water in Japan contained from 0.1 to 1 ppb with a mean of 0.91 ppb vanadium (Sugawara et al., 1956). Kleinkopf (1960) found vanadium contents of as much as 2.1 ppb, with a mean of 0.112 ppb, in 440 lake waters in Maine.

The median vanadium content of 100 municipal water supplies in the United States was reported as < 4.3 ppb, with a range from nondetectable levels to 70 ppb (Durfor and Becker, 1964). Water supplies in the Southwest are generally higher in vanadium content than those of the eastern states. Thirteen springs and wells in the vanadium-bearing Morrison formation of the Colorado Plateau had a maximum of 300 ppb vanadium (Phoenix, 1959).

Although present-day spectrographic methods using preconcentration and purified carbon are adequate for vanadium determinations in water, recent studies show that the size of filter is an important consideration in collecting raw water (Table 25) and that many reported analyses include concentrations of metals or clay.

A study of trace element data in the rivers and lakes of

TABLE 24   Vanadium Content of a Peaty Podzol[a] with a Thin Iron Pan[b]

Depth, in.	Soil Horizon	Vanadium Content, ppm	
		Total	Extractable (2.5% acetic acid)
—	Surface vegetation	1	—
1–2	Mull	20	0.04
4–8	Humus	10	0.05
10–13	A_1A_2	150	0.72
13[c]	B_1 pan	200	0.05
13–17	B_2	100	0.07
24–28	B_3	100	0.18

[a]A multilayer acidic soil with light-colored leached upper layers and dark-brown lower layers where depleted materials accumulate.
[b]A hard cement-like layer within or just beneath surface soil.
[c]< 1 in. thick.
SOURCE: Swaine and Mitchell (1960).

TABLE 25 Vanadium Content of Stream Waters in Missouri during Flood Runoff (ppb)

Type of Underlying Geologic Material	Stream	Filtered (0.1 μ)	Unfiltered
Glacial drift	Medicine Creek	<12.0	140
	Shoal Creek	< 7.0	60
Pennsylvanian sandstone, shales and limestones	Cedar Creek	<.0.9	110
	Big Creek	< 2	900
Mississippian limestones	Shoal Creek	< 2.0	67
	Spring River	< 2.0	74
Cambro-Ordovician dolomites	Osage Fork	< 2.0	33
	Jack Fork	< 5.0	22

NOTE: The vanadium contents of the unfiltered samples reflect the total vanadium concentration in the stream water and also the concentration of vanadium in the suspended sediments.
SOURCE: Feder, (1973).

the United States (Kopp and Kroner, 1968) showed only 3.4 percent frequency of detection of vanadium, with the greatest frequency (9 percent) in the Colorado River Basin. The water was filtered through a 0.45-μ filter. The maximum contents appear to reflect contaminations from various sources (Table 26). Uranium mills located on major rivers may be the largest source of vanadium in drinking water. Although there may be general differences in vanadium concentrations in the waters of the major sections of the country, it is doubtful, with the data available at the present time, that maps can be drawn to compare natural geochemical differences with disease patterns.

Pollution Sources

Petroleum and other naturally occurring hydrocarbons such as asphaltite contain appreciable quantities of vanadium. Gerrild and Lantz (1969) give vanadium analyses of 75 crude oil samples from sand units of Pliocene age in

TABLE 27 Vanadium in Nonurban Air Samples (μg/m^3)

Percentile	Vanadium Content
10th	<0.0005
20th	<0.0005
50th (median)	0.0005
80th	0.01
90th	0.02

SOURCE: Environmental Protection Agency (1972).

southern California. The values range up to 11 percent in the ash and average about 5 percent. An asphaltite deposit in Peru, containing vanadium sulfide and vanadium oxides, assays as high as 25 percent vanadium.

The combustion of petroleum products such as in electrical generating plants may be contributing detectable amounts of vanadium to the atmosphere. The Office of Air Programs of the Environmental Protection Agency (1972) reported the frequency distributions of vanadium in the air at 112 nonurban stations through the United States in 1966 and at 120 nonurban stations in 1967. The distributions for both years were similar to the values given in Table 27. Only one of the 232 measurements in nonurban areas exceeded 0.1 μg/m^3. In both 1966 and 1967, however, the measurements at eight urban stations exceeded this value.

Shacklette and Connor (1973) have used Spanish moss (*Tillandsia usneoides*) to measure the relative amounts of vanadium and other elements in the air throughout the southeastern United States. Spanish moss is an epiphyte and derives all its nutrients and other constituents directly from the air, either as gases, solutes in rainwater, or airborne particulate matter. The frequency distribution of vanadium in the ash of the Spanish moss is shown in Table 28. The region sampled includes almost all parts of the United States where Spanish moss grows. There is a general tendency for vanadium to be more highly concentrated in samples from the eastern part of this region

TABLE 26 Maximum Vanadium Content of River Waters and Their Probable Sources of Contamination (ppb)

Maximum Observed Content	Source	Location
16	Pulp Mill	Kanawha River, West Virginia
63	Oil refineries	Maumee River, Toledo, Ohio
54	Chemical industries	Kanawha River, West Virginia
38	Acid mine drainage	Ohio River, Toronto, Ohio
~15	Raw and treated sewage	Many rivers
40	Hydroelectric plant	Shenandoah River, Virginia
217	Uranium mill	San Juan River, Shiprock, New Mexico
300	Uranium mill	Colorado River, Loma, Colorado
184	Denver sewage and recycled irrigation waters	South Platte, Julesburg, Colorado
158	Unknown	Missouri River, Big Horn, Montana
500	Unknown	Nichols Hill, Oklahoma

SOURCE: Kopp and Kroner (1968).

TABLE 28 Vanadium in Ash of Spanish Moss (ppm)

Percentile	Vanadium Content
10th	50
20th	60
50th (median)	83
80th	130
90th	180

SOURCE: Shacklette and Connor (1973).

(Figure 17), a distribution that Shacklette and Connor (personal communication, 1973) feel may reflect, at least in part, the refining of Venezuelan crude oils in this part of the country. Venezuelan crude has been reported to contain among the highest amounts of vanadium of all crude oils measured from various countries (Committee on Biologic Effects of Atmospheric Pollutants, 1974).

Coal-fired power plants do not appear to introduce much vanadium into the environment (Cannon and Anderson, 1972). Analyses of native vegetation collected at the same sampling sites before and after the firing of the Four Corners Power Plant in northwest New Mexico show no significant difference with either time or distance from the power station. The plants, which were largely *Atriplex* species, averaged 3.8 ppm vanadium in 1962 and 3 ppm in 1973.

Local areas of soil pollution with vanadium may arise from aerosol particulates containing vanadium residues

that are emitted from metal refineries (Mitchell, 1964). The acetic acid-extractable vanadium in soils downwind of metal refineries in England contained as much as 1 ppm vanadium as compared to 0.2 ppm extractable vanadium in protected soils.

As fuel oils are desulfurized, the vanadium content is lowered, suggesting that this source may be reduced in the future. With the present level of vanadium released by human activity, and the relative nontoxicity of vanadium, it does not appear at this time that vanadium pollution is a significant hazard to human health.

ANALYTICAL METHODOLOGY

Until recently, vanadium analysis was mainly confined to geological and toxicological samples that contained several parts per million. Because vanadium has been established as an essential element in laboratory animals, analysis of biological samples containing a few parts per billion has become necessary. Accurate techniques such as colorimetry and emission spectrometry have been used for many years for the analysis of vanadium at levels above a few parts per million. Much more sensitive and accurate methods such as neutron-activation analysis and the catalytic method (Welch and Allaway, 1972) have been developed recently for the analysis of vanadium in biological samples down to levels of a few parts per billion.

Although cheaper, more accurate, and more sensitive

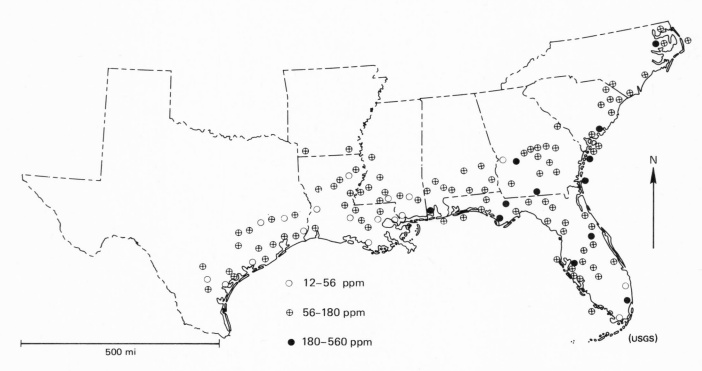

12–56 ppm

56–180 ppm

180–560 ppm

500 mi

(USGS)

FIGURE 17 Vanadium concentrations in the ash of Spanish moss samples throughout its occurrence in the southeastern United States (Shacklette and Connor, 1973).

methods are still needed, standardization of sampling techniques, analytical techniques, and laboratory methodology are needed more urgently. The analytical data in this report are in many cases questionable and difficult to interpret because of the lack of such standardization or the inability to detect vanadium at the parts-per-billion level.

PLANTS—UPTAKE AND ESSENTIALITY

Beneficial effects of vanadium for the growth of three genera of one-celled plants have been reported. Arnon and Wessel (1953) found that vanadium stimulated the growth of the alga *Scenedesmus obliquus*. Bertrand (1942) showed that as little as 0.0043 ppm vanadium in the nutrient solution produced a 21 percent increase in weight of *Aspergillus niger*, and Horner *et al.* (1942) reported that vanadium accelerates nitrogen fixation by *Azotobacter*.

Vanadium is a specific catalyst of nitrogen fixation, and it can partially replace molybdenum in this function. Present evidence suggests that this effect results from vanadium incorporation into the nitrogenase complex with consequent stabilization of the enzyme and more effective utilization of molybdenum in molybdenum-starved cells (Benemann *et al.*, 1972).

Welch and Huffman (1973) have investigated the essentiality of vanadium in higher plants by growing lettuce and tomato plants in purified nutrient solutions (< 0.04 ppb vanadium) with and without the addition of 50 ppb vanadium as NH_4VO_3. No visible deficiency symptoms attributable to vanadium deficiency developed, nor was there a significant increase in yield with added vanadium. The tops of the lettuce plants contained < 2–18 ppb vanadium when grown in the purified solutions, and the tops of the tomato plants contained 117–418 ppb when grown in the solutions containing added 50 ppb vanadium. Greater concentrations were observed in the roots. If, indeed, vanadium is essential to higher plants such as lettuce and tomato, the amount of available vanadium required in the growth media appears to be < 0.04 ppb.

Although small amounts of vanadium appear to be stimulating to some plants, large amounts are toxic. Gericke and von Rennenkampff (1940) show that 0.1–1.0 ppm vanadium is favorable and 10–1,250 ppm is toxic, depending on the chemical form in which vanadium was added to soil and on the tolerance of the particular species. Pratt (1966) found vanadium to be toxic to certain plants grown in solution culture at levels of 0.5 ppm vanadium (as NH_4VO_3) in the nutrient solution.

Warington (1957) reported that elevated vanadium caused a preliminary deepening of color in the shoots of soybean and flax plants, followed by apical iron-deficiency chlorosis. Because reddening in plants had been observed around the uranium–vanadium deposits in Utah and Colorado, Cannon (1963) ran experiments with sorghum and *Astragalus preussi* (a selenium-

accumulating legume) in nutrient solutions containing 1, 10, and 100 ppm vanadium as ammonium metavanadate. No effect was noted in the growth of germinated seeds of sorghum at 1 ppm; reddening, first of the lower stems and later of the leaf tips, was noted at 10 ppm, and stunting and death occurred after 2 weeks in 100 ppm solution. *Astragalus preussi*, which is highly tolerant of vanadium around uranium–vanadium deposits, was, on the other hand, unaffected by 100 ppm and grew 20-in. roots in 6 weeks' time.

Analyses of trees and shrubs, forage, and edible plants from soils of "normal" vanadium content, as collected from the literature and from unpublished sources, are shown in Table 29.

Vanadium in plants varies among the different parts of a plant, at different soil depths, and with the seasons. Vanadium content is generally greater in roots (particularly in legumes) than in the aerial parts of the plant. Bertrand (1942) found dry-weight values for vanadium ranging from 0.152–4.2 ppm in the aerial parts, 0.1–12.14 ppm in the roots, and 0.01–1.2 ppm in the seeds of 62 species. Mitchell (1954) analyzed six species of moorland forage growing on gneiss at four periods in three different years and reported < 0.1 ppm vanadium in dry weight, except for *Calluna* (heather), which contained 0.8 ppm. Thirty-five near-surface juniper roots collected in the Thompson uranium–vanadium district in Utah (Cannon, 1964) averaged 110 ppm vanadium in the ash; branch tips of the same trees averaged only 55 ppm (ash wt). A peeled juniper root collected from the ore zone at a depth of 9 ft contained 2,200 ppm vanadium in the ash; the same root near the surface contained only 78 ppm.

Parts of four trees [*Populus tremuloides* (aspen), *Acer glabrum* (maple), *Pinus ponderosa* (pine), and *Pseudotsuga taxifolia* (Douglas fir)] growing in forest soil over unmineralized schist were sampled seasonally by the U.S. Geological Survey. Vanadium was not detected in any part of aspen at any season. A concentration of 8.9 ppm vanadium was found in the maple roots and 2 ppm in the leaves during autumn, although none was detected in these parts at other seasons or in other parts of the tree in any season. Vanadium in Douglas fir and pine is given in Table 30.

The lowest concentrations of vanadium in plants are generally found in the seeds. Welch and Cary (1975) found a range of < 6.5–16.2 ppb in several varieties of wheat grain in six different parts of the country—at least one order of magnitude lower than in other parts of the plant (Table 31).

Great variation in vanadium uptake occurs among different plant species (Table 32). Söremark (1967), who analyzed 10 samples each of 16 fruits and vegetables, found parsley and radishes to contain considerably more vanadium than other vegetables. He was scarcely able to detect vanadium in pears, carrots, beets, or peas.

Plants collected from river alluvium, on which most of the produce is grown in the southwestern states, contain more vanadium than plants from residual soils of the Southeast (H. L. Cannon, unpublished data). Söremark

TABLE 29   Vanadium in Plants (dry weight)

Area and Source	Trees and Shrubs			Forage Grasses and Herbs			Vegetables and Fruits		
	No. Samples	Mean, ppm	Range, ppm	No. Samples	Mean, ppm	Range, ppm	No. Samples	Mean, ppm	Range, ppm
Stockholm, Sweden (Söremark, 1967)	—			—			160	0.43	<0.00001–4.52
Washington County, Maryland (USGS unpub. data)	38 (tips)	1.3[a]	<1.3–12.8	42	0.39	0.1–0.9	95	<2.0	<2–10.3
San Juan County, New Mexico (USGS unpub. data)	100 (tips)	1.8	0.4–13.6	18	<0.8	<0.8–1.2	55	<1.0	<1.0–2.1
Canandaigua, New York (USGS unpub. data)	—			—			14	<2.4	<0.7–10.5
Pinal County, Arizona (Sievers and Cannon, 1974)	4 (tips)	1.89	<1.6–2.7	5	1.3[a]	<1.1–2.6	11	0.4[a]	<0.2–1.1
Scotland (Mitchell, 1954)	—			14	0.07	0.04–0.10	—		
Georgia (Shacklette et al., 1970)	339 (tips)	0.13	0.11–1.3	—			315	<0.47	<0.35–158.8
	339 (leaves)	<0.36	<0.15–3.0						
New Jersey (Prince, 1957)	—			—			10 (corn)	0.60	0.37–1.05
New Jersey (Hanna and Grant, 1962)	43 (leaves)	1.2	0.06–4.8	—			—		

[a]Approximate values.

(1967) found lower contents in the same vegetables when grown in New Hampshire and Rhode Island rather than in Sweden. Mitchell *et al.* (1957) found no significant difference in vanadium uptake between well-drained and poorly drained soils, although concentrations in the roots were not measured.

Vanadium-tolerant species may accumulate large amounts of vanadium when grown in soils rich in available vanadium (Table 33). Bertrand (1943), who analyzed many species of fungi, found *Amanita muscaria* (the hallucinogenic mushroom) to be a true accumulator of vanadium, ranging from 61 to 181 ppm in seven samples collected from different areas.

The data in Table 32 suggest that vanadium concentrations in most edible plants are very low compared to

TABLE 30   Vanadium Content of Plant Parts, Single Samples Collected in the Fall (dry weight) Compared to Soil (ppm)

	Ponderosa Pine	Douglas Fir
Young needles	N.D.[a]	N.D.[a]
Older needles	0.41	0.51
First-year twigs	0.46	0.42
Older twigs	0.96	0.55
Cones	5.0	2.9
Wood	0.65	0.40
Roots	1.2	2.0
Root bark	5.2	1.9
Soil		
Humus layer	70	
A zone	100	
B zone	150	

[a] N.D. = not detected.
SOURCE: H. L. Cannon, U.S. Geological Survey, unpublished data, 1972.

many other trace elements. [For additional information on vanadium in foods, see Committee on Biologic Effects of Atmospheric Pollutants, 1974.]

ANIMALS AND HUMANS

Nutritional Essentiality

Hopkins and Mohr (1971a) reported that vanadium is an essential nutrient for animals in that a vanadium deficiency significantly reduced wing- and tail-feather growth in chicks consuming a diet containing less than 10 ppb vanadium. Reduced body growth has been reported by Strasia (1971) and by Schwarz and Milne (1971) in rats consuming diets low in vanadium. F. Nielsen (personal communication) has made a similar observation in chicks consuming a diet containing 30–35 ppb vanadium. Underwood (1971) gives an excellent review of the earlier literature.

The reproductive performance of rats consuming a diet of less than 10 ppb vanadium was reduced slightly in third-generation females and markedly reduced in the

TABLE 31   Concentrations of Vanadium in Wheat Grain (dry weight)

Location	No. of varieties	Median, ppb	Range, ppb
Bozeman, Montana	5	< 6.5	< 6.5–8.2
Manhattan, Kansas	5	< 6.5	—
St. Paul, Minnesota	4	< 6.5	< 6.5–11.6
Denton, Texas	5	8.6	< 6.5–10.5
Temple, Texas	5	< 6.5	—
Bushland, Texas	3	10.9	9.4–16.2

SOURCE: Welch and Cary, 1975.

TABLE 32 Vanadium in Some Fruits and Vegetables[a] and in Various Parts of Some Common Pasture Species[b]

Plant Species	Mean Vanadium Content, ppm (dry weight)
Fruits and Vegetables[a]	
Parsley	4.52
Radishes	1.26
Dill	0.84
Lettuce	0.58
Cucumbers	0.056
Wild strawberries	0.041
Strawberries	0.031
Red whortleberries	0.0102
Apples	0.0086
Potatoes	0.0064
Cauliflower	0.00109
Tomatoes	0.00053
Pears	<0.00001
Carrots	<0.00001
Common beets	<0.00001
Peas, frozen	<0.00001
Pasture Species[b]	
Cocksfoot, head	0.11
leaf	0.23
stem	0.05
Meadow fescue,	
head	0.06
leaf	0.22
stem	0.05
Perennial ryegrass,	
head	0.07
leaf	0.27
stem	0.04
Timothy, head	0.09
leaf	0.16
stem	0.02
Red clover, head	0.17
leaf and petiole	0.27
stem	0.08

[a] SOURCE: Söremark (1967).
[b] SOURCE: Fleming (1963).

fourth generation (Hopkins and Mohr, 1974). The five vanadium-supplemented fourth-generation females produced five, four, and five litters following three matings, while the five vanadium-deficient fourth-generation females produced one litter following the first mating, and three litters following the second. Two were pregnant upon autopsy following the third mating. The total number of pups from vanadium-deficient females following the first two matings was 22 as compared to 65 following the first two matings for the vanadium-supplemented controls. Thirty-two percent mortality occurred among the vanadium-deficient pups, whereas only 1.5 percent of the pups from supplemented mothers died. In a second experiment with a different strain of rat (BHE), similar results in reproduction and mortality were obtained.

Strasia (1971) has reported that rats fed a diet containing less than 0.1 ppm vanadium had a significantly in-creased packed cell volume of blood, and increased blood and bone iron levels, when compared to groups receiving 0.5, 2.5, and 5 ppm supplemental vanadium.

Söremark et al. (1962) reported that the highest uptake of subcutaneously injected radiovanadium in young rats was found in areas of rapid mineralization in the dentine and bone. Söremark and Ullberg (1962) reported that radiovanadium injected into adult mice intravenously was taken up by the teeth, bones, and fetus. The mineralization zones of bones and teeth showed an especially high uptake. Although Underwood (1971) concluded that the evidence for a beneficial effect of vanadium on dental caries was not conclusive, localization of vanadium in the tooth structure supports the view that this element might play a role in dental caries. Nielsen and Ollerich (1973) found that the tibia in vanadium-deficient chicks had an increased epiphyseal plate:primary spongiosa weight ratio. Histologically, an alteration was seen in the organization of the cells, in the amount of matrix in the zones of proliferation and maturation. This deficiency neither significantly changed the hexosamine levels, nor the uptake and distribution of $^{35}SO_4^{2-}$ in the epiphyseal plate or primary spongiosa, indicating that vanadium deficiency does not affect mucopolysaccharide metabolism (Nielsen, personal communication).

Bernheim and Bernheim (1939) reported that added vanadium increased markedly the oxidation of phospholipids *in vitro* by washed liver suspensions. More recently, numerous reports have been made that pharmacological levels of vanadium lowered tissue cholesterol levels (Underwood, 1971). Curran and Burch (1967) were able to relate these lowered levels to the ability of vanadium to inhibit cholesterol biosynthesis. Inhibition takes place in squalene synthetase (a microsomal enzyme system). Unfortunately, the ability of vanadium to lower blood cholesterol was not observed in older animals or in older humans where elevated cholesterol levels are a health problem.

Altered blood-lipid levels were found in vanadium-deficient chicks. Initially, it was reported that vanadium-deficient chicks consuming a vanadium-deficient diet for 4 weeks had lowered plasma cholesterol levels when compared to the supplemented animals (Hopkins and Mohr, 1971a). When a similar experiment was continued for a longer time, the plasma cholesterol levels of vanadium-deficient chicks again were significantly lower than the supplemented controls at 28 days. At 49 days, however, the plasma cholesterol from the deficient chicks had become significantly higher (Hopkins and Mohr, 1971b). The average cholesterol values were 249 and 224 mg/100 ml for the deficient and control groups, respectively, at 49 days. Nielsen and Ollerich (1973) have reported data that support the view that vanadium-deficient chicks have altered plasma cholesterol levels, but their observations indicate a significantly increased cholesterol level at 4 weeks of age.

In addition to vanadium-related effects on plasma cholesterol, Hopkins and Mohr (1974) found that vanadium-deficient chicks had nearly twice the amount of

TABLE 33 Unusually High Concentrations of Vanadium in Plants Grown in Soils Containing High Concentrations of Vanadium (dry weight)

Species	Plant Part	Vanadium, ppm	Area	Source
Unmineralized ground				
Carum (parsley) (10 samples)	Aerial	4.5	Sweden	Söremark, 1967
Amanita muscaria (mushroom)	Cap	181	France	Bertrand, 1943
Near vanadium deposits				
Triticum aestivum (wheat)	Aerial	5	United States	Monier-Williams, 1950
Allium macropetalum (wild onion)	Entire	133	Utah	Cannon, 1964
Oenothera caespitosa (primrose)	Aerial	38	New Mexico	USGS, unpub.
Astragalus confertiflorus (locoweed)	Aerial	144	Utah	Cannon, 1964
Astragalus preussi (locoweed)	Aerial	67	Utah	Cannon, 1964
Aster venustus (woody aster)	Aerial	21	Utah	Cannon, 1964
Castilleja angustifolia (paintbrush)	Aerial	37	Utah	Cannon, 1964
Chrysothamnus viscidiflorus (rabbitbrush)	Branch tips	37	Utah	Cannon, 1964
Eriogonum inflatum (desert trumpet)	Aerial	15	Utah	Cannon, 1964
Lepidium montanum (pepperweed)	Aerial	11	Utah	Cannon, 1964
Mielichhoferia (copper moss)	Entire	154	Alaska	Shacklette, 1967

plasma triglyceride (48.7 mg/100 ml) as did supplemented controls (25.4 mg/100 ml) at 28 days of age. Although there are some inconsistencies, it appears that vanadium at both pharmacological and physiological levels eventually decreases blood-lipid levels.

The uptake of vanadium from the gastrointestinal tract is restricted (Hopkins and Mohr, 1971a), and what is taken up is for the most part rapidly excreted in the urine. Most of the ingested vanadium is unabsorbed and excreted in the feces. Vanadium does not appear to accumulate in the body to any appreciable extent.

Because of limited data, the vanadium level required by an organism to maintain health can only be estimated at this time. Work cited above indicates that the requirement may be somewhat greater than 100 ppb vanadium.

Berg (1966) reports that the toxicity of vanadium compounds is greater with a purified diet than with a natural diet, indicating that natural rations and diets tend to reduce absorption so that a dietary level of over 100 ppb vanadium may be in order.

If nutritional needs are at this or a slightly higher level, then what vanadium levels are consumed from natural feeds and foods? Söremark (1967) reported that milk generally contained less than 0.1 ppb on a wet-weight basis, whereas liver, fish, and meat contained a few but not over 10 ppb. These limited data and those in Tables 29 and 32 show a wide variation in vanadium content, with many

feed and food ingredients containing levels of vanadium far below the estimated dietary requirement of greater than 100 ppb. Water sources were not taken into account in these studies and would probably provide additional available vanadium, at least in some areas of the United States. If inadequate dietary vanadium is a problem, the most probable effect would show up in lipid metabolism in animals and man. Although not of practical importance to the animal industry, it is conceivable that altered blood-lipid levels in some individuals result from a vanadium deficiency. Obviously, much more accurate data are needed before any firm conclusions can be drawn. Enough data are available, however, to indicate that the requirement level for vanadium and the levels in feeds and foods do not appear to overlap to a degree that will allow us to take for granted adequate vanadium nutrition. The rough calculation that the requirement might be greater than 100 ppb vanadium and that feed and food levels are generally less than 100 ppb is not reassuring.

TOXICITY

Studies of the toxicity of vanadium in man show that this element is relatively nontoxic orally. Dimond *et al.* (1963) gave vanadium orally, as ammonium vanadyltartrate, to six subjects for 6–10 weeks in amounts ranging from 4.5 to

18 mg of vanadium a day with no toxic effects other than some cramps and diarrhea at the larger dose levels. Schroeder *et al.* (1963) fed patients 4.5 mg/day as the oxytartarovanadate for 16 months with no signs of intolerance. Toxicologic studies with rats (Franke and Moxon, 1937) and chicks (Romoser *et al.*, 1961; Nelson *et al.*, 1962; Berg, 1963) show that, depending on the form and type of diet, 13–35 ppm vanadium must be in the diet before gross toxic effects are seen, such as depressed growth and diarrhea. Higher levels (50–200 ppm) may induce death.

Occasional cases of industrial toxicity have been reported as a result of unprotected workers inhaling airborne vanadium in certain industries. Respiratory disorders and eczematous skin lesions are the usual symptoms seen. There is no evidence of vanadium-induced effects on the health of individuals breathing air with increased levels of vanadium, such as is found in some large industrial cities.

RECOMMENDATIONS

From the evidence reviewed, there appears to be no relation between vanadium intake from natural sources and human disease, probably because of the relatively low toxicity of vanadium when taken orally and the lack of information concerning possible borderline nutritional deficiency. Because of this lack of information, the following recommendations are made:

1. Faster, cheaper, and more sensitive methods of vanadium analysis of biological samples at levels around 10 ppb must be found.
2. Determine what forms of vanadium are taken up by plants from soil.
3. Determine accurately the vanadium levels in foods, forage, and water.
4. Determine what forage and food plants accumulate vanadium, and determine the bioavailability of vanadium in water and foods.
5. Determine the metabolic role of vanadium in animals.
6. Determine the essentiality and quantitative requirement for vanadium in humans.
7. Determine the role, if any, of vanadium deficiency in human diseases such as cardiovascular disease, including whether the "hard-water effect" that is inversely correlated with cardiovascular diseases may be due to the vanadium content of the drinking water.
8. Determine whether areas of deficiency exist in the United States that can be correlated with disease patterns.

REFERENCES

Arnon, D. I., and G. Wessell. 1953. Vanadium as an essential element for green plants. Nature 172:1039–1040.
Benemann, J. R., C. E. McKenna, K. F. Lie, T. G. Taylor, and M. D. Kamen. 1972. The vanadium effect in nitrogen fixation by *Azotobacter*. Biochim. Biophys. Acta 264:25–38.
Berg, L. R. 1963. Evidence of vanadium toxicity resulting from the use of commercial phosphorus supplements in chick rations. Poult. Sci. 42:766–769.
Berg, L. R. 1966. Effect of diet composition on vanadium toxicity for the chick. Poult. Sci. 45:1346–1352.
Bernheim, F., and M. L. C. Bernheim. 1939. The action of vanadium on the oxidation of phospholipids by certain tissues. J. Biol. Chem. 127:353–360.
Bertrand, D. 1942. Dosage du vanadium chez les vegetaux (Content of vanadium in plants). Bull. Soc. Chim. (Paris) Ser. 5, 9:121–124, 128–135.
Bertrand, D. 1943. Le vanadium chez les champignons et plus specialement chez les amanits. Bull. Soc. Chim. Biol. 25:194–197.
Bowen, H. J. M. 1966. Trace elements in biochemistry. Academic Press, New York. 241 pp.
Cannon, H. L. 1963. The biogeochemistry of vanadium. Soil Sci. 96(3):196–204.
Cannon, H. L. 1964. Geochemistry of rocks and related soils and vegetation in the Yellow Cat area, Grand County, Utah. U.S. Geol. Surv. Bull. 1176. U.S. Government Printing Office, Washington, D.C. 127 pp.
Cannon, H. L., and B. M. Anderson. 1972. Trace element content of the soils and vegetation in the vicinity of the Four Corners Power Plant. Pt. III of Southwest energy study. Report of the Coal Resources Work Group. U.S. Geol. Surv. Open-File Rep., Denver.
Committee on Biologic Effects of Atmospheric Pollutants. 1974. Vanadium. National Academy of Sciences, Washington, D.C. 117 pp.
Curran, G. L., and R. E. Burch. 1967. Biological and health effects of vanadium. *In* Proc. 1st Annu. Conf. Trace Subst. Environ. Health, July 10–11, 1967, D. D. Hemphill [ed.]. University of Missouri, Columbia. pp. 96–104.
Dimond, E. G., J. Caravaca, and A. Benchimol. 1963. Vanadium. Excretion, toxicity, and lipid effect in man. Am. J. Clin. Nutr. 12:49–53.
Durfor, C. N., and Edith Becker. 1964. Public water supplies of the 100 largest cities in the United States, 1962. U.S. Geol. Surv. Water Supply Pap. 1812. U.S. Government Printing Office, Washington, D.C. 364 pp.
Environmental Protection Agency. 1972. Air quality data for 1968 from the national air surveillance networks and contributing state and local networks. Pub. No. APTD-0978. Office of Air Programs.
Feder, G. L. 1973. Geochemical survey of water. *In* Geochemical Survey of Missouri: plans and progress for seventh six-month period (July–December, 1972). U.S. Geol. Surv. Open-File Rep. 1982. Branch of Regional Geochemistry, USGS, Denver and Reston. pp. 50–57.
Fischer, R. P. 1959. Vanadium and uranium in rocks and ore deposits. *In* Geochemistry and mineralogy of the Colorado Plateau uranium ores, R. M. Garrels and E. S. Larsen [eds.], 3d ed. U.S. Geol. Surv. Prof. Pap. 320. U.S. Government Printing Office, Washington, D.C. pp. 219–230.
Fischer, R. P. 1973. Vanadium. *In* United States mineral resources, D. A. Brobst and W. P. Pratt [eds.]. U.S. Geol. Surv. Prof. Pap. 820. U.S. Government Printing Office, Washington, D.C. pp. 679–688.
Fleming, G. A. 1963. Distribution of major and trace elements in some common pasture species. J. Sci. Food Agric. 14:203–208.
Franke, K. W., and A. L. Moxon. 1937. The toxicity of orally ingested arsenic, selenium, tellurium, vanadium, and molybdenum. J. Pharmacol. Exp. Ther. 61:89–102.

Gericke, S., and E. von Rennenkampff. 1940. Untersuchungen über die Wirkung des V auf das Pflanzenwachstum (The effect of vanadium on plant growth). Bodenk. Pflanzenernähr. 18:305–315.

Gerrild, P. M., and R. J. Lantz. 1969. Chemical analyses of 75 crude oil samples from Pliocene sand units, Elk Hills oil field, California. U.S. Geol. Surv. Open-File Rep. Reston, Denver, and Menlo Park. 5 pp. and 14 data sheets.

Goldschmidt, V. M. 1954. Geochemistry. The Clarendon Press, Oxford. pp. 485–499.

Hanna, W. J., and C. L. Grant. 1962. Spectrochemical analysis of the foliage of certain trees and ornamentals for 23 elements. Bull. Torrey Bot. Club 89:293–302.

Hopkins, L. L., Jr., and H. E. Mohr. 1971a. The biological essentiality of vanadium. In Newer trace elements in nutrition, W. Mertz and W. E. Cornatzer [eds.]. Marcel Dekker, New York. pp. 195–213.

Hopkins, L. L., Jr., and H. E. Mohr. 1971b. Effect of vanadium deficiency on plasma cholesterol of chicks. Fed. Proc. 30:462. (Abstract)

Hopkins, L. L., Jr., and H. E. Mohr. 1974. Vanadium as an essential nutrient. Fed. Proc. 33:1773–1775.

Horner, C. K., S. Burk, F. E. Allison, and M. S. Sherman. 1942. Nitrogen fixation by Azotobacter as influenced by molybdenum and vanadium. J. Agric. Res. 65:173–193.

Jaffé, F. C. 1961. Phosphate rock of the western United States. Miner. Ind. Bull. (Colo. Sch. Mines) 4(5):11.

Kaiser, E. P., B. F. Herring, and J. C. Rabbitt. 1954. Minor elements in some rocks, ores, and smelter products. U.S. Geol. Surv. TEI-415. U.S. Atomic Energy Commission Technical Information Service, Oak Ridge, Tennessee.

Kleinkopf, M. D. 1960. Spectrographic determination of trace elements in lake waters of northern Maine. Geol. Soc. Am. Bull. 71:1231–1242.

Kopp, J. F., and R. C. Kroner. 1968. Trace metals in waters of the United States: A five-year summary of trace metals in rivers and lakes of the United States (October 1, 1962–September 30, 1967). U.S. Department of Interior, Federal Water Pollution Control Administration, Cincinnati, Ohio. 32 pp. and 16 appendixes.

Kuroda, Kazuo. 1939. Vanadium, chromium, and molybdenum contents of the hot springs of Japan. Bull. Chem. Soc. Japan 14:307–310.

Lovering, T. S., and E. N. Goddard. 1950. Geology and ore deposits of the Front Range, Colorado. U.S. Geol. Surv. Prof. Pap. 223. U.S. Government Printing Office, Washington, D.C. 319 pp.

McKelvey, V. E. 1949. Geological studies of the western phosphate field. Trans. Am. Inst. Min. Metall. Eng. 184:270–279.

Mason, B. 1958. Principles of geochemistry, 2d ed. John Wiley & Sons, New York.

Mitchell, R. L. 1954. Trace elements in some constituent species of moorland grazing. J. Br. Grassl. Soc. 9(4):301–311.

Mitchell, R. L. 1964. Trace elements in soil. In Chemistry of the soil, F. E. Bear [ed.]. Reinhold, New York. pp. 320–368.

Mitchell, R. L. 1971. Trace elements in soil. In Trace elements in soils and crops. Tech. Bull. 21. Ministry of Agriculture, Fisheries, and Food, London. pp. 8–20.

Mitchell, R. L., J. W. S. Reith, and I. M. Johnston. 1957. Trace element uptake in relation to soil content. J. Sci. Food Agric. 8(Suppl.):51–59.

Monier-Williams, G. W. 1950. Trace elements in foods. John Wiley & Sons, New York. 511 pp.

Nelson, T. S., M. B. Gillis, and H. T. Peeler. 1962. Studies of the effect of vanadium on chick growth. Poult. Sci. 41:519–522.

Ng, Siew Kee, and C. Bloomfield. 1962. The effect of flooding and aeration on the mobility of certain trace elements in soils. Plant Soil 16:108–135.

Nielsen, F. H., and D. A. Ollerich. 1973. Studies on a vanadium deficiency in chicks. Fed. Proc. 32:929. (Abstract)

Phoenix, D. A. 1959. Occurrences and chemical character of ground water in the Morrison Formation. In Geochemistry and mineralogy of Colorado Plateau uranium ores, R. M. Garrels and E. S. Larson [eds.], 3d ed. U.S. Geol. Surv. Prof. Pap. 320. U.S. Government Printing Office, Washington, D.C. pp. 55–64.

Pratt, P. F. 1966. Vanadium. In Diagnostic criteria for plants and soils, H. G. Chapman [ed.]. Division of Agricultural Sciences, University of California, Riverside. pp. 480–483.

Prince, A. L. 1957. Trace element delivering capacity of 10 New Jersey soil types as measured by spectrographic analyses of soils and mature corn leaves. Soil Sci. 84:413–418.

Rankama, K., and Th. G. Sahama. 1950. Geochemistry. The University of Chicago Press. 912 pp.

Romoser, G. L., W. A. Dudley, L. J. Machlin, and L. Loveless. 1961. Toxicity of vanadium and chromium for the growing chick. Poult. Sci. 40:1171–1173.

Schroeder, H. A., J. J. Balassa, and I. H. Tipton. 1963. Abnormal trace metals in man—Vanadium. J. Chron. Dis. 16:1047–1071.

Schwarz, K., and D. B. Milne. 1971. Growth effects of vanadium in the rat. Science 174:426–428.

Shacklette, H. T. 1967. Copper mosses as indicators of metal concentrations. U.S. Geol. Surv. Bull. 1198-G. U.S. Government Printing Office, Washington, D.C. 18 pp.

Shacklette, H. T., and J. J. Connor. 1973. Airborne chemical elements in Spanish moss. U.S. Geol. Surv. Prof. Pap. 574-E. U.S. Government Printing Office, Washington, D.C. 46 pp.

Shacklette, H. T., H. I. Sauer, and A. T. Miesch. 1970. Geochemical environments and cardiovascular mortality rates in Georgia. U.S. Geol. Surv. Prof. Pap. 574-C. U.S. Government Printing Office, Washington, D.C. 39 pp.

Shacklette, H. T., J. C. Hamilton, J. G. Boerngen, and J. M. Bowles. 1971. Elemental composition of surficial materials in the conterminous United States. U.S. Geol. Surv. Prof. Pap. 574-D. U.S. Government Printing Office, Washington, D.C. 71 pp.

Sievers, M. L., and H. L. Cannon. 1974. Disease patterns of Pima Indians of the Gila River Indian Reservation of Arizona in relation to the geochemical environment. In Proc. 7th Annu. Conf. Trace Subst. Environ. Health, June 12–14, 1973, D. D. Hemphill [ed.]. University of Missouri, Columbia. pp. 57–61.

Söremark, R. 1967. Vanadium in some biological specimens. J. Nutr. 92:183–190.

Söremark, R., and S. Ullberg. 1962. The use of radioisotopes in animal biology and medical sciences, N. Fried [ed.], Vol. I. Academic Press, New York. pp. 103

Söremark, R., S. Ullberg, and L. E. Appelgren. 1962. Autoradiographic localization of vanadium pentoxide ($^{48}V_2O_5$) in developing teeth and bones of rats. Acta Odont. Scand. 20:225–232.

Strasia, C. A. 1971. Vanadium: Essentiality and toxicity in the laboratory rat. (Ph.D. Thesis, Purdue Univ.) University Microfilms, Ann Arbor, Michigan.

Sugawara, K., H. Naito, and S. Yamada. 1956. Geochemistry of vanadium in natural waters. J. Earth Sci. (Nagoya Univ.) 4:44–61. (In English)

Swaine, D. J. 1955. The trace-element content of soils. Commonw. Bur. Soil Sci. Tech. Comm. No. 48. Commonwealth Agricultural Bureaux, England. pp. 117–120.

Swaine, D. J., and R. L. Mitchell. 1960. Trace-element distribution in soil profiles. J. Soil Sci. 11:347–368.

Tourtelot, H. A., L. G. Schultz, and J. R. Gill. 1960. Stratigraphic variations in mineralogy and chemical composition of the Pierre Shale in South Dakota and adjacent parts of North Dakota, Nebraska, Wyoming, and Montana. *In* Geological Survey Research 1960. U.S. Geol. Survey Prof. Pap. 400-B. U.S. Government Printing Office, Washington, D.C. pp. 447–452.

Turekian, K. K., and K. M. Wedepohl. 1961. Distribution of the elements in some major units of the earth's crust. Geol. Soc. Am. Bull. 72(2):175–192.

Underwood, E. J. 1971. Trace elements in human and animal nutrition, 3d ed. Academic Press, New York. 543 pp.

Vine, J. D., and E. B. Tourtelot. 1969. Geochemical investigations of some black shales and associated rocks. U.S. Geol. Surv. Bull. 1314-A. U.S. Government Printing Office, Washington, D.C. 43 pp.

Vinogradov, A. P. 1959. The geochemistry of rare and dispersed chemical elements in soils, 2d ed. Consultants Bureau, Inc., New York. 209 pp.

Warington, K. 1957. Investigations regarding the nature of the interaction between iron and molybdenum or vanadium in nutrient solutions with or without a growing plant. Ann. Appl. Biol. 44:535–546.

Welch, R. M. 1973. Vanadium uptake by plants. Absorption kinetics and the effects of pH metabolic inhibitors, and other anions and cations. Plant Physiol. 51:828–832.

Welch, R. M., and W. H. Allaway. 1972. Vanadium determination in biological materials at nanogram levels by a catalytic method. Anal. Chem. 44:1645–1647.

Welch, R. M., and E. E. Cary. 1975. Concentration of chromium, nickel, and vanadium in plant material. Agric. Food Chem. 23:479–482.

Welch, R. M., and E. W. D. Huffman, Jr. 1973. Vanadium and plant nutrition. The growth of lettuce (*Lactuca sativa* L.) and tomato (*Lycopersicon esculentum* Mill.). Plants in nutrient solutions low in vanadium. Plant Physiol. 52:183–185.

Zubovic, P. 1966. Distribution of minor elements in coals of the Appalachian region. U.S. Geol. Surv. Bull. 1117-C. U.S. Government Printing Office, Washington, D.C. 37 pp.

Zubovic, P., T. Stadnichenko, and N. B. Sheffey. 1961. Geochemistry of minor elements in coals of the northern Great Plains coal province. U.S. Geol. Surv. Bull. 1117-A. U.S. Government Printing Office, Washington, D.C. 58 pp.

Zubovic, P., T. Stadnichenko, and N. B. Sheffey. 1964. Distribution of minor elements in coal beds of the eastern interior region. U.S. Geol. Surv. Bull. 1117-B. U.S. Government Printing Office, Washington, D.C. 41 pp.

Zubovic, P., N. B. Sheffey, and T. Stadnichenko. 1967. Distribution of minor elements in some coals in the western and southwestern regions of the interior coal province. U.S. Geol. Surv. Bull. 1117-D. U.S. Government Printing Office, Washington, D.C. 33 pp.

PART TWO

Geochemical Environment and Man

Before being able to look at the relationship between the geochemical environment and human health and disease with much more confidence than at present, we need to know more about such problem areas as the nature and significance of soil imbalances and interactions between trace elements. If an imbalance occurs on agricultural soils, it can result in potentially harmful levels of toxic elements being supplied through the food chain. One major aspect of the imbalance problem (as well as many others) is the complex nature of the interactions capable of occurring between trace elements, under various conditions. The need for greater understanding of this critical topic is clearly indicated.

In evaluating the relationships of the geochemical environment to health and disease, we must search for and study specific population groups whose selection will eliminate as many variables as possible. The availability of such groups is illustrated by the studies underway by the World Health Organization, the U.S. Geological Survey, and the University of Missouri. An example of a closely contained and genetically homogeneous population group available in the United States is given in the discussion of disease patterns of southwestern Indian groups that have experienced limited environmental variability yet show differences in health and disease patterns.

One of the most effective ways to discern relationships between the geochemical environment and health and disease is to examine the appropriate data in map form. This clearly identifies spatial correlations that *may* indicate cause–effect relationships. Special consideration has therefore been given to the problems involved in displaying data in map form.

Throughout the following discussions a common theme is woven: the need for a standardized method of presenting analytical data that will facilitate both its interpretation and its comparison with data presented

109

by other investigators. This is especially pertinent when the goal is to show the possible relationships of geochemical data to health and disease patterns. Studies of these types must obviously be geographical in nature and ignore political subdivision boundaries. Similarly, common units must be used in reporting analytical data if we are to be able to compare the results of various studies; for example, agreement on the use of dry-weight versus wet-weight determinations must be made. Furthermore, analytical procedures must be standardized if our data are to be truly comparable.

A significant outgrowth of the Workshop is the development of a plan for the establishment of a National Environmental Specimen Index System, which will provide a means of access to environmental samples, ensure their future availability for retrospective studies, and aid in the establishment of standardized analytical procedures.

X

Interactions of
Trace Elements

WILLIAM H. ALLAWAY, *Chairman*

Orville A. Levander, Gennard Matrone, Harold H. Sandstead

INTERACTIONS IN SOILS AND PLANTS

In general, interactions among elements in the soil–plant system are of two types: those that affect the solubility or stability of specific elements in the soil and those involving processes at the root surfaces or inside the roots and aerial parts of the plants.

The first type includes precipitation reactions in which ions of two or more elements combine to form a compound so insoluble that neither element is available to plants. A good example of this type of interaction is the formation of calcium fluoride (CaF_2) in soils. In other cases, an element, usually a cation, may be tightly bound in an absorption site on an inorganic soil particle or organic material. Nutrient elements bound in this way are generally of low availability to plants.

Many of the interactions among trace elements in soils are strongly affected by the soil pH. Acid soils frequently have high concentrations of iron, aluminum, and manganese, and alkaline soils frequently have high levels of calcium and carbonate.

The oxidation state of certain micronutrients may also be affected by pH. At high pH, selenium tends to be oxidized to selenate, which is soluble, whereas low pH favors the formation of selenite or selenide, which are much less soluble. A detailed description of the effects of pH on micronutrient uptake by plants is given by Hodgson (1963).

The second type of element interaction—processes at root surfaces or inside plants—involves the uptake of elements by roots and their translocation inside the plant.

Interactions among elements are not the only factors affecting these processes, however, and the genetics of the plant exert a marked control over uptake or exclusion and translocation within the plant. Epstein (1972) describes the mechanics of ion accumulation from growth media, whether solution or soil.

The major process of ion accumulation by plant roots is an energy-dependent, carrier-mediated, transfer of ions from the external solution to a more concentrated solution in the conducting cells of the roots. Passage through at least one membrane is involved. One of the most common interactions between ions in the uptake process stems from competition between two different ionic species for the same carrier molecules. The exact nature of the carrier molecules and the bond between carrier and ion is largely unknown.

Finally, a few of the interactions, involving some of the elements studied at the Asilomar and Capon Springs workshops, that may affect the nutritional quality of plants are described below. These must be considered as only a few examples of a very large number of potentially important interactions.

The literature on trace element interactions in plants is voluminous. The interactions of 34 different elements are described by Chapman (1966); micronutrient interactions have been reviewed by Olsen (1972).

Zinc–Phosphorus

Plants growing in soils high in available phosphorus are often deficient in zinc, particularly in their top parts. An excellent review of the Zn–PO₄ interaction is provided by Olsen (1972).

Zinc–Cadmium

Under certain conditions, a high level of available zinc in the substrate will depress plant uptake of cadmium, and the use of zinc-rich fertilizers may be useful in minimizing cadmium concentrations in plants. This interaction has been investigated by Lagerwerff and Specht (1971).

Fluorine–Calcium–Aluminum

Large amounts of fluorine have been added to soils as accessory components of phosphate fertilizers without any observable damage to plants and with only a slight increase in the fluorine concentration in the plants. This probably reflects the formation of insoluble CaF_2 or insoluble aluminum fluorosilicates in the soil. Soil reactions affecting the uptake of fluorine by plants are reviewed by Brewer (1966).

Molybdenum–Soil pH

Most soils in the United States that support molybdenum-deficient plants are acidic soils. Liming soils appear to correct molybdenum deficiency in plants. Soils in the United States that support plants with molybdenum concentrations so high as to cause molybdenum toxicity in animals tend to be wet, neutral, or alkaline and are generally high in organic matter. Well-drained soils formed on molybdenum-rich rocks do not in general induce high concentrations of molybdenum in plants. The relation of molybdenum toxicity in livestock to soil chemistry is described by Kubota and Allaway (1972).

Nickel–Iron

The effect of the nickel–iron interaction on plant growth has been studied primarily in an attempt to improve plant growth on nickel-rich soils formed on serpentine rocks. High levels of available nickel in the substrate can result in chlorotic leaves, but this effect can be reversed by raising the level of available iron. This interaction, which is probably an interference by nickel of iron metabolism in the plant, is summarized by Crooke (1955).

Selenium–Sulfur

Although early investigations of selenium toxicity—experiments conducted under greenhouse conditions—indicated that the addition of soluble sulfate to soils would decrease plant uptake of selenium, attempts to use this interaction to prevent selenium toxicity in animals have not been successful in field trials. In fact, soils in the United States that support selenium-toxic plants nearly always are naturally high in soluble sulfates. The addition of sulfur-containing fertilizers to soils that produce plants with very low concentrations of selenium may cause even lower levels of plant selenium. Some investigators have concluded that use of sulfur-containing fertilizers in some areas has tended to increase the incidence of selenium-deficiency diseases in livestock. Literature on selenium–sulfur interactions in soils and plants has been reviewed by Allaway (1970).

INTERACTIONS IN HUMANS

The importance of element interaction in plant or animal metabolism has been apparent to nutritionists for many years. The potential importance in human health is now receiving increased attention, although specific knowledge relating to humans is limited. Studies of experimental animals have shown that magnesium metabolism is intimately linked to potassium and calcium metabolism (Grace and O'Dell, 1970a,b). These observations support findings in experimental and naturally occurring magnesium deficiency in humans (Shils, 1969; Seelig, 1971); such deficiency may interfere with potassium and calcium homeostasis. Clinical equivalents have been noted in patients with intestinal malabsorption syndrome and/or protein–calorie malnutrition. Less well appreciated is the relationship involved in thiazide diuretic-induced potassium and magnesium depletion. Resistant hypokalemia (deficiency of potassium in the blood) and the associated increased sensitivity to digitalis that occur in some patients have been found to be responsive to therapy with magnesium (see Chapter III). Magnesium depletion and associated abnormalities of potassium homeostasis have also been noted in patients with alcoholic liver disease.

The potassium loss consequent upon severe magnesium deprivation is thought to reflect magnesium's interaction with phosphate in the energy-dependent mechanism that maintains intracellular concentrations of potassium and sodium (Grace and O'Dell, 1970b).

Alterations in calcium homeostasis in magnesium-deficient humans are less well understood. The role of magnesium in the metabolism of parathyroid hormone, calcitonin, and 1,25-dihydroxycholecalciferol (a metabolite of vitamin D_3) is incompletely understood (Seelig, 1971), although these hormones may have a similar influence on magnesium metabolism to that which they have on calcium. In general, serum calcium concentrations decrease in human magnesium deficiency, probably because a magnesium deficiency inhibits the release of parathyroid hormone by the parathyroid gland.

Interactions between magnesium and phosphate at the cellular level have been documented in vitro (Wacker, 1969). Magnesium and phosphate also interact in the intestinal milieu in the presence of fat to form insoluble precipitates that are excreted in the stool. This intraluminal interaction of magnesium, phosphate, and fat is important in patients with malabsorption of fat and appears to be a major cause of magnesium deficiency in such patients.

Another example of an interaction between magnesium and calcium is the experimental prevention, by magnesium, of oxalate stones in rats with oxaluria due to vitamin B_6 deficiency (Gershoff and Andrus, 1961). Such studies and those of others (Anonymous, 1966) have prompted limited clinical trials, using magnesium therapy, to prevent stone formation in humans with some success; see also Prien and Gershoff (1974). Other elements that appear to interact in humans or animals include iron, copper, zinc, manganese, strontium, fluorine, iodine, cadmium, lead, mercury, chromium, selenium, vanadium, arsenic, cobalt, lithium, and molybdenum. Copper has long been known to be essential for the metabolism of iron in experimental animals (Seelig, 1972). A similar relation has been shown to be present in humans through the studies of infants with dietary copper deficiency (Cordano *et al.*, 1964). Such infants demonstrate impaired iron utilization, hypochromic microcytic anemia, and leukopenia.

It has been suggested that, as a consequence of its presence in the ferroxidase enzyme, ceruloplasmin, copper may participate in the binding of iron in the plasma to transferrin. Iron (Fe^{2+}), when released from tissue stores to the plasma and oxidized to Fe^{3+}, by ferroxidase, can complex with transferrin (Frieden, 1970). Transferrin transports the iron to the hematopoietic tissues or to other storage locations.

Copper and zinc interact in rats, in that a decreased intake of copper relative to zinc results in an increased level of serum cholesterol (Klevay, 1973). Epidemiologic evidence suggests that a similar phenomenon occurs in man (Klevay, 1975).

Observations in experimental animals have prompted the hypothesis that molybdenum plays a role in iron metabolism (Seelig, 1972). Molybdenum in xanthine oxidase (a molybdoflavin enzyme involved in electron transport) may participate in the reduction of ferritin–Fe^{3+} to ferritin–Fe^{2+}. It has been proposed that the Fe^{2+} is then released from the tissues to the plasma (Seelig, 1972). This interrelationship is at present unproven *in vivo*, and some of the reported studies appear to be in conflict.

Iron may also interact with certain clays in the alkaline intestinal milieu. Through cation exchange, the iron is bound to the clay and is thus rendered unavailable for intestinal absorption. This phenomenon appears to be clincially important in Turkey (Minnich *et al.*, 1968) and Iran (Halsted, 1968), where iron deficiency anemia associated with geophagia is common.

Dietary zinc may also exchange with the cations of ingested clay and thus become unavailable for intestinal absorption. Although this phenomenon has been studied in less detail than the absorption of iron, it is thought to be a factor in the pathogenesis of the zinc-responsive dwarfism that occurs among Iranian village children (Halsted, 1968).

Zinc can also interact with calcium and phytate in alkaline media to form an insoluble complex (O'Dell, 1969). The formation of this complex is thought to be one of the etiologic factors in the genesis of the zinc-responsive dwarfism (Sandstead *et al.*, 1967) that, accord-

ing to Reinhold *et al.* (1976), occurs in populations subsisting on high-fiber unleavened bread prepared from low-extraction wheat flour. (Low extraction is when less of the wheat berry is made into flour; 70 percent is low, 85 percent is high.) A second phenomenon that appears to impair the intestinal absorption of zinc is the formation of an insoluble complex between plant fiber and zinc. In contrast to its adverse effects on zinc absorption in the presence of phytate, calcium apparently will prevent the inhibitory effect of phytate on iron absorption. This is thought to occur through a competition between calcium and iron for binding sites on the phytate molecule (Apte and Venkatachalam, 1962; Haghshenass *et al.*, 1972).

Zinc-responsive growth failure also can occur in young individuals with intestinal malabsorption syndrome (Caggiano *et al.*, 1969; Sandstead *et al.*, 1976). The formation of an insoluble complex by zinc, phosphate, and fat in the alkaline environment of the small intestine is thought to account in part for zinc deficiences in such individuals.

Unconfirmed autopsy studies suggest there may be a competitive interaction between zinc and cadmium in humans (Voors *et al.*, 1973), such as that found in experimental animals (Friberg *et al.*, 1971), which indicates that zinc will protect animals from some of the toxic effects of cadmium. The studies by Voors *et al.* (1973) suggest an increase in the ratio of cadmium to zinc can result in injury to the renal cortex or to arterial walls, possibly resulting in hypertension and atherosclerosis.

The interaction of strontium, calcium, and phosphate in humans is of significance because of the presence of ^{90}Sr in the environment. In general, calcium absorption, excretion, and bone deposition is favored over that of strontium. However, in contrast to calcium, the movement of strontium through the body is apparently not homeostatically controlled. Instead, strontium uptake and retention appears to reflect the strontium:calcium ratio in the diet (Comar and Wasserman, 1970). Interactions between calcium and strontium are discussed in more detail in Chapter VII.

Fluorine interacts with calcium and phosphate in bone tissue and teeth (Shaw, 1967). The beneficial interaction of fluorine, calcium, and phosphate in growing teeth (i.e., caries resistance), when fluorine concentrations in drinking water are approximately 1.0–1.5 ppm, has been extensively documented. At concentrations of 2–4 ppm fluorine, chalky white discoloration and mottling of the teeth occur, although such teeth are also quite resistant to caries. Individuals consuming water with 4.0–5.8 ppm fluorine have been shown to have a lower incidence of degenerative bone disease (osteoporosis). Severe pathologic abnormalities may occur when fluoride concentrations in drinking water exceed 8–10 ppm, such as osteosclerosis, calcification of tendons and ligaments, deforming arthritis, myelopathies, and renal failure (Singh *et al.*, 1963).

An indirect interaction between lead and iodine has been observed both in experimental animals (Sandstead, 1967) and humans (Sandstead *et al.*, 1969). Lead has been shown to impair the uptake of iodine by the thyroid gland,

and, in animals, to decrease the *in vivo* protein binding of iodine. The significance of this phenomenon in human health is unknown. Also, lead inhibits the activity of δ-amino-levulinate-dehydratase, a zinc-dependent enzyme (Finelli *et al.*, 1974).

Selenium appears to protect against the lethal effects of mercuric chloride (Groth *et al.*, 1973) and methylmercury (Ganther *et al.*, 1973) in animals. It appears that the presence of selenium in fish that are at the end of a food chain, and thus accumulate methylmercury, tends to protect individuals who consume the fish from the toxic effects of methylmercury.

An interaction of lithium, sodium, and potassium appears to occur in the nervous system on the basis of *in vitro* physiological studies. Lithium can apparently compete with potassium for certain enzyme sites (Schou, 1957).

RECOMMENDATIONS FOR RESEARCH

1. Metabolic studies in humans to assess the interactions mentioned here in detail.

2. Metabolic studies in humans to evaluate other potential interactions that have been suggested by research on other animals and plants.

3. Research priorities for element interactions will depend in large measure upon geographical location. Some examples could be:

- the relation of selenium to mercury
- the effect of the zinc:copper ratio
- the zinc–cadmium interaction
- the effect of alkaline soils upon trace element availability
- the interference with iodine assimilation by calcium

4. In broader context, the multiple interrelationships suggested between water hardness and cardiovascular disorders, renal calculi, and other health problems are worthy of careful study. Accordingly, the Subcommittee has initiated the following panel studies: the Panel on the Geochemistry of Water in Relation to Cardiovascular Disease, the Panel on the Geochemical Environment and Urolithiasis, the Panel on Aging and the Geochemical Environment, and the Panel on the Trace Element Geochemistry of Coal Resource Development Related to Health.

REFERENCES

Allaway, W. H. 1970. Sulphur–selenium relationships in soils and plants. Sulphur Institute J. Washington, D.C. 6(3):3–5.

Anonymous. 1966. Human renal calculus formation and magnesium. Nutr. Rev. 24:43.

Apte, S. V., and P. S. Venkatachalam. 1962. Iron absorption in human volunteers using high phytate cereal diets. Indian J. Med. Res. 50:516.

Brewer, F. F. 1966. Fluorine. *In* Diagnostic criteria for plants and soils, H. D. Chapman [ed.]. Division of Agricultural Sciences, University of California, Riverside. pp. 180–196.

Caggiano, V., R. Schnitzler, W. Straus, R. K. Baker, A. C. Carter, A. S. Josephson, and S. Wallach. 1969. Zinc deficiency in a patient with retarded growth, hypogonadism, hypogamma-globulinemia and chronic infection. Am. J. Med. Sci. 257:305.

Chapman, H. D. [ed.] 1966. Diagnostic criteria for plants and soils. Division of Agricultural Sciences, University of California, Riverside. 793 pp.

Comar, C. L., and R. H. Wasserman. 1964. Strontium. *In* Mineral metabolism 2A, C. L. Comar and F. Bronner [eds.]. Academic Press, New York and London. 649 pp.

Cordano, A., J. M. Baertl, and G. G. Graham. 1964. Copper deficiency in infants. Pediatrics 34:324.

Crooke, W. M. 1955. Further aspects of the relationship between nickel toxicity and iron supply. Ann. Appl. Biol. 43:465–476.

Epstein, E. 1972. Mineral nutrition of plants: Principles and perspectives. John Wiley & Sons, New York. 412 pp.

Finelli, V. N., L. Murthy, W. B. Peirano, and H. G. Petering. 1974. Delta amino levulinate dehydratase, a zinc dependent enzyme. Biochem. Biophys. Res. Commun. 60(4):1418.

Friberg, L., M. Piscator, and G. Nordberg. 1971. Cadmium in the environment. U.S. Dep. Commerce Tech. Rep. APTD-0681. U.S. Department of Commerce, Washington, D.C.

Frieden, E. 1970. Ceruloplasmin, a link between copper and iron metabolism. Nutr. Rev. 28:87.

Ganther, H. E., P. A. Wagner, M. L. Sunde, and W. G. Hoekstra. 1973. Protective effects of selenium against heavy metal toxicities. *In* Proc. 6th Annu. Conf. Trace Subst. Environ. Health, June 13–15, 1972, D. D. Hemphill [ed.]. University of Missouri, Columbia. pp. 247–252.

Gershoff, S., and S. Andrus. 1961. Dietary magnesium, calcium and vitamin B6 and experimental nephropathies in rats: Calcium oxalate calculi, apatite nephrocalcinosis. J. Nutr. 73:308.

Grace, N. D., and B. L. O'Dell. 1970a. Interrelationships of dietary magnesium and potassium in the guinea pig. J. Nutr. 100:37.

Grace, N. D., and B. L. O'Dell. 1970b. Effect of magnesium deficiency on the distribution of water and cations in the muscle of the guinea pig. J. Nutr. 100:45.

Groth, D. H., L. Vignati, L. Lowry, G. Mackay, and H. E. Stokinger. 1973. Mutual antagonistic and synergistic effects of inorganic selenium and mercury salts in chronic experiments. *In* Proc. 6th Annu. Conf. Trace Subst. Environ. Health, June 13–15, 1972, D. D. Hemphill [ed.]. University of Missouri, Columbia. pp. 187–189.

Haghshenass, M., M. Mahloudji, J. G. Reinhold, and N. Mohammadi. 1972. Iron deficiency anemia in an Iranian population associated with high intakes of iron. Am. J. Clin. Nutr. 25:1143.

Halsted, J. A. 1968. Geophagia in man, its nature and nutritional effects. Am. J. Clin. Nutr. 21:1384.

Hodgson, J. F. 1963. Chemistry of the micronutrient elements in soils. Adv. Agron. 15:119–159.

Klevay, L. M. 1973. Hypercholesterolemia in rats produced by an increase in the ratio of zinc to copper ingested. Am. J. Clin. Nutr. 26:1060.

Klevay, L. M. 1975. Coronary heart disease: The zinc/copper hypothesis. Am. J. Clin. Nutr. 28:764–774.

Kubota, J., and W. H. Allaway. 1972. Geographic distribution of trace element problems. *In* Micronutrients in agriculture, J. J. Mortvedt, P. M. Giordano, and W. L. Lindsay [eds.]. Soil Science Society of America, Madison, Wisconsin. pp. 525–554.

Lagerwerff, J. V., and A. W. Specht. 1971. Occurrence of environmental cadmium and zinc, and their uptake by plants. *In* Proc. 4th Annu. Conf. Trace Subst. Environ. Health, June 23–25, 1970, D. D. Hemphill [ed.]. University of Missouri, Columbia. pp. 85–93.

Minnich, V., A. Okguoğlu, Y. Tarcon, A. Arcasay, S. Cin, O. Yörükoğlu, F. Renda, and B. Demirağ. 1968. Pica in Turkey. II. Effect of clay upon iron absorption. Am. J. Clin. Nutr. 21:78.

O'Dell, B. L. 1969. Effect of dietary components upon zinc availability: A review of original data. Am. J. Clin. Nutr. 22:1315.

Olsen, S. R. 1972. Micronutrient interactions. *In* Micronutrients in agriculture, J. J. Mortvedt, P. M. Giordano, and W. L. Lindsay [eds.]. Soil Science Society of America, Madison, Wisconsin. pp. 243–264.

Prien, E. L., Sr., and S. F. Gershoff. 1974. Magnesium oxide–pyridoxine therapy for recurrent calcium oxalate calculi. J. Urol. 112:509–512.

Reinhold, J. G., B. Faradji, P. Abadi, and F. Ismail-Beigi. 1976. Binding of zinc to fiber and other solids of wholemeal bread; with a preliminary examination of the effects of cellulose consumption upon the metabolism of zinc, calcium, and phosphorus in man. *In* Trace elements in human health and disease, Vol. I: copper and zinc, A. S. Prasad [ed]. Academic Press, New York. [2 vols.].

Sandstead, H. H. 1967. Effect of chronic lead intoxication on *in vivo* [131]I uptake by the rat thyroid. Proc. Soc. Exp. Biol. Med. 124:18.

Sandstead, H. H., K. P. Vo-Khactu, and N. Solomon. 1976. Conditioned zinc deficiencies. *In* Trace elements in human health and disease, Vol. I: Copper and zinc, A. S. Prasad [ed.]. Academic Press, New York. [2 vols.]

Sandstead, H. H., A. S. Prasad, A. R. Schulert, Z. Farid, A. Miale, Jr., S. Bassilly, and W. J. Darby. 1967. Human zinc deficiency, endocrine manifestations and response to treatment. Am. J. Clin. Nutr. 20:422–442.

Sandstead, H. H., E. G. Stant, and A. B. Brill. 1969. Lead intoxication and the thyroid. Arch. Int. Med. 123:632.

Schou, M. 1957. Biology and pharmacology of the lithium ion. Pharmacol. Rev. 9:17–58.

Seelig, M. S. 1971. Human requirements of magnesium, factors that increase needs. *In* First Symposium International sur le Deficit Magnesique en Pathologie Humaine, J. Durlach [ed.]. Vittel, France. p. 11.

Seelig, M. S. 1972. Relationships of copper and molybdenum to iron metabolism. Am. J. Clin. Nutr. 25:1022.

Shaw, J. H. 1967. Present knowledge of fluorine. *In* Present knowledge in nutrition. The Nutrition Foundation, New York, p. 130.

Shils, M. E. 1969. Experimental human magnesium depletion. Medicine 48:61–85.

Singh, A., S. S. Jolly, B. C. Bansal, and C. C. Mathur. 1963. Endemic fluorosis. Medicine 42:229.

Voors, A. W., M. S. Shuman, and P. N. Gallagher. 1973. Atherosclerosis and hypertension in relation to some trace elements in tissues. *In* Proc. 6th Annu. Conf. Trace Subst. Environ. Health, June 13–15, 1972, D. D. Hemphill [ed.]. University of Missouri, Columbia. pp. 215–222.

Wacker, W. E. C. 1969. The biochemistry of magnesium. Ann. N.Y. Acad. Sci. 162:717.

XI

Consequences of Soil Imbalances

JOE KUBOTA, *Chairman*

Kenneth C. Beeson, Thomas D. Hinesly, Everett A. Jenne,
Willard L. Lindsay, Perry R. Stout

An ideal balance of plant nutrients for each plant species does not exist and is not required in nature, for each plant species has a mechanism that regulates the absorption and uptake of those nutrients necessary for its growth and reproduction. Thus legumes have a high requirement for calcium and, consequently, a high concentration of that element in their tissue. On seleniferous soils certain species of *Astragalus* can take up large quantities of selenium, while native grasses accumulate only traces of the element (Beath *et al.*, 1934). Other plants may accumulate cobalt or zinc (Beeson *et al.*, 1955), barium (Robinson *et al.*, 1938), or many of the heavy metals important to man (Cannon, 1955) from soils in which these elements are very low. The selective uptake that accumulator plants exhibit can pose special situations—not related to soil imbalances—some favorable and some unfavorable to the animal or man.

For most plant species grown for food or feed, the ratio or balance of the nutrient elements is not critical until an aberrant supply of one or more nutrients occurs. For example, a low supply of one of the nutrients (Ca, P, Mg, N, Fe, Zn, B, Cu, Mn, or Mo) may result in a chlorotic plant, a poor yield, or no yield at all. A toxic supply of an element (Na, Ni, Cr, B, Al, Zn, or Mn) may retard the growth or eliminate entirely certain species of plants as occurs on the Conowingo Barrens, Maryland (McMurtrey and Robinson, 1938). Antagonistic or synergistic interactions among the elements (Ca–Mg, P–Cu, P–Zn, Fe–several heavy metals, S–Se, and B–Ca) may have important effects on the growth of certain plant species.

There are many soil properties that may modify the availability of the nutrient elements to plants. An important example is soil acidity. A low pH may enhance the availability of iron or manganese but limit that of selenium or molybdenum. In alkaline soils, the reverse is true. Organic matter may promote the uptake of iron but retard the uptake of copper. Phosphorus is less available in alkaline, arid soils than in moist soils.

Reviews of various aspects of micronutrient element movement to plants, mechanisms, and mineral interactions that affect their uptake by plants, and their translocation within plants, are available (Dinauer, 1972).

NATURAL IMBALANCES THAT AFFECT CROP PRODUCION AND QUALITY

In the United States, the areas of otherwise arable soils, where naturally occurring element imbalances result in toxicity and retarded growth of plants, are not very large. Staker and Cummings (1941) reported that excessive concentrations of zinc in certain New York peats were responsible for poor growth and quality of several vegetables. Cannon (1955, 1969) reported that, in addition to zinc, relatively large concentrations of lead and cadmium occurred in these peats.

Soils developed on serpentine rock in Oregon contain toxic quantities of nickel and chromium (Cannon, 1969), and it has long been known that the infertility of the soils of the Conowingo Barrens, Maryland—also developed on

116

serpentine rock—is due to the high levels of available chromium and nickel in those soils (McMurtrey and Robinson, 1938). Cannon (1969) has noted numerous minor areas in the United States where the soils contain above-average quantities of chromium, nickel, titanium, lead, copper, vanadium, lithium, and molybdenum. Neither high levels of selenium (Beeson, 1961) nor molybdenum (Kubota *et al.*, 1961) in arable soils appears to have any adverse effects on the growth or appearance of malnutrition in crops. In nutrient solutions, however, Hurd-Karrer (1934) reported that selenium was toxic to wheat plants if no sulfur was present in the media.

Imbalances due to a lower than normal supply of one or more nutrient elements in the soil are, of course, quite common in arable soils throughout the world. Such deficiencies may limit only the yield of the crop, or they may result in an abnormal appearance typical of a particular deficiency. However, a subnormal yield or growth of a normal-appearing plant is not necessarily an indication of a poor nutritional quality. In fact, one limiting nutrient may result in a plant taking up and storing other nutrient elements in abundance. For example, the concentration of iron in plants grown with adequate phosphorus or nitrogen may be lower than in plants grown in a less-fertile soil. This is known as the dilution effect (Beeson *et al.*, 1948). The concentrations of all nutrient elements in the plant fluctuate in a more or less normal range. As the lower level of that range for one nutrient element is reached, the growth of the plant is retarded while the uptake of other nutrients continues. For the same reason, the addition of a plant nutrient to a soil, where a low supply of the nutrient is limiting growth, may not result in an increase in the concentration of that element in the plant until a level in excess of the plant's need is reached in the soil or a deficiency of some other nutrient occurs.

Micronutrient element deficiencies or toxicities—or imbalances—often have geographical distribution patterns (Beeson, 1945; Berger, 1972; Kubota and Allaway, 1972) that also reflect the distribution of certain soils and their capabilities for crop production. Hence, more information about trace elements is generally available where nutritional imbalances occur than where they do not.

IMBALANCES THAT AFFECT THE NUTRITIVE QUALITY OF CROPS

Some of the essential micronutrients required by plants include manganese, iron, copper, zinc, and boron. Provided that the diet of man or animals is composed of plant sources recognized as containing adequate quantities of these elements, there need not be great concern. However, nutritionally important differences are seen. For example, Kentucky bluegrass (*Poa pratensis* L.) has adequate copper as compared to Para grass (*Panicum barbinode* Trin.) (Beeson *et al.*, 1947). Growing in the same soil, the latter grass may fail to supply sufficient copper for the animal, while bluegrass will supply an ample quantity.

Plants serve as a buffering mechanism in the food chain, particularly to excessive movement of many trace elements, such as lead, cadmium, and mercury, to the human food supply. The selectivity and tolerance of plants for other trace elements not required—selenium, cobalt, iodine, and fluorine—may result, however, in a serious imbalance of the micronutrients required in the diet by man and animals.

Imbalances of selenium in the soil clearly demonstrate that an element—not required by the plant—can affect animals, and possibly man, in two ways: (a) through an excess in the plant resulting in a toxicity in the animal or (b) through a very low concentration in the plant and failure to supply the needs of the animal. Areas of high levels of available selenium have long been recognized (Lakin, 1961; Beeson, 1961), while more recent studies have identified the low or deficient areas of soils and of alfalfa growing thereon (Kubota *et al.*, 1967). In general, the highly leached soils of eastern, northeastern, and far northwestern regions of the United States are the low-selenium areas, while the alkaline soils of the Great Plains and southwestern United States have ample to toxic supplies of available selenium.

Cobalt, like selenium and iodine, has no direct effect on plant growth (Beeson, 1950), but the plant is the principal source of the element for the ruminant animal. Hence, a low concentration of available cobalt in the soil results in an imbalance in the plant of nutrients that are required by the animal. Areas of cobalt deficiency in forages have been determined by Kubota (1968).

Iodine imbalances, the first of those recognized in the nutrient supply of elements for man and animals, may not be directly related to the supply of the element in the soil or its parent rock since a principal primary source of iodine is that carried inland by moisture from the seas (Vought *et al.*, 1970). Imbalances of both iodine and fluorine with respect to both man and animals are also clearly related to the supplies in water (Underwood, 1971; Shacklette and Cuthbert, 1967).

MAN-CAUSED IMBALANCES IN SOILS

The problems associated with the disposal of municipal effluents and sludges by utilizing them as sources of plant nutrients and soil moisture have been a continuing subject of study as evidenced by the appearance of over 50 papers in the last several years. The principal concern of agronomists and environmentalists is the high probability that land disposal of large quantities of sewage sludge on some soils may result in heavy contamination of not only the soil but also the groundwaters with potentially toxic elements. In a recent study, Bradford and his co-workers (1975) reported that vegetable crops took up excessive quantities of copper, molybdenum, nickel, cobalt, lead, and cadmium from an extract of municipal sludge applied to sand cultures in the greenhouse. Reduced yields and problems of phytotoxicity occur (Cunningham *et al.*, 1975; Kirkham, 1975; Lu *et al.*, 1975). These reports and others in the literature clearly point to the need, however, for more information on the range of concentrations of the potentially toxic elements in the edible portion of crops where normal yields are obtained from the application of contaminated waste materials to soils.

Surface strip mine activities may create soil imbalance problems that are reflected in plants grown on reclaimed land. The immediate concern is the rapid establishment of vegetative cover so that surface erosion and stream pollution are minimized. Long-term objectives are the ultimate return of the land to productive use through the reclamation process. Thus, in the western United States, surface enrichment of an element like selenium may result when underlying coal-bearing Tertiary rocks are overturned in mining operations and become media for plant growth. Some rocks of Tertiary age have been found to be high in selenium (Lakin, 1961), and such material if returned to the land surface may be a source of selenium in readily available forms for plant uptake. Selenium accumulator plants are native to the area, and their invasion on reclaimed mine spoils would accentuate the impact of readily available selenium. The importance to agriculture—and particularly to food production—of the recovery of mine wastes has yet to be evaluated.

Practices for increasing crop yields can be important factors in creating imbalances. The removal of topsoil to facilitate gravity irrigation may expose subsoils that are low in available zinc (Grunes *et al.*, 1961). The application of phosphate fertilizer or other nutrients under an intensive agriculture can create an imbalance with the micronutrients if proper supplementation is not included with them in the fertilizer program.

An interesting effect of heavy application of phosphates to tobacco soils that might be associated with lung cancer has been suggested by Martell (1975). Many phosphate rocks are good sources of uranium, hence ^{226}Ra. It appears that ^{222}Rn is present in the gaseous phase of tobacco soils and is adsorbed on the trichoma of the tobacco leaf. When the leaf used as smoking tobacco is ignited, the high temperature results in the formation of insoluble smoke particles containing ^{210}Po in the particulate matter of the smoke, which settles on the bronchial tissue of the lung.

SUMMARY

Naturally occurring soil imbalances of the mineral elements—both nutrient elements and other available elements—are readily recognized in relation to subnormal yields and the appearance of toxicity or deficiency symptoms in plants. These problems are resolved through adjustments in the balance of nutrients in the soil to meet the specific needs of a plant species. Where this is not feasible or economical, a crop suited or tolerant to a particular imbalance can be substituted.

An imbalance of mineral elements can occur for both man and animals when an essential nutrient not required by a food plant is in poor supply in the soil, or in those situations where a plant can take up and tolerate in its tissue a high concentration of an element that may be toxic to man or animals.

Man-caused imbalances in soils—in particular, heavy metal toxicities arising from excessive applications of sewage sludge, phosphate, or other fertilizers—must be avoided, since the long-range effects may result in abandoning a soil for all crops sensitive to such elements.

The potentially dangerous effects of toxic elements in normal appearing food crops are still to be evaluated.

REFERENCES

Beath, O. A., J. H. Draize, and C. S. Gilbert. 1934. Plants poisonous to livestock. Wyo. Agric. Exp. Stn. Bull. 200:1–84.

Beeson, K. C. 1945. The occurrence of mineral nutritional diseases of plants and animals in the United States. Soil Sci. 60(1):9–13.

Beeson, K. C. 1950. Cobalt, occurrence in soils and forages in relation to a nutritional disorder in ruminants. A review of the literature. Agric. Inf. Bull. No. 7. U.S. Department of Agriculture, Washington, D.C. 44 pp.

Beeson, K. C. 1961. Occurrence and significance of selenium in plants. Agric. Handb. USDA 200:34–40.

Beeson, K. C., L. Gray, and M. B. Adams. 1947. The absorption of mineral elements by forage plants. I. The phosphorus, cobalt, manganese and copper content of some common grasses. J. Am. Soc. Agron. 39:356–362.

Beeson, K. C., L. Gray, and K. C. Hamner. 1948. The absorption of mineral elements by forage plants. II. The effect of fertilizer elements and liming materials on the content of mineral nutrients in soybean leaves. J. Am. Soc. Agron. 40:553–562.

Beeson, K. C., V. A. Lazar, and S. G. Boyce. 1955. Some plant accumulators of the micronutrient elements. Ecology 36:155–156.

Berger, K. C. 1972. Micronutrient deficiencies in the United States. J. Agric. Food Chem. 10:178–181.

Bradford, G. R., A. L. Page, L. J. Lund, and W. Olmstead. 1975. Trace element concentrations of sewage treatment plant effluents and sludges; their interactions with soils and uptake by plants. J. Environ. Qual. 4(1):123–127.

Cannon, H. L. 1955. Geochemical relations of zinc bearing peat to the Lockport dolomite, Orleans County, New York. U.S. Geol. Surv. Bull. 1000-D:119–185.

Cannon, H. L. 1969. Trace element excesses and deficiencies in some geochemical provinces of the United States. *In* Proc. 3d Annu. Conf. Trace Subst. Environ. Health, D. D. Hemphill [ed.]. University of Missouri. Columbia. pp. 21–43.

Cunningham, J. D., D. R. Keeney, and J. A. Ryan. 1975. Yield and metal composition of corn and rye grown on sewage sludge amended soils. J. Environ. Qual. 4(4):448–454.

Dinauer, R. C. [ed.] 1972. Micronutrients in agriculture. Soil Science Society of America, Madison, Wisconsin. 666 pp.

Grunes, D. L., L. C. Boawn, C. W. Carlson, and F. G. Viets, Jr. 1961. Land leveling, it may cause zinc deficiency. Farm Res. 21:4–7.

Hurd-Karrer, A. M. 1934. Selenium injury to wheat plants and its inhibition by sulfur. J. Agric. Res. 49:343–357.

Kirkham, M. B. 1975. Uptake of cadmium and zinc from sludge

by barley grown under four different sludge irrigation regimes. J. Environ. Qual. 4(3):423–426.

Kubota, J. 1968. Distribution of cobalt deficiency in grazing animals in relation to soils and forage plants of the United States. Soil Sci. 106:122–130.

Kubota, J. and W. H. Allaway. 1972. Geographic distribution of trace element problems. *In* Micronutrients in agriculture, R. C. Dinauer [ed.]. Soil Science Society of America, Madison, Wisconsin. pp. 525–554.

Kubota, J., W. H. Allaway, D. L. Carter, E. E. Cary, and V. A. Lazar. 1967. Selenium in crops in the United States in relation to selenium-responsive diseases of animals. Agric. Food Chem. 15(3):448–453.

Kubota, J., V. A. Lazar, L. N. Langan, and K. C. Beeson. 1961. The relationship of soil to molybdenum toxicity in cattle in Nevada. Soil Sci. Soc. Am. Proc. 25(3):227–232.

Lakin, H. W. 1961. Geochemistry of selenium in relation to agriculture. *In* Selenium in agriculture Agr. Handbook 200. U.S. Department of Agriculture, Washington, D.C. pp. 3–12.

Lu, Po-Yung, R. L. Mitchell, R. Furman, R. Vogel, and J. Hassett. 1975. Model ecosystem studies of lead and cadmium and of urban sewage sludge containing these elements. J. Environ. Qual. 4(4):505–509.

Martell, E. A. 1975. Tobacco radioactivity and cancer in smokers. Am. Scientist 63:404–412.

McMurtrey, J. E., and W. O. Robinson. 1938. Neglected soil constituents that affect plant and animal development. *In* Yearbook of agriculture, Gove Hambidge [ed.]. U.S. Department of Agriculture, Washington, D.C. pp. 807–829.

Robinson, W. O., R. R. Whetstone, and H. G. Byers, 1938. Studies on infertile soils. II. Soils high in barium. Soil Sci. Soc. Am. Proc. 3:87–91.

Staker, E. V., and R. W. Cummings. 1941. The influence of zinc on the productivity of certain New York peat soils. Soil Sci. Soc. Am. Proc. 6:207–214.

Shacklette, H. T., and M. E. Cuthbert. 1967. Iodine content of plant groups as influenced by variation in rock and soil type. Geol. Soc. Am. Spec. Pap. 90:30–46.

Underwood, E. J. 1971. Trace elements in human and animal nutrition, 3d ed. Academic Press, New York. 543 pp.

Vought, R. L., F. A. Brown, and W. T. London. 1970. Iodine in the environment. Arch. Environ. Health 20:516–522.

XII

Disease Patterns of Southwestern Indians in Relation to Environmental Factors

MAURICE L. SIEVERS

Southwestern Indian tribes, because of their relative genetic homogeneity and isolated environment (the vast majority live on reservations), have for many years provided an unusual opportunity for the study of the effects of various hereditary and environmental factors on disease prevalence, health, and longevity. Although these tribes exhibit patterns of disease occurrence that are significantly different from those in the non-Indian population of the United States, the causes of this variation remain undetermined.

At present, however, both the genetic and environmental conditions are changing rapidly, especially as they involve outside cultural influences; intermarriage, changed diets, moves to urban areas, jobs in industry, or nontraditional living styles progressively impinge on reservation life. Among the Pima tribe of the Gila River Reservation in Arizona, for example, 93 percent of the population 15 yr old or older is putatively of pure Indian ancestry, whereas of those less than 15 yr old only 73 percent can make that claim. Thus the opportunities for study of an isolated genetically homogeneous population are rapidly diminishing and are increasingly confined to members of the older generations.

One aspect of the altered disease patterns among southwestern Indians that has not been adequately studied, however, is the possibility that these patterns may be related to the characteristics of local geochemical environments. Because the effects of high or low intake of some trace substances have now been established for humans, a study of the potential involvement of such substances in reservation disease patterns might prove fruitful. As a beginning, trace element analyses might be made of reservation soil, vegetation, water, and foodstuffs, as well as human tissues obtained at surgery or autopsy (as proposed by the Brigham Young University Environmental Center). These results could be evaluated in terms of differing prevalences of diseases among the southwestern tribes and between Indians and non-Indians, or even between herds of cattle grazing on the reservations.

Such a study should include consideration of the disease patterns discussed below. Only those conditions have been included in which the frequency, distribution, or manifestations differ significantly between southwestern Indians and non-Indians (defined as all races in the United States exclusive of American Indians). Table 34 summarizes these differences.

GASTRIC SECRETORY DISEASES

Full-blooded southwestern Indians have gastric carcinoma and pernicious anemia at least as often as non-Indians, but they have a markedly lower rate of peptic ulcer (Sievers and Marquis, 1962b). Gastric ulcer occurs 1/9, and duodenal ulcer 1/73 as often as predicted (Sievers, 1973). Southwestern Indians probably have achlorhydria more frequently than either the white or black racial groups (Sievers, 1966a, 1966b).

120

TABLE 34 Frequency of Conditions in Southwestern Indian versus Non-Indian Populations[a]

Condition	Athapascan[b]	Piman[c]
Diabetes mellitus	(±)	(++++)
Obesity	(±) to (+)	(++++)
Myocardial infarction rate/100,000/yr (over age 40 yr)	(−−) to (−−−) 102	(−) to (−−) 149 [Framingham: 450]
Hypertension	?(±)	(−)
Serum cholesterol level, mean (mg/100 ml)	(−) to (−−) 206	(−−) to (−−−) 190
Alcohol drinking (heavy)	(+++) to (++++)	(++) to (+++)
Cigarette smoking	(−−−)to (−−−−)	(−−) to (−−−)
Epidermoid lung carcinoma	(−−−−) to (±) [Navajo uranium miners = (±)]	(−−−−)
Biliary carcinoma	(++) to (+++)	(++++)
Cholelithiasis	(++) to (+++)	(++++)
Peptic ulcer	(−−) to (−−−)	(−−−−)
Gastric carcinoma	(±)	(+) to (++)
Skeletal and dental fluorosis	(±) to (++++) [Bylas Apache = (++++)]	(±) to (++++) [Gila Bend Papago = (++++)]
Congenital lesions	(++) to (+++)	(±) to (+)
Infectious diseases (esp. tuberculosis)	(++++) [Navajo = Whiteriver Apache > San Carlos Apache]	(++++) [Papago > Pima]

[a]Comparative frequency of selected conditions of Athapascan and Piman ethnic groups of American Indians relative to the prevalence rates for the non-Indian population of the United States. Although there are many others, these are the two major southwestern tribal groupings.
[b]Navajo and Apache tribes.
[c]Pima and Papago tribes.
KEY: Symbols are relative to rate in U.S. non-Indian population.

Greater than:			Less than:	
(++++)	very much		(−−−−)	very much
(+++)	much		(−−−)	much
(++)	moderately		(−−)	moderately
(+)	slightly		(−)	slightly
(±)	approximately equal		(±)	approximately equal

CHOLELITHIASIS

Cholesterol gallstones develop at a far higher rate in Indians than in non-Indians (Sievers and Marquis, 1962a; Sievers, 1966b). The prevalence for Pimas is 75 percent of women more than 25 yr old and 55 percent of men over 45 yr of age (Sampliner et al., 1970). Current evidence indicates that cholesterol gallstone formation results from production of bile supersaturated with cholesterol relative to the solubilizing bile acids (Grundy et al., 1972). A marked excretion of cholesterol in the bile occurs among Indians (Grundy et al., 1972), although their serum cholesterol levels are lower than those of the white population (Sievers, 1966b, 1968a; Sampliner et al., 1970).

CIRRHOSIS OF LIVER

Alcoholic (Laennec's) cirrhosis of the liver is more prevalent, occurs at a younger age, and involves women more often in southwestern Indians than in the U.S. white population (Sievers, 1966b, 1968b). Similar proportions of the Indians and whites drink alcohol, but the extent of heavy drinking is greater for Indians (Sievers, 1968b). Although some evidence suggests that alcohol may be

metabolized less rapidly by Indians than by whites (Fenna et al., 1971), a recent study found no racial differences in the rate of ethanol metabolism and in liver alcohol dehydrogenase (Bennion and Li, 1976).

DIABETES MELLITUS

The rate of diabetes mellitus varies considerably among Indian tribes, although most southwestern tribes have a much higher prevalence than whites. Observations at the Phoenix Indian Medical Center (PIMC) reveal that 45 percent of adult Pimas are afflicted (Sievers, 1966b). Population studies by Bennett et al. (1971) confirm the extraordinary frequency of diabetes mellitus in Pima Indians—the highest rate of occurrence ever reported, and 10–15 times that for the general population.

ATHEROSCLEROSIS AND MYOCARDIAL INFARCTION

Among all southwestern tribes, atherosclerosis is less common than in whites (Sievers, 1966b, 1967; Reichenbach, 1967). Myocardial infarction occurs at a rate (86.8/100,000) that is about a fourth of that expected (Sievers,

1967) from experience with Caucasians aged 30 yr or older (347.8/100,000) in the Framingham Study (Kannel *et al.*, 1961). Cigarette smoking (Sievers, 1968b) and hypercholesterolemia (Sievers, 1968a) are infrequent in southwestern Indians. The prevalence of myocardial infarction varies directly with the frequency of diabetes mellitus and obesity among the tribes (Sievers, 1966b).

HYPERTENSION

There is a clinical impression that high blood pressure is somewhat less frequent in many southwestern tribes than in non-Indians (Sievers, 1966b). A recent investigation by Inglefinger and Bennett (personal communication, 1973) compared results for Pima Indians with the Tecumseh Study for the white population 30 or more yr of age (Johnson *et al.*, 1965). The Pimas had significantly lower mean diastolic pressure overall, as well as lower mean systolic pressure in the nondiabetics.

MALIGNANT NEOPLASMS

Although cancer of the lung is the most common malignant neoplasm of the white population, bronchogenic (epidermoid) carcinoma is rare in southwestern Indians, except among uranium miners (Sievers, 1961, 1966b, 1972; Reichenbach, 1967). Cigarette smoking is infrequent and seldom extensive among these tribes (Sievers, 1968b). Indeed, alveolar cell carcinoma of the lung, which is not associated with smoking, occurs at the expected frequency (C. L. Self, personal communication, 1973). The breast is the site of most malignant lesions in white women, but southwestern Indians have a relatively low rate of breast cancer (Sievers, 1966b, 1972; Goldman *et al.*, 1972). A lower rate also exists for cancer of the colon, but not for malignant lesions of the uterus or stomach (Sievers, 1966b, 1972; Goldman *et al.*, 1972). Biliary carcinoma is much more frequent in southwestern Indians than in whites (Sievers and Marquis, 1962a; Sievers, 1966b, 1972; Goldman *et al.*, 1972); leukemia and Hodgkin's disease are less common, but multiple myeloma is apparently more common (Sievers, 1972).

SKELETAL AND DENTAL FLUOROSIS

The water supply for several reservations has a high fluoride concentration, and skeletal and dental fluorosis occur frequently (Morris, 1965; Sievers, 1966b, Goldman *et al.*, 1971, 1972). Usually, the deposition of excessive fluoride in teeth and bones is of no clinical significance, but one southwestern Indian has developed radiculomyopathy as a result of extremely severe skeletal fluorosis (Goldman *et al.*, 1971). Fluorosis is exhibited most often by the Papagos, but is noted only occasionally among the genetically related Pimas (Morris, 1965), although their reservations are less than 50 miles apart.

MISCELLANEOUS

Bronchial asthma is infrequent in southwestern Indians (Sievers, 1966b). Congenital lesions occur often in many tribes, particularly the Apache, Navajo, and Hopi (Sievers, 1966b; Goldman *et al.*, 1972). Albinism, polydactylism, hereditary hemorrhagic telangiectasia, congenital cardiovascular lesions, and dysplasia of the hip appear frequently (Sievers, 1966b; Goldman *et al.*, 1972). Most American Indian tribes have a very high prevalence of obesity. Among southwestern desert tribes, 65 percent of women and 39 percent of men over 15 yr old exceed their ideal weights by more than 25 percent (Sievers and Hendrikx, 1972). The Prima tribe has an especially high rate of marked obesity. Infectious diseases occur frequently in southwestern Indians; the prevalence of pulmonary tuberculosis is particularly great among the Navajos and Whiteriver Apaches (Sievers, 1966b).

RECOMMENDATION

The Athapascan (Navajo/Apache) and Piman (Pima/Papago) Indian tribes in the Southwest have, separately, relative genetic homogeneity; they differ, both from the general population and from each other, in the prevalence of several diseases. Their reservations are relatively isolated, and each has distinctive environmental characteristics. These conditions offer a unique opportunity to investigate the potential effects of geochemical environment on human health.

Therefore, a geochemical study should be conducted, including comparisons of reservations of the Athapascans and the Pimans, and adjacent analogous non-Indian or nonreservation areas, through the use of extensive analyses of selected trace substances in soil, water, vegetation, animals, diet (regardless of source), and human tissues (hair, bone, soft tissue, organs) obtained by biopsy or at autopsy.

REFERENCES

Bennion, L. J., and T.-K. Li. 1976. Alcohol metabolism in American Indians and whites. Lack of racial differences in metabolic rate and liver alcohol dehydrogenase. N. Engl. J. Med. 294: 9–13.

Bennett, P. H., T. A. Burch, and M. Miller. 1971. Diabetes mellitus in American (Pima) Indians. Lancet 2:125–128.

Fenna, D., L. Mix, O. Schaefer, and J. A. L. Gilbert. 1971. Ethanol metabolism in various racial groups. Can. Med. Assoc. J. 105:472–475.

Goldman, S. M., M. L. Sievers, and D. W. Templin. 1971. Radiculomyopathy in a southwestern Indian due to skeletal fluorosis. Ariz. Med. 28:675–677.

Goldman, S. M., M. L. Sievers, W. K. Carlile, and S. L. Cohen. 1972. Roentgen manifestations of disease in southwestern Indians. Radiology 103:303–306.

Grundy, S. M., A. L. Metzger, and R. D. Adler. 1972. Mechanisms of lithogenic bile formation in American Indian women with cholesterol gallstones. J. Clin. Invest. 51:3026–3043.

Johnson, B. C., F. H. Epstein, and M. O. Kjelsberg. 1965. Distributions and familial studies of blood pressure and serum cholesterol levels in a total community—Tecumseh, Michigan. J. Chron. Dis. 18:147–160.

Kannel, W. B., T. R. Dawber, A. Kagan, N. Revotskie, and J. Stokes III. 1961. Factors of risk in the development of coronary heart disease—Six year follow-up experience. Ann. Intern. Med. 55:33–50.

Morris, J. W. 1965. Skeletal fluorosis among Indians of the American Southwest. Am. J. Roent. 94:608–615.

Reichenbach, D. D. 1967. Autopsy incidence of diseases among southwestern American Indians. Arch. Pathol. 84:81–86.

Sampliner, R. E., P. H. Bennett, L. J. Comess, F. A. Rose, and T. A. Burch. 1970. Gallbladder disease in Pima Indians. Demonstration of high prevalence and early onset by cholecystography. N. Engl. J. Med. 283:1358–1364.

Sievers, M. L. 1961. Lung cancer among Indians of the southwestern United States. Ann. Intern. Med. 54:912–915.

Sievers, M. L. 1966a. A study of achlorhydria among southwestern American Indians. Am. J. Gastroenterol. 45:99–108.

Sievers, M. L. 1966b. Disease patterns among southwestern Indians. Public Health Rep. 81:1075–1083.

Sievers, M. L. 1967. Myocardial infarction among southwestern American Indians. Ann. Intern. Med. 67:800–807.

Sievers, M. L., 1968a. Serum cholesterol levels in southwestern American Indians. J. Chron. Dis. 21:107–115.

Sievers, M. L. 1968b. Cigarette and alcohol usage by southwestern American Indians. Am. J. Public Health 58:71–82.

Sievers, M. L. 1972. Cancer mortality in Indians. J. Am. Med. Assoc. 222:705.

Sievers, M. L. 1973. Unusual comparative frequency of gastric carcinoma, pernicious anemia, and peptic ulcer in southwestern American Indians. Gastroenterology 65:867–876.

Sievers, M. L., and M. E. Hendrikx. 1972. Long-term results. Two weight reduction programs among southwestern Indians. Health Serv. Rep. 87:530–536.

Sievers, M. L., and J. R. Marquis. 1962a. The southwestern American Indian's burden: Biliary disease. J. Am. Med. Assoc. 182:172–174.

Sievers, M. L., and J. R. Marquis. 1962b. Duodenal ulcer among southwestern American Indians. Gastroenterology 42:566–569.

XIII

The Missouri Study

DELBERT D. HEMPHILL, *Chairman*

Carl J. Marienfeld, Alfred T. Miesch, James O. Pierce, Herbert I. Sauer, Lloyd A. Selby

In June 1965, at the University of Missouri, a program was begun that consisted of eight related studies concerned with the effects of trace substances in the environment on man, domestic and wild animals, fish, higher plants, and microorganisms of the soil and water. This was an interdisciplinary study involving the participation of pediatricians, pathologists, veterinarians, wildlife biologists, geologists, soil scientists, plant scientists, chemists, sanitary engineers, geneticists, toxicologists, epidemiologists, and statisticians. The first part of this report will deal only with the human and animal aspects of the study, whereas the latter part describes the geochemical survey of Missouri that was carried out by scientists of the U.S. Geological Survey who joined the study group in 1969.

HUMAN AND ANIMAL STUDIES

Progress has been made on the study of human birth defects or congenital anomalies, particularly those malformations not considered to be hereditary in nature, and also with stillbirths. Progress has also been made on the study of congenital anomalies and stillbirths in swine. The hypothesis of this research was that the fetus is a sensitive indicator of both toxic and deficient substances in the environment and that either might well result in increased rates of congenital malformations in humans or swine and might also be responsible for increased rates of stillbirths or fetal deaths, with the latter being considered separately from live births. Research has also been fo-

cused on chronic diseases in adults, with the general hypothesis that geographic differences in age-specific death rates for the chronic diseases are due to geographic differences in trace substances and other environmental factors.

Human Birth Defects

This study began with an examination of birth certificates of children born to residents of the State of Missouri since 1946. The year 1946 was selected as the beginning date because it immediately preceded the widespread use of organic agricultural pesticides whose residues in foods and water are primary suspects as causal agents of numerous maladies in man, fish, and wildlife.

Data from these certificates were plotted on a map of Missouri, with a dot representing the residence of the mother of each malformed child; the map for 1968 is shown in Figure 18. Large metropolitan areas (Kansas City and St. Louis) were excluded from the plot, although these areas were included in determining rates of birth defects per thousand live births. Of particular concern was the possible clustering of malformations in any area of the state and the potential correlation of any such clustering with a particular environmental factor (Silberg et al., 1966).

The rate of total birth defects per thousand live births for the period 1953 through 1964 ranged from 7 to 7.8, with a mean of 7.4 (Silberg et al., 1966). The rates for the years 1946 through 1952 were similar. In 1965 the rate

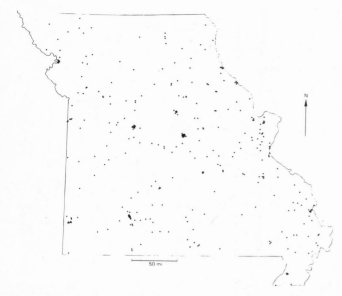

FIGURE 18 A map of Missouri showing the distribution of residences of mothers of 735 malformed children born during 1968. Each dot represents a malformed child. The large metropolitan areas of Kansas City and St. Louis are excluded. Clusters tend to reflect the location of moderate-sized cities.

increased to 9.09 and has continued to be between 9 and 10 since. The reason for the increase in rate has not been determined. However, since about 1965, teenage mothers have become more numerous, and mothers under 19 or over 35 yr of age are more likely to give birth to children with congenital anomalies. The most common types of birth defects that have been reported are impaired hearing, blindness, clubfoot, heart ailment, spinal deformity, cleft palate, cleft lip, hernia, mongolism, and dislocated hip.

Swine Birth Defects

Birth defects in swine were also studied. The reason for studying swine was that the large number of animals in the state could be compared to human population in rural areas. We suspected that pathological defects were similar (we now know this to be true), and many physiological functions of swine are similar to the same functions in man. Care is needed during pregnancy, and swine are more closely related to the local environment than are humans.

A pilot study in one county (Nodaway) was conducted between June 1, 1966 and May 31, 1967, covering two farrowing periods (Selby *et al.*, 1971a). A postcard reporting form was developed, and an orientation meeting was held for the county extension staff, related agencies, veterinarians, physicians, and selected community leaders. A list of all known swine producers was compiled, and the reporting cards were mailed to them. Approximately 40 percent of the forms were returned.

The malformation rate was found to be 4.8 per thousand

total births and 5.6 per thousand live births of the swine. Malformations were classified mostly according to the following body systems: (a) central nervous system, (b) sensory organs, (c) alimentary and respiratory system, (d) genitourinary system, (e) bones and joints, and (f) other systems, e.g., hair, skin, monster or conjoined (Siamese) twin. They were then compared with known human malformations for 1966. The frequency distribution of each type was reasonably similar in both swine and humans.

After the pilot study, the decision was made to extend the study to all counties in the state. The statewide study was conducted over a 3-yr period between October 1, 1967, and September 30, 1970, and covered six biannual report periods (October 1 to March 31, or April 1 to September 30). Results are given in Table 35. A seventh specific category (rupture) was added for this part of the study. As shown in Table 35, the defect rate is slightly different than in the pilot study, but more specific rates were calculated by body system. For each report period, the birth-defect rate was calculated for each farm. The defect rate and the X and Y coordinate location of each farm reporting malformations can be displayed on computer-drawn maps. For example, Figure 19 shows the location of those farms with birth-defect rates of more than 20 per thousand for the report period ending September 30, 1968. Such maps can be directly overlain for purposes of comparison with maps of soil, vegetation, surface drainage, geochemical elements, and other environmental factors. As in the human study, we were interested in the possible clustering of high-rate farms.

The rate of congenitally malformed pigs per thousand total births for the 3-yr period was calculated for the state and for each county (Figure 20). The stillbirth rate per thousand pigs farrowed ranged from 11.5 to 104.5 for the counties, with a state average of 59.2 (Figure 21). A summary of the data for live, stillborn, and malformed pigs is presented in Table 36.

During the course of the study, several epidemics occurred. On one farm, a group of 14 sows farrowed 149

TABLE 35 Distribution of Reported Congenital Malformation Rates in Swine, by Body System, in Missouri, for the 3-Yr Period October 1, 1967, to September 30, 1970

Body System	Rate Per 1,000 Total Births
Central nervous system	0.46
Sensory organs	0.38
Alimentary and respiratory	0.99
Genitourinary	0.42
Bones and joints	1.16
Rupture	2.77
Other	0.68
Total malformations (45,433)	6.86
Total malformed pigs (44,445)[a]	6.69

[a]Includes stillbirths. Stillbirth is not a defect; a defective pig could have been born either alive or dead.

FIGURE 19 A computer-printed map of malformed swine in Missouri for the period April 1–September 30, 1968. Each dot represents the location of a farm with a malformation rate of greater than 20 per 1,000 total births. The sparsity of dots in southeast Missouri tends to reflect the lower number of swine-producing farms in this area.

pigs, of which 59 were abnormal; the abnormalities were apparently related to the feeding of tobacco stalks to the sows, because other sows on this same farm not exposed to the tobacco stalks had no pigs with birth defects (Menges *et al.*, 1970). Another epidemic was associated with consumption of poison hemlock (*Conium maculatum*). Four sows farrowed 6 pigs with limb mal-

FIGURE 20 Rate of congenitally malformed pigs per 1,000 total births in Missouri, reported by mailed questionnaire for the 3-yr period October 1967–September 1970. State rate = 6.7; county range = 0.0–36.1.

FIGURE 21 Stillbirth rate per 1,000 pigs born in Missouri, reported by mailed questionnaire for the 3-yr period October 1967–September 1970. State rate = 59.2; county range = 11.5 to 104.5.

formations and 34 pigs with signs of central nervous system disturbance (Edmonds *et al.*, 1972). A third epidemic was possibly associated with the consumption of the fruit of the wild black cherry (*Prunus serotina*) (Selby *et al.*, 1971b). Except for these three epidemics, no causal environmental factors have yet been found to be associated with malformations in Missouri.

Human Death Rates for Chronic Diseases

In cooperation with the Missouri State Division of Health (Sauer *et al.*, 1964), mortality patterns for adults have been studied by county and other geographic areas for various chronic diseases by means of age–sex–race-specific death rates. Emphasis has been placed on the four 10-yr age groups between 35 and 74, inclusive, and also on all causes of death, because for these age groups: (a) the overwhelming majority of the deaths are caused by chronic diseases, (b) these deaths may properly be considered premature, (c) the number of deaths involved is substantial, and (d) the cause of death will tend to be more specific and more easily determined in these groups than in the elderly. Attention is focused especially on males, because among whites the male rates are about double the female death rates for the same age. To present a single death rate that would not be appreciably affected by differences in age composition from county to county, death rates for ages 35–74 were age-adjusted, or standardized, for 10-yr age groups. The U.S. total population in 1950 was used as the standard, and the direct-method procedure was used, as described by Linder and Grove (1943) and Shryock and Siegel (1971). The counties with the lowest and highest death rates for this group from all causes for the period 1959–1969 are shown in Figure 22. Fourteen counties had average annual death

TABLE 36 Number and Rate per 1,000 Total Births of Live, Stillborn, and Congenitally Malformed Pigs for Missouri, October 1, 1967, to September 30, 1970

	Some Malformations	No Malformations	Totals
Number of farms reporting	13,036	36,736	49,772
Total births	2,843,030	3,794,344	6,637,374
Live births	2,664,659	3,579,532	6,244,191
Live birth rate/1,000	937.3	943.4	940.8
Stillbirths	178,371	214,812	393,183
Stillbirth rate/1,000	62.7	56.6	59.2
Malformations	44,445	—	44,445
Malformation rate/1,000	15.6	—	6.7

rates of less than 12.5 per thousand population. The city of St. Louis and nine counties had rates greater than 16.5, and 91 counties had rates between 12.5 and 16.5. The high rates for Jasper and Iron counties have contributed heavily toward generating the hypothesis that trace substances or other environmental factors associated with mining or a history of mining in some way increase the risk of death in middle age. The "Bootheel" counties in the Mississippi Delta (southeast) portion of the state also suggested the hypothesis that environmental factors associated with floodplains contribute to high risk. The high rates for Jackson and Buchanan counties, as well as for St. Louis, add evidence to the association between population density and the risk of dying. All three of these hypotheses are currently being tested using data for the counties of the United States for several time periods and for more specific cause categories.

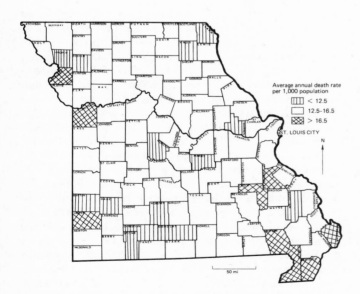

FIGURE 22 Counties in Missouri with lowest and highest death rates, all causes, white males aged 35–74 (age-adjusted), 1959–1969.

Biologic Monitoring for Toxic Heavy-Metal Contamination

An effort was also made in this study to search for species of animals that could serve as monitoring or early warning systems for the buildup of toxic heavy-metals in the environment.

Missouri has a large deer population. Moreover, deer do not range far from their place of birth, large numbers are killed each year, and the locations of the kills are recorded by Missouri Department of Conservation agents and biologists. For this study, the agents and biologists collected approximately 500 deer legs, 50 from each of 10 counties. Two counties were in northern Missouri and had no major cities or industrial development; seven counties were in southeastern Missouri and included present or recent lead-mining and smelting activities; and one county was in south-central Missouri with no mining or other industrial activity. The levels of lead in the leg bones, and particularly in the marrow, were found to reflect those in the environment (3.97 ppm for lead-mining and smelting counties against 1.72 ppm for control counties).

Preliminary results with the use of the fruit fly, *Drosophila* spp., as a monitoring species appear very promising. *Drosophila* are ubiquitous, easy to collect in adequate numbers for analysis, and appear to reflect very accurately the levels of the toxic heavy metals in their local environment. The usefulness of these species has been tested by making collections at various distances from a lead smelter in southeastern Missouri. Lead content was 1,200 ppm at 0.3 km, 175 ppm at 1.6 km, 100 ppm at 3.2 km, and 30 ppm at 11 km from the smelter.

GEOCHEMICAL SURVEY OF MISSOURI

Soon after the beginning of the human and animal studies, the U.S. Geological Survey entered into an informal cooperative study to make a geochemical survey of the state. Work began in 1969 and was completed in 1973. This survey is viewed as a pilot investigation into the problems of preparing statistically stable geochemical maps (i.e., maps not likely to change substantially with the addition of new data) of large regions. The survey included bedrock, surficial unconsolidated geologic deposits, agricultural and currently uncultivated soils, selected native and cultivated plants, groundwaters, and surface waters.

In considering the conduct of the survey and the kind of results obtained, it is useful to consider one of the principal underlying concepts employed in its design. The concept will be familiar and obvious to statisticians and statistically oriented geochemists but is strangely unfamiliar to a broad group of geologists and other scientists.

The basis of the concept is that any geochemical value obtained from a laboratory on a sample submitted for analysis will depart to some degree from the true value that it is intended to represent. There are two reasons for this: (a) a sample, by definition, is only a part of the whole that it is to represent, and its composition is never exactly the same as that of the whole; and (b) no laboratory method has perfect precision (reproducibility).

This situation may be expressed:

$$X = T + d + e$$

where:

X = the analytical value received from the laboratory
T = the true value we are trying to estimate
d = the deviation of the sample value from the true value
e = the analytical error.

Both d and e may be positive or negative quantities, and, although neither is ever known for an individual analysis, their general magnitudes may be estimated in any specific geochemical study if it is based on a suitable formal experimental design. Because knowledge of these magnitudes is absolutely imperative when describing geochemical variation from place to place over a region, all studies in the geochemical survey of Missouri were based on formal designs of this type. The interested reader will find a more detailed account in Miesch (1967).

To the extent possible, the analytical method for each constituent determined in each type of material (rocks, soils, vegetation, and water) was consistent throughout the Missouri program; none of the observed geochemical variations could have resulted from the use of different methods. Also, with a few unavoidable exceptions, the analysis in each case was done by the same analyst. As a precaution against systematic errors (drifts) that may occur in even the best of laboratories, all samples were analyzed in a sequence that was randomized with respect to the geographic locations from which they were taken. For the most part, the analytical errors encountered in the Missouri survey, as measured by the variance of e in the equation given above, were small in comparison to the errors in sampling as measured by the variation in d.

Study Design

Most of the sampling programs were based on nested designs (Krumbein and Graybill, 1965, p. 161), in which each level of the nested hierarchy consisted of areal units. The top level always represents some mapped areal unit of rock, soil, vegetation, or water; successively lower levels represent areal subdivisions of decreasing size. The areal subdivisions at all but the top level were, in all cases, selected for sampling by formal randomization procedures in an attempt to minimize sampling bias.

The reason for the choice of the nested design for the sampling program becomes apparent when we compare this plan with the conventional method of spacing sampling localities at equal intervals over the region of interest. For example, if we were to sample soils only, and if we were to be satisfied with a sampling density of one sample per square mile, almost 70,000 samples would be required. Although more detailed information would have been obtained than was obtained by nested sampling, such a program was too costly. Moreover, two other points had to be considered: (a) a fixed sampling density at the beginning of the survey might result in an overwhelming amount of redundant and costly information, and (b) both the geochemists and the epidemiologists were interested not only in the soils, but also in rocks, plants, and water. The nested sampling designs allowed description of the gross geochemical variations—the first-order variations—in all these materials with fewer than 7,000 samples. The results from nested designs can be used also to establish supplementary sampling programs with any desired balance of detail, scientific reliability, and cost.

Conclusions

Some principal conclusions and products of the geochemical survey of Missouri follow:

1. The major geochemical contrast in the bedrock and surficial geologic deposits is between the northern and southern halves of the state. The northern half of the state is underlain primarily by sandstones and shales of Pennsylvanian age, which are covered by glacial and periglacial debris. These materials are relatively rich in a large suite of major and minor elements. The southern half of the state is underlain primarily by carbonate rocks, which are generally poor in trace element content, although many of the transition elements are enriched in the clay residuum that has developed from weathering of the carbonate rocks at the surface. Nevertheless, the clay residuum contains lower concentrations of most of the alkaline, alkaline-earth, and other soluble elements than does the glacial and periglacial debris in the north.

2. Geochemical maps for most major and many minor elements in agricultural soils show good correspondence to the underlying geology. It is only through a complex multivariate analysis in which 32 constituents of the soils are examined simultaneously that a correspondence between soil composition and the conventional soil taxonomy becomes apparent, and then only at the upper level of the taxonomic hierarchy (Tidball, 1971; personal communication, 1973).

3. In soils that are at present uncultivated, the compositions of the B-horizons vary significantly with both the geology and the type of vegetation present. These soils tend to be compositionally distinct in six major areas of the state that have been defined on the basis of characteristic vegetation. These areas are also roughly coincident with physiographic (and geologic) provinces.

4. Both native vegetation and farm crops vary greatly in composition from one species to another, but compositional variations over various parts of the state are notably slight. Detailed studies of the relations between plant chemistry and the total chemistry of the supporting soil show that the correlations are exceedingly small, although large differences in soil pH may result in important changes in plant chemistry for some trace elements. Thus, the trace element intake from the consumption of plants by animals and humans depends a great deal more on what plants are eaten and in what quantity than on the geochemical environment in which they are grown. Important exceptions occur, however, where the geochemical environment is extreme, as where the soils have been polluted by industrial activity or by naturally occurring and pronounced geochemical anomalies.

5. Groundwaters and surface waters of the state vary in major- and minor-element composition among the seven major geohydrologic units, and the variation corresponds well to variations in the compositions of the enclosing bedrock and surface geology (Feder *et al.*, 1972).

6. The average composition and compositional variation have been estimated for the following units of rock, soil, vegetation, and water in Missouri:

- Nine different units of bedrock, including granite and rhyolite, sandstones, carbonates, and shales.
- Two major categories of unconsolidated geologic materials—loess and carbonate residuum.
- Agricultural soils in each of the 114 counties of the state.
- Soils, at present uncultivated, in each of the six major vegetation-type areas of the state.
- Buckbush and smooth sumac in each of the six major vegetation-type areas of the state.
- Other selected native vegetation, including white oak, post oak, willow oak, red cedar, shortleaf pine, and sweetgum, in each of the vegetation-type areas where they occur.
- Corn, soybeans, and mixed pasture grasses in each of the four major vegetation-type areas where agriculture is widespread.
- Groundwaters in each of the seven major geohydrologic units of the state.
- Surface waters in each of four major geohydrologic units of the state.

7. Many of the results listed in item 6 above have been summarized in seven semiannual progress reports in which 129 geochemical maps appear (U.S. Geological Survey, 1972a, 1972b, 1972c, 1972d, 1972e, 1972f, 1973). These maps show the gross geochemical variations over the state at a glance and are being used by the Environmental Health Center in epidemiological studies. See also more recent U.S. Geological Survey Reports by Erdman *et al.* (1976a, 1976b), Miesch *et al.* (1976), and Tidball (1976).

8. Four investigations of geochemical or environmental pollution were completed. These were directed at (a) effects of highways on the compositions of roadside soils and vegetation throughout southern Missouri (Connor *et al.*, 1971a), (b) contamination of the roadside environment near Centerville because of uncovered ore trucks (Connor *et al.*, 1971b); (c) environmental effects of a clay-pit operation in Callaway County (Ebens *et al.*, 1972; Ebens *et al.*, 1973; Case *et al.*, 1973), and (d) the effects of mining operations on some soils along the Big River in the east-central part of the state (U.S. Geological Survey, 1973).

9. A cooperative investigation involving the Environmental Health Center and the U.S. Geological Survey has shown some subtle but statistically significant relation between the chemistry of agricultural soils and the human death rates in Missouri counties (Tidball and Sauer, 1975).

Other general accounts of the geochemical survey of Missouri appear in Connor *et al.* (1972), Ebens (1973), Erdman and Shacklette (1973), Feder (1973), Miesch and Connor (1973), and Tidball (1973). Some of the specific aspects of the sampling were discussed by Feder (1972), by Miesch (1972), and by Miesch *et al.* (1976).

SUMMARY

Initially, the Missouri program focused on the development of methods and the collection of fundamental or baseline data on health effects and the geochemistry of the environment. Currently, the program is concentrating on epidemics and areas with rates at near-epidemic levels of disease, statewide maps to identify such patterns of disease, and the association of variations in disease rates with geochemical and geological patterns. Geological and other hypotheses generated to a substantial extent from Missouri data are currently being tested against national data; the results indicate a clear need for more intensive work in specific counties and areas of Missouri that are epidemiologically and geochemically unique.

REFERENCES

Case, A. A., L. A. Selby, D. P. Hutcheson, R. J. Ebens, J. A. Erdman, and G. L. Feder. 1973. Infertility and growth suppression in beef cattle associated with abnormalities in their geochemical environment. *In* Proc. 6th Annu. Conf. Trace Subst. Environ. Health, June 13–15, 1972, D. D. Hemphill [ed.]. University of Missouri, Columbia. pp. 15–21.

Connor, J. J., J. A. Erdman, J. D. Sims, and R. J. Ebens. 1971a. Roadside effects on trace element content of some rocks, soils, and plants of Missouri. *In* Proc. 4th Annu. Conf. Trace Subst. Environ. Health, June 23–25, 1970, D. D. Hemphill [ed.]. University of Missouri, Columbia, pp. 26–34.

Connor, J. J., H. T. Shacklette, and J. A. Erdman. 1971b. Extraordinary trace element accumulations in roadside cedars near

Centerville, Missouri. *In* Geological Survey Research, 1971. U.S. Geol. Surv. Prof. Pap. 750-B. U.S. Government Printing Office, Washington, D.C. pp. 151–156.

Connor, J. J., G. L. Feder, J. A. Erdman, and R. R. Tidball. 1972. Environmental geochemistry in Missouri—A multidisciplinary study. Twenty-fourth International Geological Congress, Montreal, Canada, August 20 to 30, 1972. Symposium 1. pp. 7–14.

Ebens, R. J. 1973. Geochemistry of loess and carbonate residuum in Missouri. Abstracts with Programs. Geol. Soc. Am. 5(4):312. (Abstract)

Ebens, R. J., J. A. Erdman, and G. L. Feder. 1972. Geochemical anomalies in a claypit area, Callaway Co., Missouri, where metabolic disorders in cattle were reported. Abstracts with Programs. Geol. Soc. Am. 4(5):319. (Abstract)

Ebens, R. J., J. A. Erdman, G. L. Feder, A. A. Case, and L. A. Selby. 1973. Geochemical anomalies of a claypit area, Callaway County, Mo., and related metabolic imbalance in beef cattle. U.S. Geol. Surv. Prof. Pap. 807. U.S. Government Printing Office, Washington, D.C. 24 pp.

Edmonds, L. D., L. A. Selby, and A. A. Case. 1972. Poisoning and congenital malformations associated with consumption of poison hemlock by sows. J. Am. Vet. Med. Assoc. 160:1319–1324.

Erdman, J. A., and H. T. Shacklette. 1973. Concentrations of elements in native plants and associated soils of Missouri. Abstracts with Programs. Geol. Soc. Am. 5(4):313. (Abstract)

Erdman, J. A., H. T. Shacklette, and J. R. Keith. 1976a. Geochemical survey of Missouri—elemental composition of selected native plants and associated soils from major vegetation-type areas in Missouri. U.S. Geol. Surv. Prof. Pap. 954-C. U.S. Government Printing Office, Washington, D.C. 87 pp.

Erdman, J. A., H. T. Shacklette, and J. R. Keith. 1976b. Geochemical survey of Missouri—elemental composition of corn grains, soybean seeds, pasture grasses, and associated soils from selected areas in Missouri. U.S. Geol. Surv. Prof. Pap. 954-D. U.S. Government Printing Office, Washington, D.C. 23 pp.

Feder, G. L. 1972. Problems of sampling in trace element investigations. Ann. N.Y. Acad. Sci. 199:118–123.

Feder, G. L. 1973. Geochemical survey of trace elements in waters of Missouri. Abstracts with Programs. Geol. Soc. Am. 5(4):314. (Abstract)

Feder, G. L., R. J. Ebens, and J. J. Connor. 1972. The relationship between lithology and trace element content of ground water. *In* Short Pap. 9th Annu. Water Resources Conf. American Water Resources Association, St. Louis, Missouri. p. 104.

Krumbein, W. C., and F. A. Graybill. 1965. An introduction to statistical models in geology. McGraw-Hill, New York. 475 pp.

Linder, F. E., and R. D. Grove. 1943. Vital statistics rates in the United States 1900–1940. Bureau of the Census. U.S. Government Printing Office, Washington, D.C. pp. 66–69.

Menges, R. W., L. A. Selby, C. J. Marienfeld, W. A. Aue, and D. L. Greer. 1970. A tobacco related epidemic of congenital limb deformities in swine. Environ. Res. 3:285–302.

Miesch, A. T. 1967. Theory of error in geochemical data. U.S. Geol. Surv. Prof. Pap. 574-A. U.S. Government Printing Office, Washington, D.C. 17 pp.

Miesch, A. T. 1972. Sampling problems in trace element investigations of rocks. Ann. N.Y. Acad. Sci. 199:95–104.

Miesch, A. T., P. R. Barnett, A. J. Bartel, J. I. Dinnin, G. L. Feder, T. F. Harms, C. Huffman, Jr., V. J. Janzer, H. T. Millard, Jr., H. G. Neiman, M. W. Skougstad, and J. S. Wahlberg. 1976. Geochemical survey of Missouri—methods of sampling, laboratory analysis, and statistical reduction of data. U.S. Geol.

Surv. Prof. Pap. 954-A. U.S. Government Printing Office, Washington, D.C. 39 pp.

Miesch, A. T., and J. J. Connor. 1973. Geochemical survey of Missouri—Methods and goals. Abstracts with Programs. Geol. Soc. Am. 5(4):337–338. (Abstract)

Sauer, H. I., J. E. Banta, and W. W. Marshall, Jr. 1964. Cardiovascular diseases mortality patterns among middle-aged white males in Missouri. Missouri Med. 61:921–926, 929 (November 1964).

Selby, L. A., C. J. Marienfeld, W. Heidlage, and V. E. Young. 1971a. Evaluation of a method to estimate the prevalence of congenital malformations in swine using a mailed questionnaire. Cornell Vet. LXI (2):203–213.

Selby, L. A., R. W. Menges, E. C. Houser, R. E. Flatt, and A. A. Case. 1971b. Outbreak of swine malformations associated with the wild black cherry (*Prunus serotina*). Arch. Environ. Health 22:496–501.

Shryock, H. S., and J. S. Siegel. 1971. The methods and materials of demography. U.S. Bureau of the Census. U.S. Government Printing Office, Washington, D.C. pp. 418–421.

Silberg, S. L., C. J. Marienfeld, H. Wright, and R. C. Arnold. 1966. Surveillance of congenital anomalies in Missouri, 1953–1964—A preliminary report. Arch. Environ. Health 13:641.

Tidball, R. R. 1971. Geochemical variation in Missouri soils. *In* Proc. 4th Annu. Conf. Trace Subst. Environ. Health, June 23–25, 1970, D. D. Hemphill [ed.]. University of Missouri, Columbia. pp. 15–25.

Tidball, R. R. 1973. Distribution of elements in agricultural soils in Missouri. Abstracts with Programs. Geol. Soc. Am. 5(4):358–359. (Abstract)

Tidball, R. R. 1976. Geochemical survey of Missouri—chemical variation of soils in Missouri associated with selected levels of the soil classification system. U.S. Geol. Surv. Prof. Pap. 954-B. U.S. Government Printing Office, Washington, D.C. 16 pp.

Tidball, R. R., and H. I. Sauer. 1975. Multivariate relationships between soil composition and human mortality rates in Missouri. *In* Trace element geochemistry in health and disease, J. Freedman [ed.]. Geol. Soc. Am. Spec. Pap. No. 155. Geological Society of America, Boulder, Colorado. pp. 41–59.

U.S. Geological Survey. 1972a. Environmental geochemistry: Geochemical survey of Missouri, plans and progress for first six-month period (July–December, 1969). U.S. Geol. Surv. Open-File Rep. 1658. Branch of Regional Geochemistry, USGS, Denver and Reston. 49 pp.

U.S. Geological Survey. 1972b. Environmental geochemistry: Geochemical survey of Missouri, plans and progress for second six-month period (January–June, 1970). U.S. Geol. Surv. Open-File Rep. 1659. Branch of Regional Geochemistry, USGS, Denver and Reston. 60 pp.

U.S. Geological Survey. 1972c. Environmental geochemistry: Geochemical survey of Missouri, plans and progress for third six-month period (July–December, 1970). U.S. Geol. Surv. Open-File Rep. 1660. Branch of Regional Geochemistry, USGS, Denver and Reston. 33 pp.

U.S. Geological Survey. 1972d. Environmental geochemistry: Geochemical survey of Missouri, plans and progress for fourth six-month period (January–June, 1971). U.S. Geol. Surv. Open-File Rep. 1661. Branch of Regional Geochemistry, USGS, Denver and Reston. 63 pp.

U.S. Geological Survey. 1972e. Environmental geochemistry: Geochemical survey of Missouri, plans and progress for fifth six-month period (July–December, 1971). U.S. Geol. Surv. Open-File Rep. 1706. Branch of Regional Geochemistry, USGS, Denver and Reston. 145 pp.

U.S. Geological Survey. 1972f. Environmental geochemistry: Geochemical survey of Missouri, plans and progress for sixth six-month period (January–June, 1972). U.S. Geol. Surv. Open-File Rep 1800. Branch of Regional Geochemistry, USGS, Denver and Reston. 86 pp.

U.S. Geological Survey. 1973. Environmental geochemistry: Geochemical survey of Missouri, plans and progress for seventh six-month period (July–December, 1972). U.S. Geol. Surv. Open-File Rep. 1982. Branch of Regional Geochemistry, USGS, Denver and Reston. 59 pp.

XIV

World Health Organization Studies in Geochemistry and Health

ROBERTO MASIRONI

Deficiencies or excesses in the content or availability of trace elements in rocks and soils, or in water flowing through them, is hypothesized as a possible cause of certain chronic diseases. This hypothesis is supported by the association of the incidence of goiter and dental caries with areas that are low in iodine and fluorine, respectively. Other possible associations that have been reported between geochemical and geological environments and the distribution of certain diseases include gout (Kovalsky and Yarovaya, 1966), certain forms of cancer (Burrell, 1962; Armstrong, 1972; Kmet and Mahboubi, 1972), and cerebrovascular and cardiovascular diseases (Takahashi, 1967; Shacklette et al., 1970).

The potential effects on health of the geochemical environment are also suggested by the apparent relation between hardness of water and cardiovascular diseases. With six or seven exceptions, 27 studies in eight countries (see Masironi et al., 1972; Neri et al., 1972) have shown that death rates from cardiovascular diseases are usually higher in areas served by soft water than in areas served by hard water.

Further, preliminary observations (Masironi, 1971) concerning the geographic distribution of ischemic heart disease in Europe suggest that mean national death rates increase with increasing age of the surface rocks and underlying strata in each country or group of countries (Table 37).

Because further study may help to explain why many chronic diseases are geographically distributed in characteristic ways, the World Health Organization (WHO) is coordinating three principal geochemically oriented surveys on: cardiovascular disease in several areas, esophageal cancer along the Caspian coast of Iran, and dental caries in Papua New Guinea. The information reported here concerning the cancer and dental caries studies was obtained through personal communication with J. Kmet and D. Barmes, respectively, as well as from the literature (Barmes, 1969; Barmes et al., 1970; Kmet and Mahboubi, 1972).

CARDIOVASCULAR DISEASE

In 1969, WHO, in collaboration with the International Atomic Energy Agency (IAEA), began a multidisciplinary research program on trace elements in relation to cardiovascular diseases.

This program has principally involved coordination of the work of pathology departments, epidemiological teams, and analytical laboratories in 20 countries: Argentina, Brazil, Bulgaria, Czechoslovakia, the Federal Republic of Germany, Finland, Greece, Hong Kong, Iran, Israel, Jamaica, New Zealand, Nigeria, Norway, the Philippines, Singapore, Switzerland, Papua New Guinea, the United Kingdom, and the United States. WHO coordinates the medically oriented aspects of these studies; the IAEA coordinates the analytical aspects.

After a planning meeting (WHO, 1971) at which research needs and protocols were established and activities or-

TABLE 37 Death Rates from Ischemic Heart Disease in Europe, Related to Age of Surface Rocks and Underlying Strata, 1967 (All Ages; Both Sexes; per 100,000)

Precambrian (> 600 m.y.[a] old)		Early Paleozoic (600–300 m.y. old)		Late Paleozoic (300–180 m.y. old)		Mesozoic (< 180 m.y. old)	
Sweden	320	Norway	263	Ireland	294	Italy	200
Finland	274	Northern Ireland	302	Austria	256	Yugoslavia	129
Denmark	318	England	308	Hungary	259	Bulgaria	144
Scotland	345			France	83	Greece	100
				Germany (FR)	226	Switzerland	220
				Netherlands	186		
				Belgium	180		
				Czechoslovakia	197		
				Poland	92		
				Spain	68		
				Portugal	119		
				Romania	144		
MEAN	314[b]		291		175		159[a]
± SD	± 29		± 24		± 75		± 50

[a]Million years.
[b]Difference significant at $P \leq 0.001$.
SOURCE: Masironi (1971).

ganized, collaborative studies were begun in the following five areas of research:

1. Comparison of trace elements in tissues obtained at autopsy from healthy subjects accidentally killed and from persons whose deaths were caused by myocardial infarction.

2. Comparison of the cadmium content and of cadmium: zinc ratios in the livers and the kidneys obtained at autopsy from normotensive and hypertensive subjects.

3. Epidemiologic studies of the blood pressure, electrocardiogram (EKG), and blood-cholesterol levels in nonindustrialized groups, and of their correlation with the trace element content of specimens of blood, toenails, hair, food, and water taken from those groups and their environment.

4. Determination of the trace element content of staple foods (like rice and sugar) from several countries, and correlation of the content with the degree or kind of processing used on the foods.

5. Continued study of the hardness of water supplies in relation to cardiovascular mortality or morbidity. This will be supplemented with limited studies of the relation of trace elements in rocks and soils to cardiovascular mortality.

Of the many studies concerning the water/cardiovascular disease relation mentioned above, only three, coordinated by WHO, continue in depth: one in England (Crawford *et al.*, 1971; Stitt *et al.*, 1973); one in Finland (Punsar, 1973); and one in Canada (Neri *et al.*, 1971). For these studies, WHO is acting mainly as a center for coordination and the exchange of information. WHO is also carrying out some other, more recent studies in New Guinea and in five European cities.

In England, 61 towns of comparable size were ranked according to the hardness of their water supplies. Death rates from cardiovascular diseases in the six towns with the hardest water proved to be lower than those in the six towns with the softest water. Moreover, the populations of the hard-water towns also generally were found to have relatively lower blood pressures, lower heart rates, and lower blood-cholesterol levels. Socioeconomic, dietary, and environmental factors appeared to have no influence on these associations.

In Finland, a higher cardiovascular death rate, generally higher blood pressures, and higher blood-cholesterol levels were found in eastern Finland, as compared to western Finland, and are associated with the softer water in the east. Other parameters, such as dietary habits, socioeconomic factors, and physical activity of the populations in the two areas did not explain the differences in cardiovascular pathology.

In Canada, general mortality, as well as cardiovascular mortality, has been found to be higher in soft-water areas.

The WHO study in New Guinea confirms the findings in England and Finland that higher blood pressure is associated with areas having softer water than those with harder water. The villages studied are located along the Wogupmeri and Blackwater rivers, which originate from limestone mountains and flow into the Sepik River (Figure 23). The calcium content of the river water decreases from 7.8 ppm to a minimum of 1.2 ppm and rises again to 15.9 ppm as the two tributaries approach the Sepik. When the villages are grouped according to high, medium, and low calcium content of the local riverwater, the corresponding mean systolic and diastolic blood pressures in both their male and female inhabitants were found to be significantly greater in the low-calcium group (Table 38). The difference, moreover, was not attributable to age or

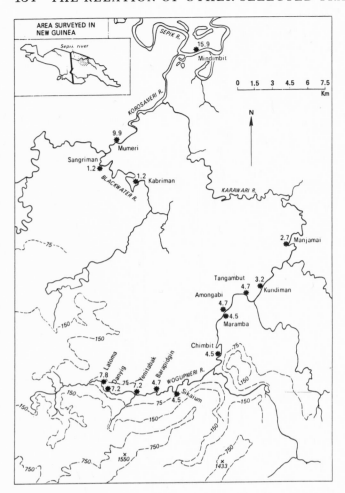

FIGURE 23 Villages surveyed in Sepik River Basin (New Guinea) and parts per million of calcium in local river water (WHO, 1973).

body build. When the mean blood pressures measured in individual villages located along the Wogupmeri River only are plotted against the calcium content of the local river water, the same trend is evident (Figure 24). These findings are of particular interest because they indicate that a cardiovascular parameter may be influenced by a chemical parameter of the local geological environment.

The WHO study in five European cities includes a standardized assessment of the rates of occurrence of, and extent of damage from, atherosclerosis, hypertension, and myocardial infarction in autopsy cases from Prague (Czechoslovakia), Malmö (Sweden), and Ryazan, Tallin, and Yalta (USSR). Full details on the epidemiological and autopsy aspects of this study are contained in the WHO report: "Atherosclerosis of the Aorta and Coronary Arteries in Five Towns" (Kagan et al., 1976). Water hardness values were also obtained from these cities. Preliminary results indicate an inverse relation between water hardness and extent of cardiovascular damage (Figure 25).

Thus, both the independent studies carried out in the past and the more recent ones that are more or less closely coordinated by WHO agree in indicating the existence of undesirable trends in the cardiocirculatory parameters of populations using soft waters, regardless of their socioeconomic conditions or geographic location.

In 1972, WHO and the U.S. Department of Health, Education, and Welfare began an informal collaboration in water/cardiovascular disease studies. A number of cities in the United States were tentatively selected that have contrasting cardiovascular mortality rates but are otherwise comparable in population size, degree of industrialization, distance from the seacoast, and altitude. Water analyses, for the bulk elements and as many trace elements as possible, will then be carried out and the results compared with the corresponding death rates from cardiovascular disease.

TABLE 38 Mean Blood Pressure in Relation to Calcium Content of River Water in Villages along Wogupmeri and Blackwater Rivers in New Guinea

Calcium concentration, range in ppm	Number of Subjects[a]	Sex	Mean Age	Mean Systolic Blood Pressure, mm Hg	Mean Diastolic Blood Pressure, mm Hg
9.6 (7.2–15.3)	15	M	50	104.5 ± 12.5[b]	67.4 ± 5.3[b]
	21	F	43	108.9 ± 12.3[b]	70.7 ± 7.9[c]
4.4 (3.2–4.7)	31	M	44	111.6 ± 13.4	69.7 ± 8.2
	53	F	40	107.6 ± 13.6	70.7 ± 7.9
1.7 (1.2–2.7)	26	M	48	115.8 ± 9.9[b]	75.1 ± 9.2[b]
	24	F	43	117.5 ± 13.8[b]	74.7 ± 7.5[c]

[a]The size of each village is only about 200 persons including children. Blood pressure was measured only in fathers and mothers.
[b]P ≤ 0.05.
[c]P ≤ 0.1.
SOURCE: WHO (1973).

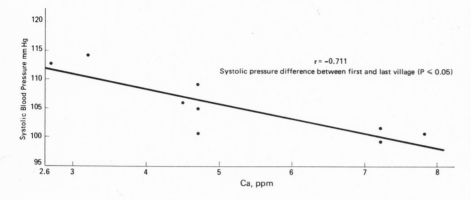

FIGURE 24 Mean systolic blood pressure in male and female adults versus calcium content in water of New Guinea villages along the Wogupmeri River (WHO, 1973).

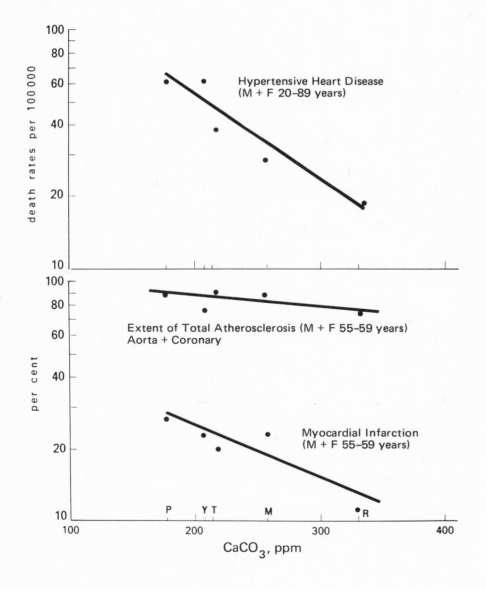

FIGURE 25 Hypertensive heart disease, atherosclerosis, and myocardial infarction in Prague (P), Yalta (Y), Tallin (T), Malmö (M), and Ryazan (R) in relation to water hardness. M + F means both sexes (WHO, 1973).

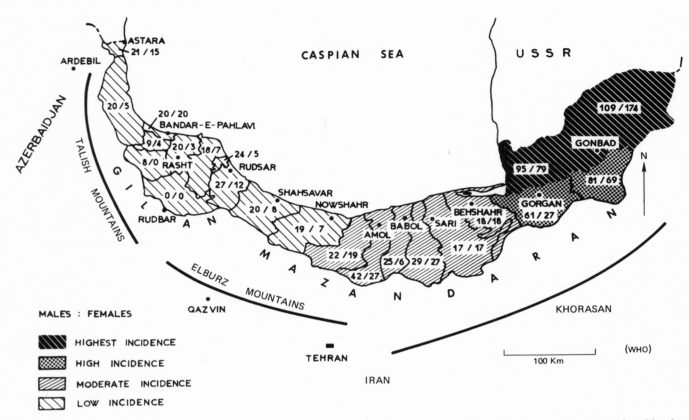

FIGURE 26 Incidence of esophageal cancer per 100,000 people per year along the Caspian Sea coast in Iran (Kmet and Mahboubi, 1972).

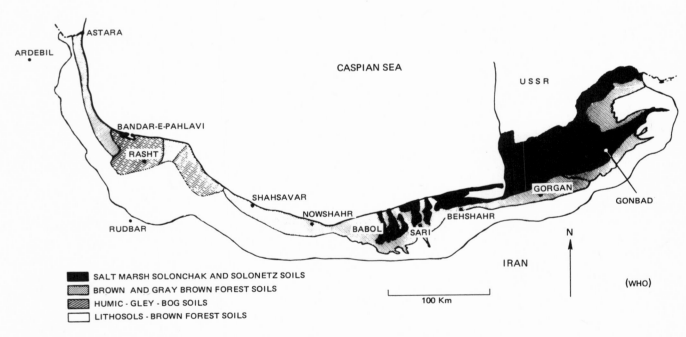

FIGURE 27 Soils of the Caspian Sea coast area of Iran (Kmet and Mahboubi, 1972).

WHO is beginning studies on the possible relation between soil composition and the prevalence of cardiovascular disorders near the Caspian coast of Iran in association with a survey of the occurrence there of esophageal cancer described below. Information on blood pressure, electrocardiogram (EKG), blood-cholesterol levels, and degree of obesity is being collected to see whether any characteristic differences will become evident between population groups in two highly contrasting geochemical environments: the saline, semidesert area to the east and the humid, subtropical area to the west. This study will also examine differences in trace elements in rocks and in their availability in plants, animals, and water in the two areas. Samples of toenails, hair, blood, water, and food are also being collected for bulk and trace element analysis.

Similar studies are being carried out in Polynesia on the Tokelau Islands that will provide additional information on the influence of the geochemical environment on human health, particularly cardiovascular health, in nonindustrialized populations living in close contact with their natural environment, consuming food grown and produced locally, and drinking untreated water.

ESOPHAGEAL CANCER

Esophageal cancer is also being studied in populations along the coast of the Caspian Sea by the International Agency for Research on Cancer (a branch of WHO) in close collaboration with the Iranian public health authorities. This coast is characterized by one of the highest incidences in the world of esophageal cancer in its eastern area and very low incidence in its western area (Figure 26). As mentioned above, the geochemical environments of these two areas are sharply contrasted.

An environmental survey, started in late 1968, includes the evaluation of climate, soil and water characteristics, flora and fauna, disease patterns in man and animals, socioeconomic characteristics of the population groups, dietary habits, and genetic markers. This information is being correlated with the incidence of esophageal cancer. Of the parameters analyzed so far, the soil data show a striking association with the epidemiology of the disease. The western low-incidence area is characterized by brown and gray-brown forest soils and humic-gley-bog soils (Figure 27). The eastern high-incidence area is characterized by saline, alkaline soils of the solonchak and solonetz type, and loess. These soils are usually unsuitable for general agricultural use, so that browsing livestock, particularly sheep and goats, represent the main use of the land in this region (Kmet and Mahboubi, 1972).

DENTAL CARIES

A study of dental caries is being carried out in primitive village communities in Papua New Guinea by the WHO Dental Health Unit in collaboration with Australian public health authorities and investigators and with the U.S. National Institutes of Health. This geochemically and nutritionally oriented survey was prompted by the observation that caries prevalence in Papua New Guinea is generally low, and that even at such reduced levels distinct differences exist among the various population groups. Several of the villages studied are actually caries-free. These isolated population groups in Papua New Guinea, living in a close relationship with their natural environment, provide an ideal opportunity for carrying out comprehensive dietary, environmental, and trace element surveys to elucidate the potential relation between the local geochemical environment and dental health.

Although the benefits of fluorine to dental health are well known, it cannot by itself completely prevent dental caries. It is possible that other elements, such as molybdenum, vanadium, or strontium, may act in a complementary or a synergistic way with fluorine, while others, such as selenium and lead, may perhaps act antagonistically. To test these hypotheses, studies of the prevalence of dental caries and the trace element composition of food, water, and soil have been carried out in a number of villages in Papua New Guinea.

Barmes (1969) and Barmes *et al.* (1970) conclude that:

1. A strong inverse association exists between the prevalence of dental caries and the existence of concentrations in village soils and water of alkali and alkaline-earth elements, especially strontium, barium, potassium, magnesium, calcium, and lithium. There is a possible direct association of caries incidence with the presence of lead and copper in the soil and water.

2. Inverse associations were found between the incidence of caries and the concentrations of vanadium, molybdenum, magnesium, aluminum, titanium, and phosphorus, and direct associations with concentrations of lead, copper, chromium, zinc, and selenium in the stable foodstuffs—sage, sweet potato, and Chinese taro.

3. The differences in the prevalence of caries among the villages cannot be explained in terms of the fluorine content of food, soil, and water.

Follow-up studies are now underway to test these associations.

REFERENCES

Armstrong, R. W. 1972. Is there a particular kind of soil or geologic environment that predisposes to cancer? Ann. N.Y. Acad. Sci. 199:239–248.

Barmes, D. E. 1969. Caries etiology in Sepik villages: Trace element, micronutrient and macronutrient content of soil and food. Caries Res. 3:44–59.

Barmes, D. E., B. L. Adkins, and R. G. Schamschula. 1970. Etiology of caries in Papua–New Guinea: Associations in soil, food and water. Bull. WHO 43:769–784.

Burrell, R. J. W. 1962. Esophageal cancer among Bantu in the Transkei. J. Natl. Cancer Inst. 28:495–514.

Crawford, M. D., M. J. Gardner, and J. N. Morris. 1971. Changes in water hardness and local death rates. Lancet 2:327–329.

Kagan, A. R., N. H. Sternby, K. Uemura, R. Vaněček, A. M. Vihert, A. M. Lifšic, E. E. Matova, Z. Záhoř, and V. S. Ždanov. 1976. Atherosclerosis of the aorta and coronary arteries in five towns. Bull. WHO 53(5–6):485–645.

Kmet, J., and E. Mahboubi. 1972. Esophageal cancer in the Caspian littoral of Iran: Initial studies. Science 175:846–853.

Kovalsky, V. V., and G. A. Yarovaya. 1966. Biogeochemical provinces rich in molybdenum. Agrokhimiya 8:68–91.

Masironi, R. 1971. Associations between geology, types of soil, and cardiovascular diseases. WHO Internal Document CVD/WP/70.16. World Health Organization, Geneva.

Masironi, R., A. T. Miesch, M. D. Crawford, and E. Hamilton. 1972. Geochemical environments, trace elements and cardiovascular diseases. Bull. WHO 47:139–150.

Neri, L. C., D. Hewitt, and J. S. Mandel. 1971. Risk of sudden death in soft water areas. Am. J. Epidemiol. 94:101–104.

Neri, L. C., D. Hewitt, and G. B. Schreiber. 1972. Water mineralization and health: A critical synopsis of literature. WHO Internal Document CVD/WP/TE/72.12. World Health Organization, Geneva.

Punsar, S. 1973. Cardiovascular mortality and quality of drinking water. WHO Internal Document CVD/WP/TE/72.11. World Health Organization, Geneva.

Shacklette, H. T., H. I. Sauer, and A. T. Miesch. 1970. Geochemical environments and cardiovascular mortality rates in Georgia. U.S. Geol. Surv. Prof. Pap. 574-C. U.S. Government Printing Office, Washington, D.C. 39 pp.

Stitt, F. W., M. D. Crawford, D. C. Clayton, and J. N. Morris. 1973. Clinical and biochemical indicators of cardiovascular disease among men living in hard and soft water areas. Lancet 1(7795):122–126.

Takahashi, E. 1967. Geographic distribution of mortality rates from cerebrovascular diseases in European countries. Tohoku J. Exp. Med. 92:345–378.

World Health Organization. 1971. Report of a WHO meeting of investigators on trace elements in relation to cardiovascular diseases. WHO Internal Document CVD/71.2. World Health Organization, Geneva.

World Health Organization. 1973. Report of second meeting of investigators on trace elements in relation to cardiovascular diseases. WHO Internal Document CVD/73.4. World Health Organization, Geneva.

XV

Display of Data in Map Form— Problems of Compatibility

HOWARD C. HOPPS

GENERAL CONSIDERATIONS

Generally speaking, our approach to understanding the causes of disease has been far too simplistic. We act as though most diseases had a single cause, whereas, in truth, multiple causes are almost always at work. Disease, from the broad viewpoint of cause and effect, is not an entity; it is the result of a complex interplay among host factors and disease–agent factors. The host factors are, in part, genetic, but it is the environmental elements that play the greater overall role in determining what diseases will occur, where, when, and in what form (Hopps, 1971a, 1972a; May, 1958, 1961).

How does this view of disease relate to the display of data in map form? Simply stated:

1. From a pragmatic view, environmental factors that affect both cause and course of disease are more important than genetic factors because they are more easily controlled (as of today, at any rate).

2. For the most part, environmental factors are neither evenly nor uniformly distributed; moreover, there are reasons why they occur where they do.

3. Spatial distribution of the level of environmental factors is best portrayed in map form because:

    a. Maps can show in a very effective, three-dimensional way which disease is where, and when; and they can help to answer the questions "Why?" and "What can be done about it?"

    b. An enormous amount of nonmedical information is already available in map form—such as population density, transportation routes, climate, land usage, and soil types—ready for use with appropriate medical information that has been comparably assembled and displayed.

    c. Maps can present medical information in a manner that allows:

        i. analysis (with other mapped factors) by pattern matching—a very powerful analytic approach, particularly useful in generating new ideas about cause and effect relationships;

        ii. cluster analysis—particularly useful in testing preformed hypotheses; and

        iii. evaluation of time and distance factors—with respect to such problems as deciding how many and what kind of controls (of pollution, for example) should be exerted, and where, to meet specific needs.

4. Maps, particularly isopleth (contour-type) ones, are very useful in trend analysis, so that, given an adequate data base, the items considered under (3) above can be projected for the future as well as determined for the present.

Types of Maps

There are three principal types of maps: *point maps*, on which dots are used, largely restricting the information to a "present" or "not present" statement; *choropleth maps* (shading-type), on which there is an unfortunate effect of

139

averaging the factor in question and expressing that average value over an area much larger than a point; and *isopleth maps* (contour-type), which has the great advantage of freely expressing quantitative aspects in a manner that dramatically indicates both the values of the functions and their rates of change from one area to another. Isopleth maps are by far the best if one is attempting to relate patterns of disease distribution and environmental characteristics.

This very brief and simplistic discussion can easily be extended by consulting a reference work on cartography, such as Robinson and Sale's *Elements of Cartography* (1969).

Obviously, it is important to know where a particular disease occurs if one is to work toward preventing or controlling it. But merely a yes or no for place of occurrence (such as dots on a map or cross-hatching of affected districts) is not enough. Moreover, because disease is no respecter of political boundaries, the use of choropleth (shading-type) maps that give the average values of disease data for political regions, such as countries or states, may be quite misleading. For example, the incidence of many diseases in New York State is dominated by their rate of occurrence in New York City.

If choropleth maps are to be used, the size of the unit area should be sufficiently small to represent a reasonably homogeneous set of the data depicted. For example, when mapping large regions of the United States (e.g., the Northwest), county units might be suitable; state units would be far too large for most purposes (Hopps, 1972b). Isopleth maps are much more effective than point (dot-type) or choropleth maps in displaying quantitative aspects of data, including rate of change. They are more difficult to produce, however.

Producing an isopleth map for a given simple function of two variables that are restricted to some rectangular domain is a reasonably straightforward task. In the case of contour mapping of disease, however, serious problems arise, because one is not given a simple function to contour, but the average value of a function at selected points, irregularly distributed in the domain. A major hazard is unrecognized error resulting from inappropriate interpolation (Jenks and Caspall, 1971).

Of greatest value are maps that show, quantitatively, the distribution of indices (i.e., data complexes) that relate directly and comprehensively to the problem at hand. A given data complex (which, in effect, represents the solution of an equation) can reflect the factors or parameters known to affect the cause and course of the disease, including the effects of various treatments. For example, a single data complex might incorporate age, sex, race, body build and weight, occupation, social and economic status, specific aspects of past medical and social history, blood pressure, level of serum cholesterol, and specified electrocardiographic characteristics.

This is not a new concept, of course. Distribution patterns of disease are, for example, commonly presented in terms of levels that are above or below the expected. However, indices have not been used to the extent that

they could and should be, now that we have at our disposal powerful new mathematical methods such as regression on principal components, linear probability models, and cross-spectrum analyses, along with the computer technology that makes the application of such complex mathematical analyses feasible. Moreover, computer science and technology have also developed to the point at which it is now possible to produce distribution maps quickly and with relative ease (Tarrant, 1970; Peucker, 1972). Thus, we have potentially many more maps to work with than ever before. But the production of maps by computer is not a simple matter. One of the major problems has to do with converting narrative and tabular data to a form that is suitable for computer operation and that will allow direct map output (Hopps *et al.*, 1969; Hopps and Gabrieli, 1970; Hopps, 1971b).

SPECIFIC CONSIDERATIONS

Aside from the content of the map, three important factors to be considered from a technical standpoint are: projection, scale, and symbolism.

Projection

There are well over 100 different projections, and, since each of them has its unique set of advantages and disadvantages, the user must weigh the pros and cons to arrive at an appropriate choice for his specific purpose. The ideal map accomplishes equality of areas, truth in shape, and accuracy in distance. This happy combination can only be found in global maps; when the sphere is flattened, distortion is inevitable. There are four principal ways to flatten the sphere, so to speak, but many variations within each—thus, the hundred-plus projections. The four basic projections are described briefly below:

1. Cylindrical projections, the Mercator form being best known, are good for very large areas, such as the world, but distort size badly as one gets beyond the 30°–40° latitudes, North or South (assuming that the middle of the cylinder corresponds to the equator).

2. Orthographic projections present an image of any chosen half of the sphere (as a maximum) as though parallel rays of light were projected through this portion. If an entire hemisphere is displayed, the map will be a perfect circle and distortion will be minimal at the center and maximal at the periphery. Of course, areas smaller than half the globe can be mapped in this manner, and, the smaller the area, the less the marginal distortion.

3. Conic projections represent a compromise of truth in shape, area, and distance that is quite tolerable for relatively large areas, especially if they are wider than they are long (see Figure 28).

4. Azimuthal projections are made on a tangent plane with the origin point at the opposite surface of the globe or at its center. An outstanding advantage of this projec-

tion is that all distances extending from the center point are true to scale.

In producing maps that will help relate the geochemical environment to conditions of health and disease, the task is not merely to select the projection that best shows disease distribution or geochemical characteristics, but to choose a projection that is suitable and that also fits closely the many existing maps that will be useful for comparison (e.g., maps depicting land usage, population density, and so forth).

Responses from inquiries to a number of major agencies and private groups that present data in map form indicate that the Albers Conic Equal-Area Projection is widely used and may well be the best choice for our purposes. It is better suited to the United States than its close rival, the Lambert Conformal Projection, because the two parallels "cut" by the cone (representing lines of true values) are 4° farther apart than those on the Lambert projection. Thus, the Albers projection produces somewhat less distortion in the northernmost and southernmost parts of our 48 contiguous states, viewed as a composite, as is shown in Figure 28.

Unfortunately, there is no easy way of transposing data presented in a map of one projection to a map of significantly different projection, particularly if a large area is involved. With small areas, differences are often insignificant.

Scale

As with projection, there is no absolute best selection; one must match scale with objective. Resolution is usually the deciding factor. Here, however, the problem is considerably simpler, because many of the important map producers have more-or-less standardized on a limited number of scales, to match the size of area and the resolution they are interested in. Moreover, changing the scale is easily accomplished by photographic enlargement or reduction.

In general, large areas such as regions of the United States and continents, as shown in atlases, range from 1:40,000,000 (1 in. ≈ 640 mi) to 1:4,000,000 (1 in. ≈ 64 mi); topographic maps range from 1:1,000,000 (1 in ≈ 16 mi) to 1:10,000 (1 in. ≈ 0.16 mi); and cadastral maps, such as one showing a township plat, have a scale of 1:10,000 or larger. Since scale is expressed as a fraction (1:10,000 = 1/10,000), the smaller the denominator the larger the scale and, given the same area, the larger the map. For example, taking the scales used in the *National Atlas*, a 1:34,000,000 map of the contiguous 48 states measures 4 × 6 in. A map of the same region at 1:17,000,000 measures 8 × 12 in., and at 1:7,500,000, 17 × 26 in. Table 39 shows scales and projections of base maps used by the U.S. Geological Survey.

The U.S. plan to convert to the metric system is already having an impact on map production. The Geological Survey announced in March 1973 that the first of its series of all-metric maps would cover parts of Alaska. These maps will be at a scale of 1:25,000 (4 cm = 1 km).

Symbolism

Symbolization in maps is, in essence, the means of coding the information to be presented. The goal is to select a system that displays the information in an esthetically pleasing manner and in a form that facilitates rapid, comprehensive retrieval by the map user. There are many, many ways to portray the distribution of things, and each time a new map is designed this poses a new challenge. The symbolism that is used becomes an inherent part of the map, the importance of which can hardly be overemphasized. Many maps fail to deliver their message merely because of inappropriate symbolism. Sometimes the individual symbols are appropriate, but the map fails because too many of them are used, leading to clutter and confusion.

Data can be represented in a map in three basically different ways, and these may be combined:

1. Point symbols (Figure 29), such as actual dots, numbers, letters—even small pictures. These can be considered as zero-dimensional symbols since, in practice, they approximate a geometric point.

2. Area symbols (Figure 30), such as various intensities of gray, or various colors, or patterns. These can be considered as two-dimensional symbols since, in practice, they approximate a geometric planar area.

3. Line symbols (Figure 31) include contour lines, which can be considered as three-dimensional symbols since a set of contour lines, in practice, approximates a geometric curved surface, either real or imaginary. Line symbols are also used to represent essentially two-dimensional things such as roads, rivers, or boundaries. In the form of directional arrows, they may also indicate flow.

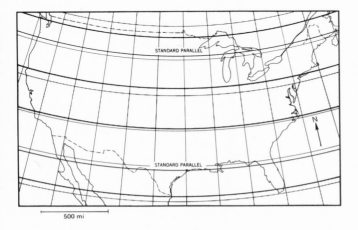

FIGURE 28 The Albers Conic Projection. Values of lines of equal maximum angular deformation are 1°. Reproduced by permission from Robinson and Sale, *Elements of Cartography*, 3rd ed., John Wiley & Sons, New York, 1969.

TABLE 39 Scales and Projections of Base Maps in the U.S. Geological Survey

Series	Scale	1-in. Equals	Size	Area, sq mi	Projection
7½ min	1:24,000	2,000 ft	7½ × 7½ min	49–70	Polyconic
15 min	1:62,500	≈ 1 mi	15 × 15 min	197–282	Polyconic
1:250,000	1:250,000	≈ 4 mi	1° × 2° 1° × 3°	458–8,669	Universal Transverse Mercator
State	1:500,000	≈ 8 mi	Variable		Lambert Conformal Conic
State	1:1,000,000	≈ 16 mi	Variable		Lambert Conformal Conic
IMW (International Map of the World)	1:1,000,000	≈ 16 mi	4° × 6°	73,734–102,759	Modified Polyconic
U.S.	1:2,500,000	≈ 40 mi	2 sheets 36 × 78 in.		Albers Equal Area
U.S.	1:3,168,000	≈ 50 mi	1 sheet 39 × 61		Albers Equal Area
U.S.	1:5,000,000	≈ 80 mi	1 sheet 24 × 38		Albers Equal Area
U.S.	1:7,000,000	≈ 110 mi	1 sheet 15 × 27		Albers Equal Area
U.S.	1:17,000,000	≈ 268 mi	1 sheet 8 × 12		Albers Equal Area
U.S.	1:34,000,000	≈ 536 mi	1 sheet 4 × 6		Albers Equal Area

SOURCE: Douglas Kinney, U.S. Geological Survey, unpublished, 1973.

Each of these basic classes provides for an enormous variety of symbols. The illustrations graciously provided by Professor Robinson demonstrate this point very well.

Millions of dollars are being spent to accumulate health and disease-related data, but much of this is not now being used effectively, if at all. Clearly, it is not enough that the data exist. They must be brought together, manipulated to identify significant relationships, and the results displayed in a form that will effectively communicate their significance. Maps, because they present relationships among three or more variables (the usual graph deals only with two), are ideal for presenting large amounts of complex information in a way that shows clearly the distribution of components in terms of quantity, location, and time.

Moreover, maps are readily adaptable to whatever level of sophistication is necessary to fit the quality and quantity of available data. Depending on the data, the information that disease maps present can range from a simple yes or no answer with respect to the presence or absence of a particular health characteristic or disease, through quantitative expressions of the amount and character of the disease at various locations, to a demonstration of the complex interrelationships among disease rates and a wide variety of environmental factors.

In this era of increasing public concern with environmental influences, many important judgments must be made in the absence of sufficient information. The data that are available must be collated and put in a clear and understandable form for use by those who must make the decisions "now." The display of pertinent data in map form is an important means toward that end.

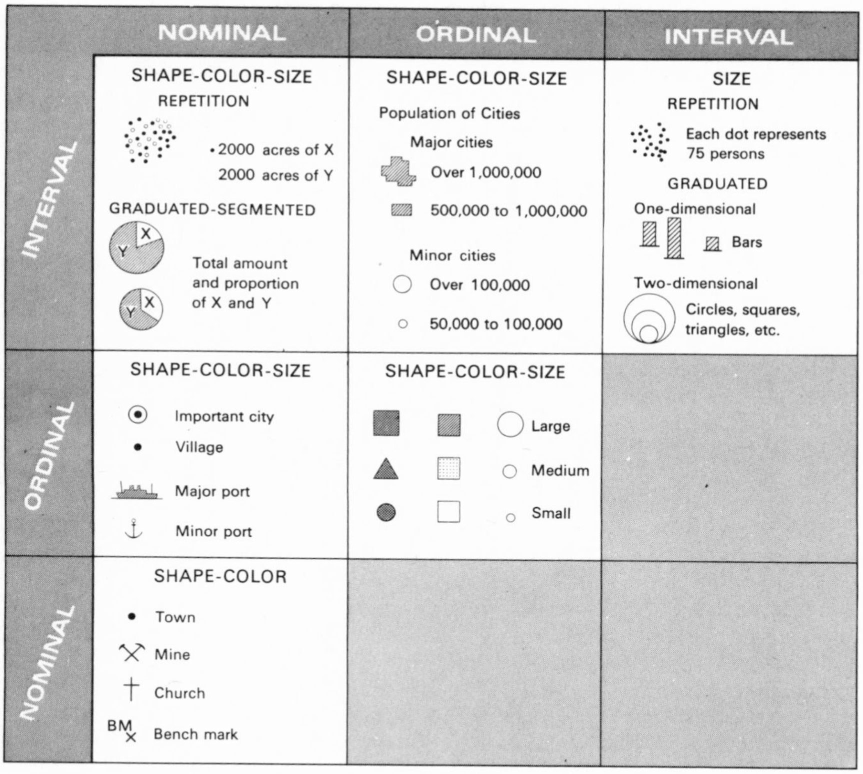

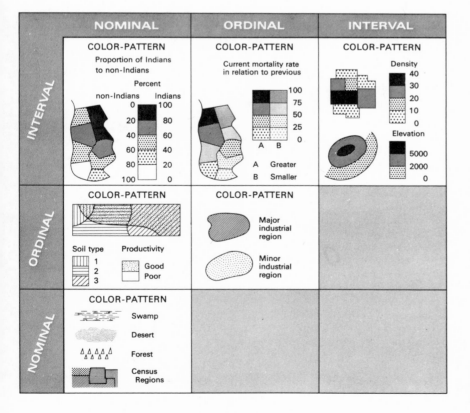

	NOMINAL	ORDINAL	INTERVAL
INTERVAL	**SHAPE-COLOR-SIZE** REPETITION •2000 acres of X 2000 acres of Y GRADUATED-SEGMENTED Total amount and proportion of X and Y	**SHAPE-COLOR-SIZE** Population of Cities Major cities Over 1,000,000 500,000 to 1,000,000 Minor cities ◯ Over 100,000 ○ 50,000 to 100,000	**SIZE** REPETITION Each dot represents 75 persons GRADUATED One-dimensional Bars Two-dimensional Circles, squares, triangles, etc.
ORDINAL	**SHAPE-COLOR-SIZE** ⊙ Important city • Village Major port ⚓ Minor port	**SHAPE-COLOR-SIZE** ◯ Large △ ▢ ○ Medium ● ▢ ○ Small	
NOMINAL	**SHAPE-COLOR** • Town ✕ Mine ✝ Church BM✕ Bench mark		

FIGURE 29 Illustrative outline of the kinds of *point* symbols and the ways that they can be used. Reproduced by permission from Robinson and Sale, *Elements of Cartography*, 3d ed., John Wiley & Sons, New York, 1969.

	NOMINAL	ORDINAL	INTERVAL
INTERVAL	**COLOR-PATTERN** Proportion of Indians to non-Indians Percent non-Indians / Indians 0 / 100 20 / 80 40 / 60 60 / 40 80 / 20 100 / 0	**COLOR-PATTERN** Current mortality rate in relation to previous 100 75 50 25 0 A B A Greater B Smaller	**COLOR-PATTERN** Density 40 30 20 10 0 Elevation 5000 2000 0
ORDINAL	**COLOR-PATTERN** Soil type / Productivity 1 2 Good 3 Poor	**COLOR-PATTERN** Major industrial region Minor industrial region	
NOMINAL	**COLOR-PATTERN** Swamp Desert Forest Census Regions		

FIGURE 30 Illustrative outline of the kinds of *area* symbols and the ways that they can be used. Reproduced by permission from Robinson and Sale, *Elements of Cartography*, 3d ed., John Wiley & Sons, New York, 1969.

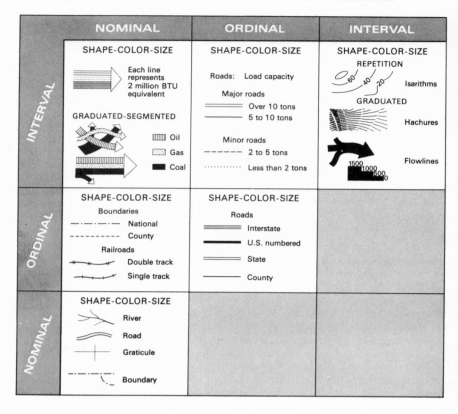

FIGURE 31 Illustrative outline of the kinds of *line* symbols and the ways that they can be used. Reproduced by permission from Robinson and Sale, *Elements of Cartography*, 3d ed., John Wiley & Sons, New York, 1969.

REFERENCES

Hopps, H. C. 1971a. Geographic pathology and the medical implications of environmental geochemistry. *In* Environmental geochemistry in health and disease, H. L. Cannon and H. C. Hopps [eds.]. Geol. Soc. Am. Mem. No. 123. Geological Society of America, Boulder, Colorado. pp. 1–11.

Hopps, H. C. 1971b. Cause–effect relationships in disease—A problem in data manipulation. J. Clin. Comput. 12:1–3.

Hopps, H. C. 1972a. Ecology of disease in relation to environmental trace elements—particularly iron. Geol. Soc. Am. Bull. 83. Geological Society of America, Boulder, Colorado. pp. 797–804.

Hopps, H. C. 1972b. Display of data with emphasis on the map form. Ann. N.Y. Acad. Sci. 199:325–334.

Hopps, H. C., R. J. Cuffey, J. Morenoff, W. L. Richmond, and J. D. H. Sidley. 1969. Computerized mapping of disease and environmental data. U.S. Government Printing Office, Washington, D.C. 431 pp.

Hopps, H. C., and E. R. Gabrieli. 1970. Automated data processing in renal diseases. *In* Laboratory diagnosis of kidney diseases, F. W. Sunderman and F. W. Sunderman, Jr. [eds.]. Warren H. Green, St. Louis. pp. 362–368.

Jenks, G. F., and F. C. Caspall. 1971. Error on choroplethic maps; definition, measurement, reduction. Ann. Assoc. Am. Geogr. 61:217–244.

May, J. M. 1958. The ecology of human disease. MD Publications, New York. 327 pp.

May, J. M. 1961. Studies in disease ecology. Hafner, New York. 613 pp.

Peucker, T. K. 1972. Computer cartography. Commission on College Geography, Assoc. Am. Geogr. Resour. Pap. 17, Washington, D.C. 75 pp.

Robinson, A. H., and Randall D. Sale. 1969. Elements of cartography, 3d ed. John Wiley & Sons, New York. 415 pp.

Tarrant, J. R. [comp.]. 1970. Computer in geography. *In* GEO abstracts. School of Environmental Sciences, University of East Anglia, Norwich NDR 33C, England. p. 76.

XVI

A National Environmental Specimen Index System

BOBBY G. WIXSON, *Chairman*

William Fulkerson, Philip D. LaFleur, James O. Pierce, Wilbur D. Shults,
Ivan C. Smith, Robert I. Van Hook, Darryl von Lehmden, Douglas Worf

INTRODUCTION AND OBJECTIVES

A National Environmental Specimen Index System (NESIS) would fill several important service needs as a comprehensive, centralized data bank of information on the nation's rapidly expanding collection of environmental specimens, including in particular those obtained with public funds. The major function of NESIS would be to serve as a guide to accessible collections of stored environmental samples available for further study. The range of services envisioned for such a system would include complete descriptive information and analytical results on the indexed samples, as well as recommendations for uniform sampling, measurement, and storage criteria. The system would also extend well beyond normal curatorial activities by developing suggestions for improving the usefulness of the indexed specimens to both research scientists and environmentally oriented regulatory officials. It would, in addition, foster coordination of specimen collection and banking activities and cooperation both nationally and internationally.

The concept of a NESIS is not intended to create a new agency, to generate data, or to establish a massive centralized sample bank, but to develop a mechanism, at nominal cost, to provide an indexing system and a central means of access to selected environmental specimens in existing single-purpose collections and to future materials banked under its recommended procedures. NESIS also could suggest a quality-control program for the accessible materials to help ensure that items needed for present as well as future environmental studies are acquired, validated, and properly preserved.

The idea of an environmental specimen index system first emerged at the October 18–19, 1972, meeting of the Subcommittee on the Geochemical Environment in Relation to Health and Disease (GERHD). Consideration of a national environmental sample index was proposed publicly at an Environmental Protection Agency (EPA) symposium held in February 1973. In May 1973, the topic was included at the GERHD Capon Springs Workshop (Wixson and Jennett, 1975). Participants in the Work Group that considered this topic were selected from interested sample-bank and information-handling specialists from the National Science Foundation (NSF), EPA, and the National Bureau of Standards (NBS), as well as from Oak Ridge National Laboratory (ORNL), Midwest Research Institute (MRI), and the University of Missouri. Workshop participants and the organizations represented saw in this study a valuable and fiscally reasonable tool to aid environmental research efforts.

As an outgrowth of the Workshop's deliberations, EPA and NSF jointly funded a preliminary study, in close collaboration with NBS, aimed at developing such an indexing system (National Bureau of Standards, 1975). This study, conducted at ORNL, resulted in a draft report by Van Hook and Huber (1975), "National Environmental Specimen Bank Survey," whose purpose was to identify repositories in the United States that store materials collected for either research or monitoring activities. The survey's aims were to identify where collections are located, who maintains them, what comprises the collections, what analyses have been performed, and how the sample collections have been preserved and stored, as well as to determine the accessibility of the stored materials and associated data to both research and regulatory personnel. The survey was designed to include information on collections of atmospheric, geological, human-tissue, animal-tissue, plant-tissue, and water samples.

Beyond the initial survey, the ORNL effort is designed to organize a planning committee to evaluate the feasibility of such an indexing system, to develop a 5-yr plan for carrying out the work, and to develop a program to establish criteria for sample collection, preparation, storage, and analysis.

The most important function of the system would be to ensure the future availability of banked specimens necessary for retrospective studies on the occurrence of selected substances and of their changes through time. The NESIS concept would thus help to identify and evaluate time-dependent changes in the environment and would provide a means of measuring properties not previously sought and of reevaluating previously determined analytical values with respect to newer techniques.

More complete knowledge and availability of environmental samples is necessary, because at present many investigators are not always aware of existing samples collected, analyzed, and stored in the normal course of research that may be useful to them; in addition a significant fraction of the data generated is probably not generally available or made known to any substantial portion of the scientific community. Furthermore, such individual collections undoubtedly reflect a troublesome variety of collection methods and conditions of storage.

NESIS would eliminate many of these problems, while increasing the quantity of samples and information available for general use. NESIS could help to compile descriptions of existing specimen collections and assist in establishing procedures for future sample collection, handling, storage, and analysis. Such a program could also identify needs for augmenting existing collections and suggest specific objectives for future needs.

The objectives of NESIS are to:

• Provide an index to historical specimens for the measurement of properties, including contaminants not previously investigated, long-term evaluation of analytical methods, and measurements of time-dependent changes.
• Establish a centralized data and information storage and retrieval system containing all information appropriate to the selected specimens. NESIS should be open to all qualified parties.
• Establish uniform criteria for sampling, storing, and measuring the properties of various types of specimens needed to be stored.
• Provide information useful for the assessment and revision of current environmental policies.
• Establish a framework for national coordination of current and future specimen banks and collection activities.

• Foster cooperation and establish working arrangements for the international exchange of information and specimens.

Operating Procedures

NESIS could function as a central coordinating group able to draw on many organizations and agencies. Specifically, it would need the active participation both of groups currently building specimen collections and of researchers involved with general environmental monitoring. This participation might best be ensured by the establishment of a steering committee composed of representatives of all major participating groups. The steering committee could establish NESIS and act as a board of directors for the operational system. To meet the needs of the wide variety of contributors, the committee membership should reflect a diversity of disciplines, including, at least, analytical chemistry, geology and geochemistry, meteorology, soil science, ecology, biology, medicine, and statistics.

The system could be operationally managed by a central staff supervised by a full-time executive officer. The executive officer should be responsible to the steering committee, and the staff should handle requests and maintain regular contacts with the various environmental specimen banks and collections associated with NESIS.

Among the existing sample banks that should be affiliated with NESIS are:

• The National Tissue Specimen Bank (NTSB), a part of the Community Health and Environmental Surveillance System of EPA. NTSB is conducting a series of standardized epidemiological studies, designed to evaluate existing environmental standards and to improve the knowledge and understanding of human exposure to toxic metals and persistent man-made chemical compounds. Pilot studies, involving the collection of human-tissue specimens from biopsies and autopsies, are also under way with the objective of systematically determining baseline pollutant levels in different organs as a function of age.
• The Atomic Energy Commission Soil Bank, now a part of the Energy Research and Development Administration (ERDA). It has collected soil samples, both nationally and worldwide, over the past 15 yr as a means of monitoring changes in strontium-90 levels.
• The National Air Surveillance Network (NASN) of the EPA Environmental Research Center. NASN, established in 1953, conducts high-volume air sampling for Total Suspended Particulate (TSP) in 237 urban and 20 nonurban areas throughout the United States. Half of the filter (on which the sample collects) is used for immediate study, and the other half is placed in storage. This TSP air-sampling network has provided data for several studies, most notably concerning the occurrence of particulate polycyclic organic matter in the atmosphere.

NESIS would maintain and disseminate a complete description of all samples located within the system through a computerized information storage and retrieval system. Descriptive information would include such things as the types and numbers of specimens, their handling and storage history, the results of analyses, the locations from which they were taken, and other descriptive items.

NESIS could designate certain significant specimens in various sample-bank collections as "NESIS specimens" as a means of ensuring their availability for special use in the national interest. In addition, cooperating banks might, as the need arose for specific samples not otherwise banked in NESIS, entrust or make available through NESIS certain fractions of their own, non-NESIS collections.

NESIS would identify the types of samples that should be banked but are not. Use could also be made of the extensive and diverse U.S. environmental monitoring activities; government environmental research grants might provide for a certain fraction of specimens collected to be banked for future use.

With an eye to present and future needs, NESIS would encourage preservation of a wide variety of specimens. Those currently regarded as essential include:

- Geochemical specimens, such as air, soil, sediments, and water.
- Ecological specimens, such as plants and animals.
- Animal tissues, such as specimens of domestic animal tissue used for food.
- Human tissues obtained from biopsy and autopsy.
- Foods, such as those from the Food and Drug Administration's (FDA) Market Basket Surveys and other high-usage commodities.
- Miscellaneous specimens, such as energy sources, waste materials, and effluents.

Exhaustive analysis of all specimens prior to deposit seems unnecessary. However, a sequence of testing procedures is required to ensure that the needed prestorage tests are performed and that less-destructive testing precede more-destructive testing. Other measurements could be made as needed. All operations should be designed and conducted to ensure high quality in the preservation and measurement of all specimens in the system. Although not directly involved in field sampling, NESIS might review and help to plan long-term sampling strategies of various monitoring agencies and groups. NESIS might also seek to encourage the maintenance in perpetuity of sites around the country to serve as environmental study parks for the accurate monitoring of environmental quality.

CONCLUSIONS

A National Environmental Specimen Index System (NESIS) could provide a much-needed central source of information on past, present, and future environmental specimens, repositories, and associated monitoring techniques. The following sequence of three tasks is envisioned for NESIS.

Task I

Establish a NESIS Steering Committee, composed of representatives of groups or organizations providing funds, collecting specimens, operating monitoring programs, or using its services—organizations such as the Environmental Protection Agency, the U.S. Department of Agriculture, the Energy Research and Development Administration, the National Science Foundation, the Food and Drug Administration, the U.S. Department of the Interior, appropriate academic institutions (particularly land-grant and sea-grant universities), and the Council on Environmental Quality.

The chairman and members of such a steering committee should be expected to devote substantial time to this activity, and sufficient funds should be provided for their travel to attend meetings.

The Steering Committee should:

- Develop the organization and management structure of NESIS.
- Identify the types of specimens and information to be stored in the system.
- Develop interim procedures for sampling, sample handling, and storage of specimens to be included in the system.
- Plan the data handling, storage, and retrieval system.
- Publish approved analytical techniques, and support research in analytical procedures.
- Serve as a clearinghouse for information on environmental data banks.
- Promote research in environmental geochemistry.

Task II

Inventory and assess the value of existing specimen collections that are potential candidates for participation in NESIS. The findings should be published and should include reviews and recommendations concerning:

- Present methods of collection and storage.
- Availability of banked specimens for future analysis.
- Reliability of existing analytical data.
- Assessment of the value of the specimens for future reference.

Task III

Implement the NESIS concept, including:

- Defining sampling procedures and strategies.
- Developing preparative and storage procedures that minimize chemical or biological degradation.

Because of the multiagency interest, existing cooperation and participation, and the diverse types of environmental samples desired, the National Science Foundation logically should be considered as the initial coordinating and funding agency. The Environmental Protection Agency would be the most logical organization to establish such a system, because of its present involvement in various sample banks. However, other agencies, organizations and individual users also have expressed a desire to participate by cooperation, coordination, and funding, to ensure the development of a viable National Environmental Specimen Index System.

REFERENCES

National Bureau of Standards. 1975. Banking the environment. Dimensions (U.S. Department of Commerce, Washington, D.C.) 59:7, 147–149.

Van Hook, R. I., and E. E. Huber. 1975. National environmental specimen bank survey. Oak Ridge National Laboratory, Oak Ridge, Tennessee. 207 pp.

Wixson, B. G., and J. C. Jennett. 1975. The new lead belt in the forested Ozarks of Missouri. Environ. Sci. Technol. 9:1128–1133.

Capon Springs
Workshop
Participants

JERRY K. AIKAWA, University of Colorado Medical Center, Denver

WILLIAM H. ALLAWAY, Plant, Soil, and Nutrition Laboratory, Agricultural Research Service, Ithaca, N.Y.

KENNETH C. BEESON, Consulting Soil Scientist, Sun City, Ariz.

HELEN L. CANNON, Branch of Exploration Research, U.S. Geological Survey, Denver, Colo.

EDITH M. CARLISLE, School of Public Health, University of California, Los Angeles

RICHARD A. CARRIGAN, Environmental Systems and Resources, National Science Foundation, Washington, D.C.

GEORGE K. DAVIS, Department of Sponsored Research, University of Florida, Gainesville

WILLIAM FULKERSON, ORNL–NSF Environmental Program, Oak Ridge National Laboratory, Tenn.

WALLACE R. GRIFFITTS, Geologic Division, U.S. Geological Survey, Denver, Colo.

DAVID H. GROTH, Toxicology Branch, National Institute for Occupational Safety and Health, Cincinnati, Ohio

DELBERT D. HEMPHILL, Department of Horticulture, University of Missouri, Columbia

THOMAS D. HINESLY, Program Planning Group, Department of the Army, Washington, D.C.

LEON L. HOPKINS, JR., Human Nutrition Research Division, Agricultural Research Service, Fort Collins, Colo.

HOWARD C. HOPPS, Department of Pathology, University of Missouri Medical Center, Columbia

DONALD J. HORVATH, Division of Animal and Veterinary Sciences, West Virginia University, Morgantown

EVERETT A. JENNE, Water Resources Division, U.S. Geological Survey, Menlo Park, Calif.

JOE KUBOTA, Plant, Soil, and Nutrition Laboratory, Soil Conservation Service, Ithaca, N.Y.

PHILIP D. LAFLEUR, Analytical Chemistry Division, National Bureau of Standards, Gaithersburg, Md.

ORVILLE A. LEVANDER, Nutrition Institute, Agricultural Research Service, Beltsville, Md.

WILLARD L. LINDSAY, Department of Agronomy, Colorado State University, Fort Collins

JUSTIN A. McKEAGUE, Soils Research Institute, Canadian Department of Agriculture, Ottawa, Ontario

ROBERTO MASIRONI, Cardiovascular Disease Unit, World Health Organization, Geneva

GENNARD MATRONE, Department of Biochemistry, North Carolina State University, Raleigh (now deceased)

ISMAEL MENA, Medical Research Center, Brookhaven National Laboratory, Upton, N.Y.

WALTER MERTZ, Nutrition Institute, Agricultural Research Service, Beltsville, Md.

ALFRED T. MIESCH, Branch of Regional Geochemistry, U.S. Geological Survey, Denver, Colo.

DAVID B. MILNE, Laboratory of Experimental Metabolic Diseases, Veterans Administration Hospital, Long Beach, Calif.

PAUL M. NEWBERNE, Department of Nutrition and Food Science, Massachusetts Institute of Technology, Cambridge

FORREST H. NIELSEN, Human Nutrition Laboratory, Agricultural Research Service, Grand Forks, N.D.

JAMES O. PIERCE, Environmental Trace Substances Center, University of Missouri, Columbia

HORACE T. RENO, Ferrous Metals Division, U.S. Bureau of Mines, Washington, D.C.

EVAN M. ROMNEY, School of Medicine, University of California, Los Angeles

HAROLD H. SANDSTEAD, Human Nutrition Laboratory, Agricultural Research Service, Grand Forks, N.D.

WILBUR D. SHULTS, ORNL–NSF Environmental Program, Oak Ridge National Laboratory, Tenn.

RAYMOND SIEVER, Department of Geology, Harvard University, Cambridge, Mass.

MAURICE L. SIEVERS, Phoenix Indian Medical Center, Indian Health Service, Department of Health, Education, and Welfare, Phoenix, Ariz.

MARVIN W. SKOUGSTAD, Water Resources Division, U.S. Geological Survey, Denver, Colo.

IVAN C. SMITH, Chemistry Department, Midwest Research Institute, Kansas City, Mo.

PERRY R. STOUT, Department of Soils and Plant Nutrition, University of California, Davis (now deceased)

LEE O. TIFFIN, Agricultural Environmental Quality Institute, Agricultural Research Service, Beltsville, Md.

ROBERT I. VAN HOOK, ORNL–NSF Environmental Program, Oak Ridge National Laboratory, Tenn.

PETER J. VAN SOEST, Department of Animal Science, Cornell University, Ithaca, N.Y.

DARRYL VON LEHMDEN, Source Sample and Fuel Analysis Branch, Environmental Protection Agency, Research Triangle Park, N.C.

WARREN E. C. WACKER, University Health Services, Harvard University, Cambridge, Mass.

ROBERT H. WASSERMAN, Department of Physical Biology, N.Y. State Veterinary College, Cornell University, Ithaca, N.Y.

ROSS M. WELCH, Plant, Soil, and Nutritional Laboratory, Agricultural Research Service, Ithaca, N.Y.

BOBBY G. WIXSON, Environmental Research Center, University of Missouri, Rolla

DOUGLAS L. WORF, National Environmental Research Center, Environmental Protection Agency, Research Triangle Park, N.C.

Staff

WILLIAM L. PETRIE, Division of Earth Sciences (now Assembly of Mathematical and Physical Sciences), National Research Council

JOHN L. LUDWIGSON, Editorial Consultant

JUNE R. GALKE, Division of Earth Sciences (now Assembly of Mathematical and Physical Sciences), National Research Council

Index

Boldface page numbers indicate extended discussion of topics.